Generis
PUBLISHING

INNOVATIONS IN AGRICULTURE, CHEMISTRY AND PHARMACEUTICAL BIOTECHNOLOGY FOR SUSTAINABLE ADVANCEMENT

Godwin Mtetwa
Beloved Mtetwa
Answer Mtetwa

Title: **INNOVATIONS IN AGRICULTURE, CHEMISTRY AND PHARMACEUTICAL BIOTECHNOLOGY FOR SUSTAINABLE ADVANCEMENT**

ISBN: 979-8-89966-443-4

Author: Godwin Mtetwa, Beloved Mtetwa,Answer Mtetwa, Desire Mtetwa / Evidence Mtetwa

Cover image: www.pixabay.com

Publisher: Generis Publishing
Online orders: www.generis-publishing.com
Contact email: info@generis-publishing.com

Authors

	**Godwin Mtetwa:** BSc Agriculture Management (Zimbabwe Open University), Diploma (Gwebi Agricultural College). Principal Research Officer in the Department of Agriculture Research, Innovation and Specialist Services. An author of the books *"A Step Towards Sustainable Agriculture Technology Development for Climate Resilience" and "The Fearless Hannibal Cooper."* Has also published several papers in peer reviewed International scientific journals. Currently stationed at Chiredzi Research Station in Zimbabwe and doing research on Deficit irrigation in maize, cassava, rice and wheat.
	**Beloved Mtetwa:** Attained a *MSc*. Crop Science, Midlands State University (MSU) -Zimbabwe, BSc. Horticulture (Honours), Midlands State University (MSU), Diploma-Sports Coaching and IAAF (World Athletics) Level 1 Coaching certificate. A dynamic leader with passion for innovation. Spearheads Vollig Pvt, Ltd, a pioneering solar company in Zimbabwe. He also drives growth and excellence at Glymorn Security, Cubic Bay Cleaners, and other ventures. Drives innovation in renewable energy and empowers communities through his diverse ventures. Author of Chapter, "Evaluation of planting time, pest incidence and yield for 5 (Five) Industrial hemp varieties (Cannabis Sativa. L) in Zimbabwe.
	**Desire Mtetwa:** *MPhil* in Computational Biology and Biotechnology, Chinhoyi University of Technology (CUT), BSc. Biological Sciences (Honours), Midlands State University (MSU) in Zimbabwe. A Computational Biologist and researcher specializing in genomic data science and environmental biotechnology with research focused on the molecular epidemiology and evolution of SARS-CoV-2 in Africa. Her work spans human pathogen genomics and plant-microbe interactions, using high-performance computing and R/Python-based bioinformatics. She is currently based in Czech Republic, applying integrative 'omics to address ecological and global health challenges. Author of the Chapter, "Molecular epidemiology of SARS-CoV-2 genetic variants in Africa".

	**Evidence Mtetwa:** MSc. Pharmaceutical Biotechnology, Chinhoyi University of Technology (CUT) Zimbabwe, BSc. Chemical Technology (Honours), Midlands State University (MSU). Pharmaceutical Biotechnology Specialist. Contributor to Innovations in Agriculture, Chemistry and Pharmaceutical Biotechnology for sustainable advancement. Author of the Chapter, "Fabrication of Fe-W Oxide Alginate Bean Composite for Gold Recovery".
	**Answer Mtetwa:** BSc. Applied Chemistry (Honours), National University of Science and Technology (NUST) in Zimbabwe. A meticulous chemist dedicated to quality assurance, quality control, Research and Development. He possesses certified expertise in global laboratory and quality management systems. Worked in the Agro-pharmaceutical industry and now within the mining sector, upholding the highest standards of safety, analytical excellence and operational quality. Author of the Chapter, "Preparation of antimicrobial soap using *Kigelia africana* (Sausage tree)-extracted phytochemicals".

Table of Contents

Acknowledgement

The compiled scientific papers went through various stages of review and editing before taken to different journals for further reviews, editing and publishing. All these tasks were performed judiciously and expeditiously by members of Chiredzi Research Station, Research on World Agricultural Economy, Journal of Agriculture and Horticulture Research, Octa Journal of Agricultural Sciences, International Journal of Pharmacy and Biological Sciences and Royal Society Open Science. The result is an excellent collection of Agricultural Science book, Chemistry and Pharmaceutical Biotechnology with technologies that are life changing to the society all areas of life. The technologies will strengthen the adaptive capacity of vulnerable people in Africa. I trust readers will find these papers to be a useful contribution to the development of our society in Africa.

PART 1: Agronomy and Plant Protection

1.1 Yield Performance of Sugar Bean Varieties (Phaseolus spp.) Influenced by Sowing Time Under Overhead Irrigation at Chiredzi Research Station in Natural Region V of Zimbabwe

Octa J. Biosci. Vol. 11 (1):12-26

Received: 04/01/2023

Revised: 23/05/2023

Accepted: 18/06/2023

Published: 30/06/2023

Godwin Mtetwa

*Chiredzi Research Station, P. O. Box 97, Chiredzi. Zimbabwe *corresponding author e-mail address: godwin.mtetwa@gmail.com*

1.1.1 Abstract:

The most favorable time of sowing and a promising variety are of prime importance to harness potential yield in bean production. A trial was conducted on the effect of sowing date and variety on yield of sugar beans in the Lowveld at Chiredzi Research Station Semi-arid region of Zimbabwe. Twelve different determinate varieties of sugar beans and eight indeterminate varieties were evaluated. The determinate varieties were OPS-RSI, Natal Sugar beans, AND 897, Mus 97, Mug 38, Red Canadian wonder, Michigan pea bean, Cim 9314-17, Zebra, Grey Light, White Kidney, Cim 9406-3. The indeterminate varieties were Carioka, Ren 22, Iris, MCM 2203, Nyanga Red, APN 136, Black bean and Xan 76. Results showed that the number of pods varied across months with highest numbers produced May (20). Indeterminate varieties had more pods per plant than determinate varieties. The number of seeds per plant increased significantly ($p<0.05$) with Michigan producing the highest number (68). The number of seeds per pod was highest in May (36). Michigan pea bean had the highest average number of pods per plant (17). The average number of days to flowering were highest in May (51), June (55) and July (53). Average number of days to maturity were highest in April (99), May (106), June (98) and July (97). Marketable average yield increased significantly $p<0.05$) in April (1620 kg/ha), May (1719 kg/ha) and June (1633 kg/ha). Mean maximum temperature was above optimum requirement for bean production affecting number of pods per plant, seeds per plant and ultimately economic yield. Variety APN 136 planted in May was the best in the Lowveld of Zimbabwe.

Keywords: Variety; sowing date; temperature, indeterminate, determinate

1.1.2 Introduction

Sugar bean (Phaseolus vulgaris L.) is a leguminous crop, implying that it has the capacity to fix atmospheric nitrogen and can be sown without rhizorbium inoculums at planting (Marco et al 2006). This biological aspect makes sugar beans a good fit in crop rotation and intercropping systems (Walter 2014). As a protein-rich food, sugar beans play an important role in human nutrition, especially in developing countries because it is a good source of iron and zinc (Buruchara et al., 2011). They contribute up to 33% of the dietary protein needs of humans (Vance et al., 2002). Common bean is a warm season annual legume crop grown primarily for its protein and energy-rich dry seeds. An estimated 23.1 million tons of common bean is produced annually on about 8.7 million hectares worldwide (FAO, 2014). In semi-arid regions of Zimbabwe (Natural Region 4 and 5) yield of field beans are reduced mainly by high day and night temperature (Catuchi et al 2019). In contrast, late planting date causes the reproductive growth stage of the crop to face the fall chilling in areas that have chilling temperatures, which ultimately lowers yield (Esmaeilzadeh and Aminpanah 2015). Unfortunately, the sensitivity of sugar bean to low temperature (Badowiec & Weidner, 2014) may be troublesome in temperate climate regions where the transitional drop of temperature occurs frequently during late growing season. Lowveld receives mild temperature in winter which is favorable to growth and development of the crop. Sugar beans being a C3-cycle plant, cultivation in high temperature environments results in decreased photosynthesis, mainly due to increased respiration and photorespiration (Karavidas et al. 2022). From the observations that were made at Chiredzi Research Station it was shown that some varieties of beans were affected very much by time of sowing, whilst some varieties performed relatively well regardless of time. High temperatures above 32^0C during flowering and pod filling stages contributed to poor pod filling (Madamba, 2003). Yoldas and Esiyok, (2009), reported that planting when temperatures were high, reduced plant growth cycle and flowering period. It was also shown that values of yields and number of normal seeds and pods were reduced. Heavy rains and humidity that contribute to the build-up of leaf and stem diseases reduced marketable grain. In Middle Sabi some newly settled farmer planted beans in August and yielded less than a half a ton on alluvial soils.

Beans can tolerate low temperatures and even occasional mild frost during the early growth stages but this reduces yields (Madamba, 2003). However, yields of up to 3 t/ha are possible under the same soil type and good management. In Natural Region 1, 2 and 3 in Zimbabwe farmers normally sow beans in December and January when temperatures are still high (35^{0}C). From observations, Red mottled, Iris, Michigan Pea Bean and Zebra varieties were some of the good varieties that showed good yields when planted during these hot months of the year. The information generated from this trial would help to answer some of the challenges that are being faced by farmers as far as time of sowing; variety type and the economics of production are concerned. Varieties that are not affected by time of sowing would be selected and information made available to farmers. The objective of this trial was to find the effect of sowing date on the yield and financial returns of 20 varieties of field beans.

1.1.3 Materials and Methods

1.1.3.1 Experimental procedure and treatments

The trial was planted at Chiredzi Research Station in the Southeastern Lowveld of Zimbabwe (21^0 33' S and 31^{0}30' E) at an altitude of 429 m above sea level. The trial was set up in a randomized complete block design with three replications and twenty varieties from Crop Breeding Institute were planted. The determinate varieties planted were OPS-RSI, Natal Sugar beans, AND 897, Mus 97, Mug 38, Red Canadian wonder, Michigan pea bean, Cim 9314-17, Zebra, Grey Light, White Kidney and Cim 9406-3. Indeterminate varieties were Carioka, Ren 22, Iris, MCM 2203, Nyanga Red, APN 136, Black bean and Xan 76.

1.1.3.2 Crop establishment and management

The trial was planted every first week of each month from January to December for three years. The mean maximum and minimum temperatures were 29^{0}C and 12^{0}C respectively. The soils were predominately sandy-clays, and shallow (0.6-1.0m deep) with 27-30% clay. Compound D (N:7: P:14: K:7) at a rate 300 kg/ha was applied as basal dressing and ammonium nitrate (34.5%N) at a rate of 50 kg/ha N was applied as topdressing at three weeks after planting. The land was prepared using a tractor drawn disc plough and discing done to attain fine tilth. Plot sizes were 4.2 m long and 2.7 m wide. Inter-rows spaced at 0.45 m and intra rows at 0.07 m (317,460 plants/ha) were made by hands. All varieties were supplied by Crop Breeding Institute. The seed was sprayed with Thiram at planting to prevent fungal infection using knapsack sprayers.

1.1.3.3 Measurements

The number of days to 50% flowering, 95% maturity and marketable yields were recorded. Grain yield was adjusted to 13% moisture content and was determined by grain moisture meter. Two rows on either side of the plots were discarded as guard rows leaving a net plot 7.56 m2.

1.1.3.4 Statistical analysis

The data were subjected to analysis of variance procedure as Randomized Complete Design. Analysis was done using Genstat 14th Edition. Where necessary least significant differences (L.s.d) were used to separate means at $p=0.05$.

1.1.4. Results and Discussions

1.1.4.1 Days to flowering and maturity

The number of days to flowering increased significantly $p<0.001$ from 39 days in January (summer) to 55 days in June (winter) and dropped to 43 days by December (table 1.1.1). Crop planted from March to June took significantly more days to flower. During periods of high temperature sugar beans hasten to flower. Similar results were reported (Essubalew Getachew Seyum 2014; Ndegwa et al. 2001). Consequently, the number of days to maturity were significantly delayed $p<0.001$ for April, May and June crop 99, 106 and 98 days respectively compared to 80; 76; 75; 86; 82; 93; 80 and 81 days in January; February; March; August; September; October; November and December respectively (table 1.1.2). MCM 2203 and Iris had the longest days to maturity during winter (100-114) compared to 77- 97 days during summer months. During winter (April, May and June) it took longer for the varieties to flower and mature than in summer. This was also confirmed by Yoldas and Esiyok, (2009), who reported that plant growth cycle and flowering period were shorter when temperatures were high. When temperatures are low the crop takes a longer period to develop because of the reduced heat units.

1.1.4.2 Number of pods per plant and number of seeds per plant

Significant number of pods per plant was recorded among varieties and sowing dates. There were no significant differences on number of pods per plant among Black bean, Michigan pea bean, Nyanga Red, Iris, Ren 22, Mus 97 and Carioka (table 1.1.3). The number of seeds per plant showed significant differences $p<0.001$. The variety that exhibited highest significant differences in the number of seeds per plant was Michigan pea bean. Indeterminate varieties that had significantly the highest number of seeds per plant were APN 136, Nyanga red, Iris, Ren 22, Carioka and Xan 76. Determinate varieties that showed significant high numbers of seeds per plant were Michigan pea bean and Mus 97. The month of May had significantly the highest number $p<0.001$ of pods per plant and seeds per plant (fig. 1.1.2). There were no significant differences on number of pods per plant and number of seeds per plant for the months February, March, April, June and July. From August through to January the number of pods per plant was significantly lower than February through to July. There was a decrease in the number of seeds per plant from August to December. Sowing time may also affect yield due to the temperature and rainfall that prevail at critical developmental stages, specifically flowering and pod-filling (Karavidas et al. 2022). Catuchi et al (2019) found that when night and day temperatures exceed 25^0C to 300C respectively flower buds and pod formation are adversely affected and reduce in numbers per plant. Results also concur with (Getachew et al. 2014 and Qiu Ying Zhang et al. 2010) who reported that early sowing towards cooler months produced more branches and more pods and ultimately more seeds than late sowing towards hotter months. The increase in the number of pods and seeds per plant in early planted sugar beans could also be attributed to the development of more vegetative parts in earlier sowing, which resulted in the production of a greater number of pods per plant, and the difference between the varieties could probably be genetic, hence some varieties might be more tolerant to low temperature.

1.1.4.3 Response of sugar beans to time of sowing on yield

It was clear that planting sugar beans from March up to June gave significantly high yields ($p<0.05$) on average from the 20 varieties evaluated. In Zimbabwe's Natural Region V temperatures go down from April as winter will be drawing closer (Fig 1.1.5; 1.1.7 and 1.1.9). Reduction in yield for the January and February crop might be due to high temperatures during time of flowering that occurred in February and March respectively and pod setting (Elhag et al. 2014 and Sharma et al 2014) because the crop flowered when temperatures were still higher than optimum temperature requirement. This may be ascribed to short growing season and ultimately lesser accumulation of photosynthates (Sharma et al. 2013). Stephen et al. (2014) found out that temperatures exceeding 30°C during flowering causes flower abortion in sugar beans (Fig 1.1.6; 1.1.8 and 1.1.10). Reduced yield from July through to December may be ascribed to very high temperatures during flowering periods causing flower abortion (Karavidas et al. 2022, Sharma et al 2014; Stephen et al 2014), rainfall during flowering (fig 1.1.11) that causes flower drop and bean rot at harvest and short growing period that occurs when temperatures are higher than optimum. Lower pod yield in the late planting season was due to a smaller biomass production from a shorter vegetative growth period and moreover, the decline in pod production may simply result from declining flower production as vegetative growth ceases (Molosiwa and Kgokong 2017). Sowing after July has negative consequences on yield because the reproductive stage occurs when weather conditions are less favorable (Karavidas et al. 2022). The reproductive period of sugar bean plants coincides with the highest summer temperatures and this cause abscission of many buds and flowers that results in a significant decrease in productivity (Marlene et al., 2008; Sharma et al 2013). Yield trends (fig 1.1.1) show that bean yield increased from February up to May sowing and starts to decline from June. Sugar beans should be planted from March until June as shown in (Fig 1.1.4). However, some varieties can be planted as early as February and still break even under these conditions. The varieties Xan 76 (1567 kg/ha), Mus 97 (1454.5 kg/ha), Nyanga red (1372.5 kg/ha), APN 136 (1698 kg/ha), Michigan pea bean (1718 kg/ha), Red Canadian wonder (1511.5 kg/ha) and black bean (1575 kg/ha) were conspicuous. If farmers are late in planting varieties like Iris (1770.5 kg/ha), Xan 76 (1488.5 kg/ha, CIM 9314-17 (1406.5 kg/ha), Black bean (1294.2 kg/ha) and Nyanga red (1449.5 kg/ha) can be sown in July and still break even.

The Lowveld experiences temperatures that are above the optimum temperature for sugar bean from August to January (fig. 1.1.6, 1.1.8 and 1.1.10). However, from February temperature starts to be suitable for bean production up to July. Considering viability of the crop from August to January conditions are not favorable for growing beans in the Lowveld (fig 1.1.2, 1.1.3 and 1.1.4). A combination of rainfall and high temperatures in November, December, January and respectively when flowering takes place causing flower abortion resulting in low counts of pod numbers per plant. In January yield started to rise above breakeven point but not economical (fig 1.1.4). Most of these bean varieties performed better from February to February caused flower abortion and pod rotting reducing yields of the crop that was sown in October, November, December and January. August, September and October sowing were affected mainly by high temperatures that occurred in September, October and November July. Varieties that are the best from February to July are Xan 76, Carioka, Red Canadian wonder, APN 136 and Michigan pea bean. Natal sugar bean showed a short range of time from February to April.

Table 1.1.1: Effect of time of sowing on days to 50% flowering across season

Table 1.1.2: The effect of time of sowing on days to 50% Maturity across seasons

Variety	January	February	March	April	May	June	July	August	September	October	November	December
OPS-RSI	91^{a}	73^{cd}	76^{cd}	97^{defgh}	96^{b}	94^{fgh}	95^{f-h}	86^{fg}	85^{a-c}	92^{fg}	79^{d-e}	70^{fg}
Xan 76	89^{bc}	78^{abcd}	78^{abc}	104^{bc}	112^{a}	100^{cd}	99^{a-d}	90^{cd}	84^{a-c}	90^{fg}	84^{ab}	81^{d}
Zebra	73^{f}	73^{cd}	69^{f}	90^{h}	96^{b}	91^{h}	92^{h-I}	77^{hi}	78^{fg}	80^{i}	80^{cd}	81^{d}
Grey Light	66^{g}	72^{d}	70^{f}	92^{gh}	94^{b}	94^{fgh}	93^{g-I}	76^{i}	78^{f-h}	91^{fg}	77^{ef}	70^{f}
Carioka	83^{d}	77^{abcd}	71^{ef}	100^{cdef}	111^{a}	99^{cde}	97^{c-f}	90^{cd}	84^{a-d}	95^{ef}	80^{cd}	80^{de}
Natal sugar beans	67^{g}	73^{cd}	70^{f}	91^{gh}	95^{b}	92^{gh}	93^{h-i}	76^{hi}	77^{gh}	93^{f}	77^{f}	67^{g}
Mus 97	74^{f}	75^{bcd}	76^{cd}	99^{cdef}	112^{a}	99^{cde}	99^{a-e}	91^{ab}	82^{c-e}	92^{fg}	77^{ef}	80^{de}
White Kidney	67^{g}	73^{cd}	69^{f}	91^{gh}	96^{b}	92^{gh}	91^{i}	77^{hi}	75^{h}	76^{j}	79^{de}	79^{de}
AND 897	82^{d}	82^{ab}	79^{ab}	103^{bcd}	109^{a}	100^{cd}	99^{a-e}	93^{a}	86^{a}	86^{h}	84^{a}	87^{b}
Cim 9406-3	88^{c}	79^{abcd}	78^{abc}	95^{efgh}	99^{b}	95^{fgh}	96^{e-g}	85^{g}	83^{c-e}	104^{ab}	82^{bc}	88^{b}
Ren 22	91^{ab}	77^{abcd}	75^{cd}	104^{bcd}	112^{a}	109^{a}	101^{a}	91^{bc}	83^{b-e}	96^{def}	79^{de}	82^{d}
Iris	88^{c}	77^{bcd}	79^{ab}	103^{bcd}	114^{a}	105^{b}	99^{a-e}	91^{ab}	84^{a-d}	97^{de}	80^{cd}	82^{cd}
MCM 2203	88^{c}	80^{ab}c	79^{ab}	104^{bc}	114^{a}	110^{a}	100^{a-} c	90^{cd}	84^{a-d}	94^{f}	79^{de}	85^{bc}
Nyanga Red	79^{e}	75^{bcd}	71^{ef}	100^{cdef}	111^{a}	99^{cde}	96^{d-g}	87^{ef}	83^{c-e}	99^{cd}	79^{de}	82^{d}
Cim9314-17	93^{a}	84^{a}	77^{bc}	111^{a}	110^{a}	95^{efg}	101^{ab}	88^{e}	86^{ab}	104^{ab}	80^{cd}	88^{b}
APN 136	74^{f}	77^{bcd}	76^{cd}	95^{efgh}	112^{a}	99^{cde}	98^{b-f}	86^{eg}	84^{a-d}	101^{bc}	78^{d-f}	81^{d}
Michigan Pea bean	78^{e}	76^{bcd}	74^{de}	101^{cde}	111^{a}	102^{bc}	100^{a-} c	87^{ef}	83^{c-e}	94^{f}	79^{d-f}	78^{e}
Red Canadian Wonder	74^{f}	74^{bcd}	75^{cd}	93^{fgh}	100^{b}	94^{fgh}	95^{f-h}	78^{h}	81^{d-f}	90^{fg}	81^{c}	80^{de}
Black Bean	68^{g}	75^{bcd}	76^{bcd}	98^{cdefg}	112^{a}	99^{cde}	97^{d-f}	89^{d}	81^{e-f}	92^{fg}	79^{d-e}	78^{e}
Mug 38	91^{a}	76^{bcd}	81^{a}	108^{ab}	111^{a}	96^{b}	101^{ab}	90^{cd}	86^{a}	107^{a}	80^{cd}	93^{a}
Means	**80**	**76**	**75**	**99**	**106**	**98**	**97**	**86**	**82**	**93**	**80**	**81**
LSD	**2**	**6**	**3**	**6**	**7**	**3**	**3**	**2**	**3**	**3**	**2**	**3**
CV%	**5**	**5**	**2**	**4**	**1**	**2**	**2**	**1**	**3**	**18**	**1**	**2**

Table 1.1.3: Number of pods/plant and number of seeds per plant

Variety	Number of pods/plant	Number of seeds/plant
OPS-RSI	10^{ef}	30.4^{ef}
Xan 76	13^{bcde}	52^{bc}
Zebra	10^{def}	25^{f}
Grey Light	12^{bcdef}	41^{cde}
Carioka	15^{ab}	51^{bc}
Natal sugar beans	12^{bcdef}	41^{cde}
Mus 97	14^{abc}	56^{b}
White Kidney	10^{ef}	24^{f}
AND 897	8^{f}	25^{f}
Cim 9406-3	10^{def}	25^{f}
Ren 22	14^{abc}	50^{bc}
Iris	16^{ab}	52^{bc}
MCM 2203	12^{bcdef}	41^{cde}
Nyanga Red	14^{abc}	55b
Cim9314-17	10^{def}	31^{ef}
APN 136	14^{abc}	45^{bcd}
Michigan Pea bean	17^{a}	68^{a}
Red Canadian Wonder	12^{bcdef}	35^{def}
Black Bean	15^{ab}	56^{b}
Mug 38	11^{cdef}	33^{def}
Means	12.42	41.7
LSD (0.05)	3.4	11.04
CV%	18.1	16.5
P-value	<0.001	<0.001

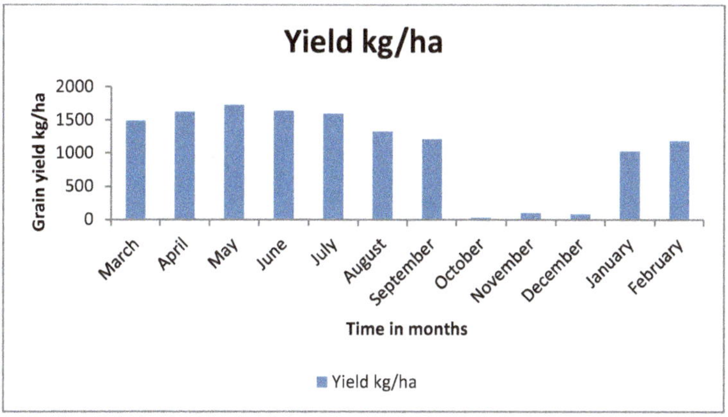

Fig. 1.1.1: Average yield performance of 20 sugar beans varieties grown at different times of the year across season.

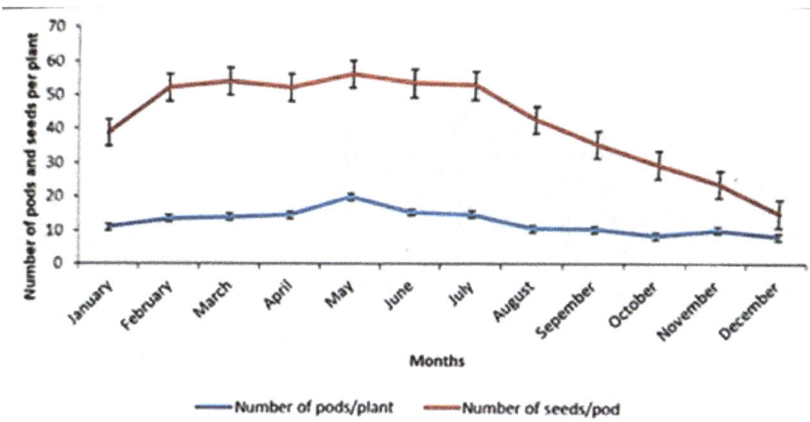

Fig. 1.1.2: Average number of pods per plant and average number of seeds per plant

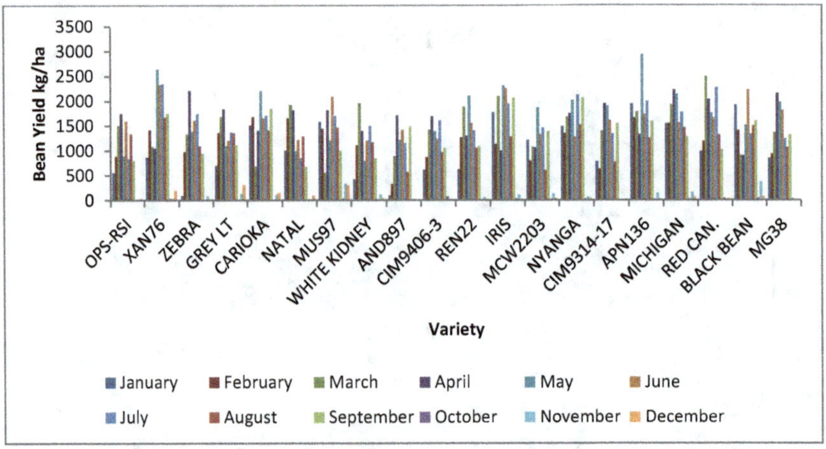

Fig 1.1.3: Performance of twenty varieties of beans in relation to sowing dates across season.

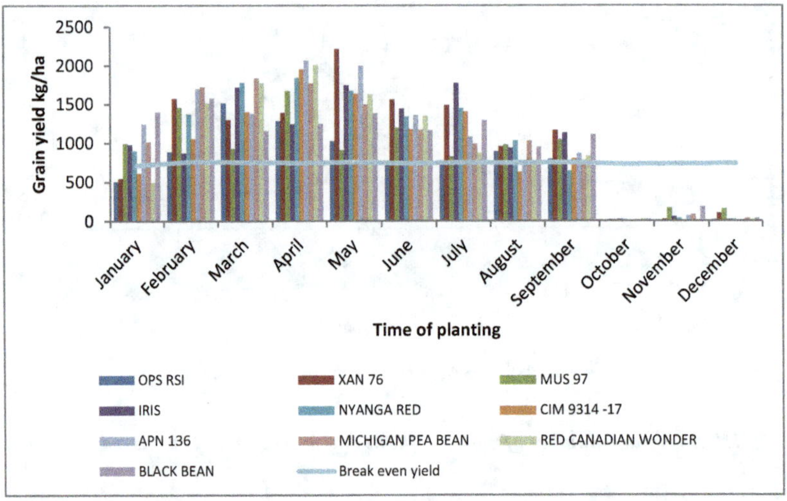

Fig 1.1.4: Break-even yield (kg/ha) of 10 best varieties of beans from the trial

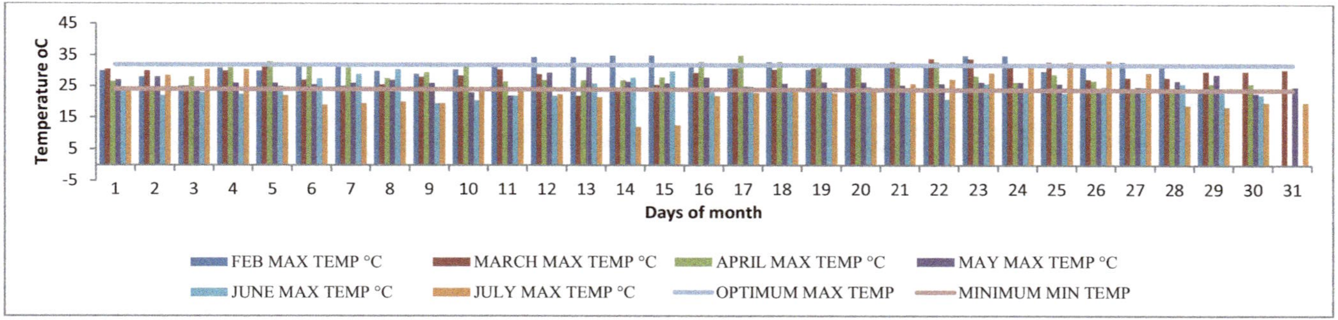

Fig 1.1.5: Maximum and minimum temperatures February to July year 1 in relation to optimum

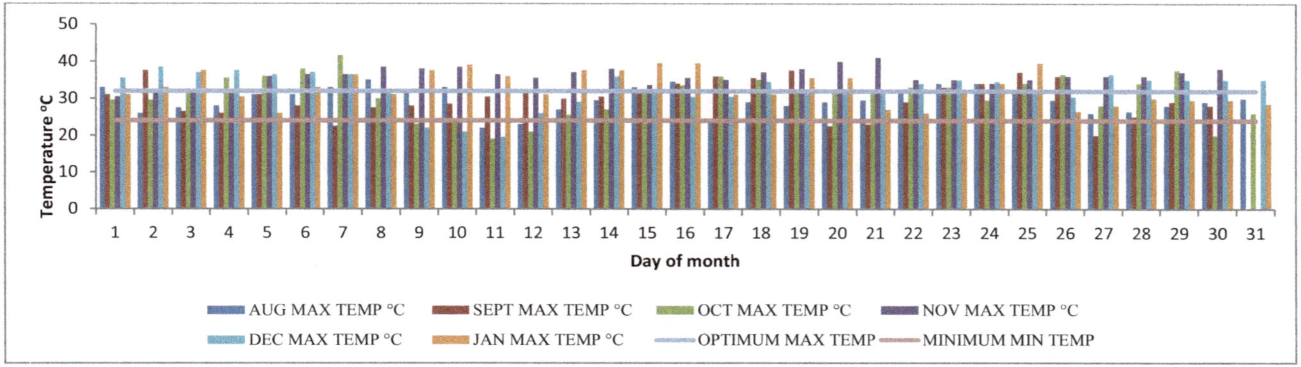

Fig 1.1.6: Minimum and maximum temperatures August to January year 1 in relation to optimum temperatures

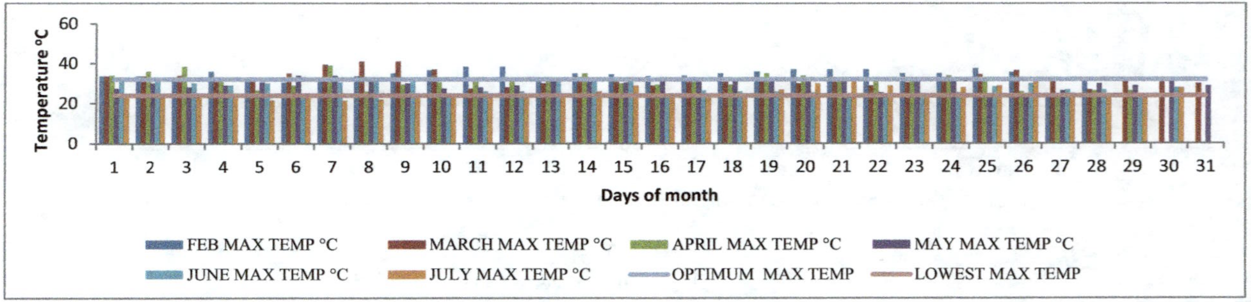

Fig 1.1.7: Minimum and maximum temperatures February to July year 2 in relation to optimum temperatures

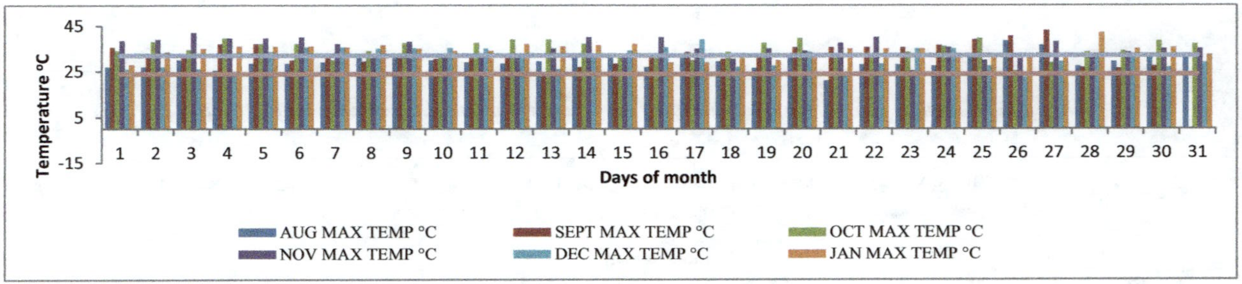

Fig 1.1.8: Minimum and maximum temperature August to January year 2 in relation to optimum temperatures

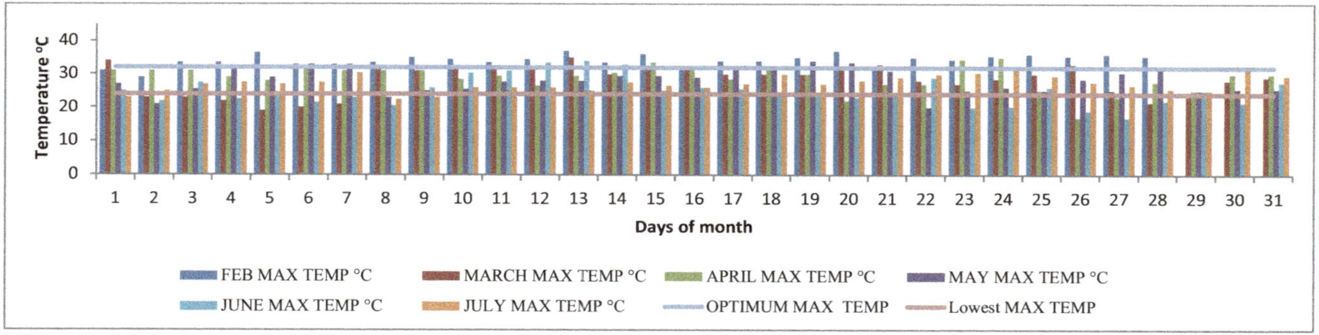

Fig 1.1.9: Minimum and maximum temperature February to July year 3 in relation to optimum temperatures

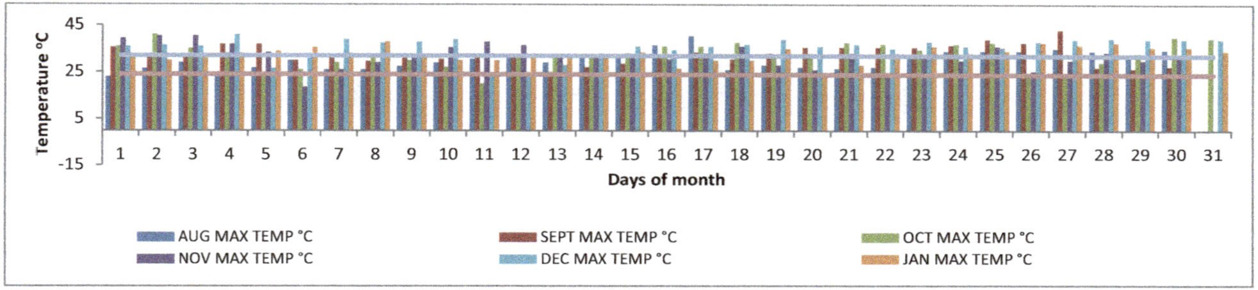

Fig 1.1.10: Minimum and maximum temperature August to January year 3 in relation to optimum temperatures

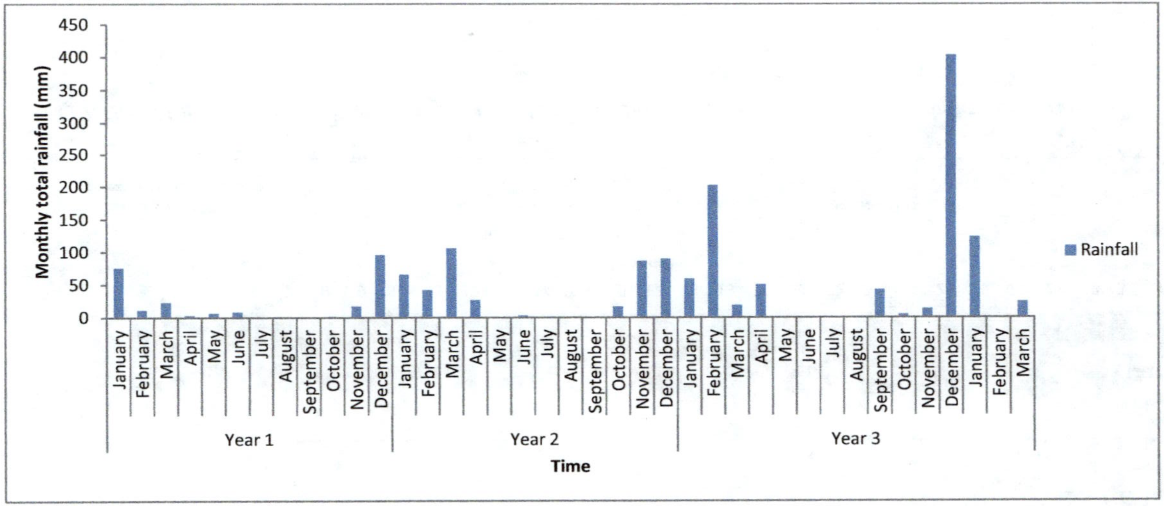

Fig 1.1.11: Monthly rainfall totals during the trial period

1.1.5 Conclusions:

Selection of best varieties, with appropriate sowing dates is a very important factor in increasing the productivity of common beans. The best sowing date was May with an average of 1718 kg/ha across twenty varieties. The best varieties are APN 136 (indeterminate), Michigan Pea Bean (determinate), Red Canadian Wonder (determinate), Xan 76 (indeterminate), Mug 38 (determinate) in that order. The most suitable time for planting was from March to July with yields ranging from 1482 to 1718kg/ha on average. However, there are some varieties that do well when sown in March and July. The highest yields of 2940 kg/ha came from APN 136 May sowing. Farmers should not sow a crop that would flower when temperatures are above optimum temperatures to avoid flower abortion. There is also need to avoid maturity of the crop during wet months to avoid diseases like bean rot and flowering during rainy months as flower shedding occurs due to heavy rain drops. Therefore, the results of this study have shown that different sowing dates and varieties had a significant positive influence on the yield of sugar bean.

Acknowledgments:

I would like to acknowledge Mr. Fambisai, for trial data collection and supervision of the trial. I would like to thank Mr. Charles Mhazo Head of the Institute for the resources that enabled us to carry out the research work. Author Contributions: Mtetwa prepared original draft, visualization, investigation and editing.

Funding: No funding agency

1.1.6 References:

1. Badoweic, A.; Weidner, S. 2014. Proteomic changes in the roots of germinating Phaseolus vulgaris seeds in response to chilling stress and post-stress recovery. J. Plant Physiol., v. 171, n. 6, p. 389-398, 2014.
2. Buruchara R, Chirwa R, Sperling L, Mukankusi C, Rubyogo J.C, Muthoni R, Abang M.M. 2011. Development and Delivery of bean varieties in Africa: The Pan-Africa bean research Alliance (PABRA) model. Afr. Crop Sci. J., v. 19, n. 4, p. 227-245, 2011.

3. Catuchi, T.A.; Guidorizzi, F.V.C.; Peres, V.J.S.; Dias, E.S.; Parmezan, G.C.; Galdi, L.V., 2019. Development and grain yield of common bean cultivars according to sowing season. Cientifica 2019 47, pp. 296–303.
4. Elhag A. Z., and Hussein A. M. 2014. Effects of Sowing Date and Plant Population on Snap Bean (Phaseolus vulgaris L.) Growth and Pod Yield in Khartoum State. Universal Journal of Agricultural Research 2(3): 115-118, 2014 http://www.hrpub.org DOI: 10.13189/ujar.2014.020306
5. Esmaeilzadeh and Aminpanah, 2015. Effects of Planting Date and Spatial Arrangement on Common bean (Phaseolus vulgaris) Yield under Weed-free and Weedy Conditions. Planta Daninha, Viçosa-MG, v. 33, n. 3, p. 425-432.
6. Essubalew Getachew Seyum 2014. Influence of plant spacing and date of sowing on yield and yield components of two snap bean (Phaseolus vulgaris L.) varieties in Jimma, Southwestern Ethiopia. Merit Research Journal of Agricultural Science and Soil Sciences (ISSN: 2350-2274) Vol. 2(7) pp. 086-095.
7. FAOSTAT (2014), FAO Statistics online database, "Production/crops – 'beans, dry', year 2014", Food and Agriculture Organization.
8. Getachew E., Tesfaye A. and Mulualem T. 2014. The effects of sowing date and spacing for yield of green bean (Phaseolus vulgaris L.) varieties at Jimma, Southwestern Ethiopia. Sky Journal of Agricultural Research Vol. 3(9), pp. 174 - 180
9. Karavidas I., Ntatsi G., Vougeleka V., Karkanis A., Ntanasi T., Saitanis C., Agathokleous E., Ropokis A., Sabatino L., Tran F., Iannetta P.P.M. and Savvas D., 2022. Agronomic Practices to Increase the Yield and Quality of Common Bean (Phaseolus vulgaris L.): A Systematic Review. Agronomy 2022, 12, 271. https://doi.org/10.3390/ agronomy12 020271
10. Madamba, C. at al, 2003. A guide to pulse production in Zimbabwe. Crop breeding Institute pp21.
11. Marco A, Rondon MA, Lehmann J, Ramirez J, Hurtado M. 2006. Biological nitrogen fixation by common beans. Biol Fert Soils DOI 10.1007/s00374-006-0152-z.
12. Marlene PB, Ana María González AM, Rodiño AP,De Ron AM, Marta SantallaM (2008). Effects of planting season and plant cultivar on growth, development, and pod production in snap bean (Phaseolus vulgarisL.) Aust. J. Agricultural Res., 59: 1121–1129.

13.Molosiwa O. O. and Kgokong S. B. 2017. Effect of planting date on tepary bean yield and yield components sown in Southern Botswana. African Journal of Agricultural ResearchVol. 13(4), pp. 137-143, 25.

14.Ndegwa AM, Muchui MN, Wachiuri SM, KARI-Thika JN (2001). Evaluation of introduced snap bean (phaseolus vulgaris l.) Varieties for adaptability and pod quality. Thika, Kenya.

15.Qiu-ying Zhang, Yan-sheng Li, Qing-luGao, Herbert SJ, Hashemi AM (2010). Influence of sowing date on phenological stages, seed growth and marketable yield of four vegetable soybean cultivars in Northeastern USA. African J. Agricultural [Res., 5(18): 2556-2562

16.Sharma, A. Singh, Y. and Sharma, K.C., 2013. Optimum sowing dates and varieties for seed productivity of Frenchbean (Phaseolus vulgaris L.) under north western Himalayas.

17.Sharma R., Jain M., Kumar S., Kumar V., 2014. Evaluation of differences among Vigna aconitifolia varieties for acquired thermotolerance Agril. Res., 3 (2) (2014), pp. 104-112

18.Stephen E, Beebe, Idupulapati M, Rao, Mura JD, Jose Polania. 2014. Common beans, biodiversity, and multiple stresses: challenges of drought resistance in tropical soils. Crop and Science Department, North Carolina State University, USA. Vance, C. P. Root-bacteria interactions: symbiotic nitrogen fixation. In: Waisel, Y; Eshel, A.; Kafkati, U. (Ed.) Plant roots: The Hidden Half, Ed 3. New York: Marcel Dekker, 2002. p. 839-867.

19.Walter T. 2014. Planting dates effects on growth, development and yield of selected agronomic crops at Makhathini Irrigation scheme.

20.Yoldas, F. and Esiyok, D. 2009. The Influence of temperature on Growth and Yield of Green Beans for Processing. International Journal of Agricultural Research. Vol. 4, Issue 3 pp. 124-130

1.2 Yield Assessment of Irrigated Summer and Winter Maize (Zea mays L.) to Soil and Foliar Applied Fertilizers under Lowveld Conditions of Zimbabwe

Researchers: **Mtetwa G., Dube T., Mutata M.*

1.2.1 Abstract

*In this era of massive intensification crop production, maintaining soils in a state of adequate fertility is among the basic tasks of crop production techniques. The role of macronutrients and micronutrients are crucial in maize production in order to achieve higher grain yields. To study the response of soil applied nutrients and multi-nutrients foliar fertilizers an experiment was conducted at Chiredzi Research Station Natural Region 5 of Zimbabwe. The objectives were to determine the response of maize to essential nutrients through soil and foliar fertilizer application and to determine the most suitable sowing season in the Lowveld of Zimbabwe. The experiment was designed in a randomized complete block design 2x3x3 factorial, with application of phosphate and potash, nitrogen and folia. Treatments were No application (control); 2% Omniboost; 2% Kelp-p-max; 75Nkg/ha; 75N+2%Omniboost; 75N+2% Kelp-p-max; 150N; 150N+2% Omniboost; 150N+2% Kelp-p-max; 80*P_2O_5 *+ 40*K_2O*; 80*P_2O_5 *+ 40*K_2O*+2% Omniboost; 80*P_2O_5*+40*K_2O*+2% Kelp-p-max; 80*P_2O_5*+40*K_2O*+75N; 80*P_2O5*+40*K_2O*+75N+2% Omniboost; 80*P_2O5*+40*K_2O*+75N+2% Kelp-p-max; 80*P_2O_5*+40*K_2O*+150N (standard); 80*P_2O_5*+40*K_2O*+ 150N+2% Omniboost and 80*P_2O_5*+40*K_2O*+150N+2% Kelp-p-max. Results indicated that plant height was influenced by season. Application of soil nitrogen influenced plant height. Winter crop showed significant 1000 seed weight differences among treatments while summer crop did not. Application of soil nitrogen influenced cob height. Significant cob lengths were shown in winter with application of soil nitrogen. Significant grain yield increases were recorded when folia were applied in combination with P, K and 150 kg/ha N. Application of P, K, 150 kg/ha N with kelp-P-max gave 34% yield advantage to 150 kg/ha N + kelp-P-max. Omni-boost appeared to have higher leaf colour score compared to Kelp-P-max when solely applied.*

Key words: foliar fertilizer, winter crop, summer crop, soil nitrogen, essential nutrients.

1.2.2 Introduction

Maize is one of the world's most important cereal crops, which has the widest distribution compared to other cereals (Erenstein O., 2022). It is the major source of carbohydrate in most meals in Zimbabwe (Maurawa et al., 2021; FAO, 2015). Nutrients have an important role in plant growth and development (Brankov et al., 2020). In most cases, when soil is optimally supplied with available nutrients, plants have a higher possibility to reach a genetic yield potential and quality (Marschner, 2012). Plant nutrition provides optimal conditions for growth and development which makes it an important part of an Integrated Pest Management strategy (Vasileidas et al 2011). Complete crop growth and development is influenced by essential nutrients. Lack of any essential nutrient, cannot complete the entire crop growing process to attain the optimum yields when all the other production inputs are not limiting because they are involved in metabolic functions required in the life cycle of the plant. Therefore, essential elements may not replace each other in the production process. The important fertilizer laws include nutrient return theories, equally important irreplaceable law and the law of the minimum nutrient. They state that crops in different ways absorb mineral nutrients from the soil thereby reducing soil nutrients (Tisdale et al., 1993). Continuous cultivation without replenishing nutrients will reduce soil nutrients. In order to maintain soil fertility, it is necessary to add used nutrients to the soil by fertilization methods. The second law is equally important, irreplaceable law which states that both large, medium, and trace elements are equally important and indispensable. The lack of certain trace elements, result in symptoms of deficiency and will result in reduced productivity (Njinga 2013). The importance of nutrient elements is the same. The third law is the law of the minimum nutrient. It states that yield of crops is decided by the minimum nutritive element (Van der Ploeg. 1999). Soil cannot supply all the nutrients required for high yielding maize production and therefore foliar application are essential. It is also important to know that not all the soil applied fertilizers become available to crops. Some are lost through leaching, erosion and the impact of pH on nutrient fixation. Therefore, there is need to apply foliar fertilizers.

Due to the fact that leaf tissue has the same morphological structure as root tissue (they originate from the meristem tissue), plants are able to rapidly absorb dissolved minerals (Martinka et al., 2014). Thus, soil deficiency in micro-elements can be successfully managed by foliar fertilizing. This could be the solution to the identified problem. Nutrient deficiency affects more than 3 billion people worldwide, causing so-called "hidden hunger" in people (Graham 2007).

With general fertilizer prices on the rise, many farmers are looking at ways to cut back on expenses and still produce satisfactory maize yields (Baligar et al., 2001). Modern technologies for maize production require soil fertilization to be combined with application of foliar fertilizers (Bruulsema et al., 2008; Chien et al., 2009). Foliar fertilization is a widely used practice to correct nutritional deficiencies in plants caused by improper supply of nutrients to roots (Shaaban, 2001). Intensive agriculture and use of high productivity cultivars led to a continuous decrease in soil micronutrient content. In addition, unfavourable soil and climate conditions cause the less availability of these nutrients to plant roots and addition of micronutrients to soil proved to be less beneficial (Hirschi, 2009). Fertiliser use in the developing world is low < 30kg per hectare and this is hardly enough to support high yields even to replace what the crops would have utilised. Two pre-formulated foliar fertilisers were used in this experiment. These were Kelp-P-max and Omni-boost. Kelp-P-Max contains a higher percentage of natural auxins and natural cytokinins, as well as some free amino acids, while Omini-boost contains Sulphur and magnesium in addition as the main element. The objectives of this study were to determine the response of maize to essential nutrients through foliar fertilizer application in both summer and winter season and to determine the effects of planting in summer and winter season using pre-formulated foliar fertilisers.

1.2.3 Materials and Methods

The trial was carried out at Chiredzi Research Station ($21^0 33'$ S and $31^0 30'$ E) at an altitude of 429 m above sea level in a 2x3x3 factorial design. Soil samples were taken before land preparation and analysed. Samples taken were put in Vikings paper pockets, air dried and sieved and sent to Chemistry and Soils Research for analysis. Soil texture was determined by the finger texturing method. Organic matter content was determined using the sulphuric acid potassium dichromate method. Soil pH was done with 1:5 Suspensions (soil: $CaCl_2$). Clay, sand and silt were determined using the sedimentation method by soil hydrometer. Mineral nitrogen was done using 0.01m KCL extract done by titrimetry. Phosphate analysis was done using the Resin extract on UV-V is spectroscopy. Potash, calcium, magnesium and sodium were done by Atomic Absorption Spectroscopy (Weaver et al. 1962). Fertilizer was applied as); Ammonium nitrate (AN) 34.5%N., Omni-boost and Kelp-P-Max.;

treatment combinations were per treatment requirement. Nitrogen, Potassium and Phosphates was applied as straight fertilizers and the essential elements applied as foliar which were pre-formulated and was applied at 5 weeks after planting. A variety SC 727 a long season and high yielding variety was used. Data collection on 1000 seed weight, cob heights, cob length, leaf colour and grain yield per plot were collected. Genstat version 14 was used for data analysis. Treatments were 80 kg/ha P_2O_5 (Single super phosphate);, 40 kg/ha K_2O (Muriate of potash as follows: Zero (nothing was applied); Omni-boost (foliar applied); Kelp-P-max (foliar applied); 75 kg N/ha; 75 kg N/ha +Omni-boost; 75 kg N/ha +Kelp-P-Max; 150 kg N/ha; 150 kg N/ha + Omni-boost; 150 kg N/ha +Kelp-P-Max; Phosphate +Potash (PK); PK+ Omni-boost; PK+ Kelp-P-Max; PK+150 kg N/ha; PK+150 kg N/ha +Omni-boost; PK+150 kg N/ha +Kelp-P-Max. The design was a 2 x 3 x3 factorial.

The two foliar fertilizers have different nutrient compositions (table 1). It is important to note that Kelp-P-Max has higher content of copper, zinc, boron, molybdenum and manganese compared to Onmiboost and vice versa for nitrogen, phosphate, and iron.

Table 1.2.1: Chemical composition of foliar fertilizers used in the experiment

Nutrient elements	**Kelp-P-Max**	**Omni-boost**	**Ammonium nitrate**	**Single superphosphate**	**Muriate of Potash**
magnesium	-	3.10%			
sulphur	-	4.20%			
nitrogen	3.50%	7.70	34.5% N		
phosphate	11.1%	39.60%		18% P_2O_5	
iron	2818mg/l	3480mg/kg			
copper	1409mg/l	40mg/kg			
zinc	1409mg/l	440mg/kg			
boron	2818mg/l	920mg/kg			
molybdenum	198mg/l	80mg/kg			
manganese	1409mg/l	720mg/kg			
free amino acids	1.74%	-			
natural auxins	2.13mg/l	-			
natural cytokinins	0.006mg/l	-			
Potash					60% K_2O

1.2.4 Results and Discussion

1.2.4.1Rainfall distribution and monthly mean maximum and minimum temperatures

Rainfall distribution for the experimental site from September to May for the two summer seasons is shown in fig. 1.2.1. The rainfall totals were 343.2 mm and 859.8 mm in 1st and 2nd year respectively. The first year depicted a very poor season in terms of rainfall amount with light rainfall right through the year and the second year had a high rainfall amount with good distribution. The amount of irrigation applied during the season was 650mm. Monthly mean maximum temperature could rise to 35°C in summer (fig. 1.2.2) while in winter it could reach 24^{0}C. The monthly mean minimum could also reach 9^{0}C while the monthly mean maximum reached 24^{0}C showing that the winter season in this area is mild and can be utilised for maize production.

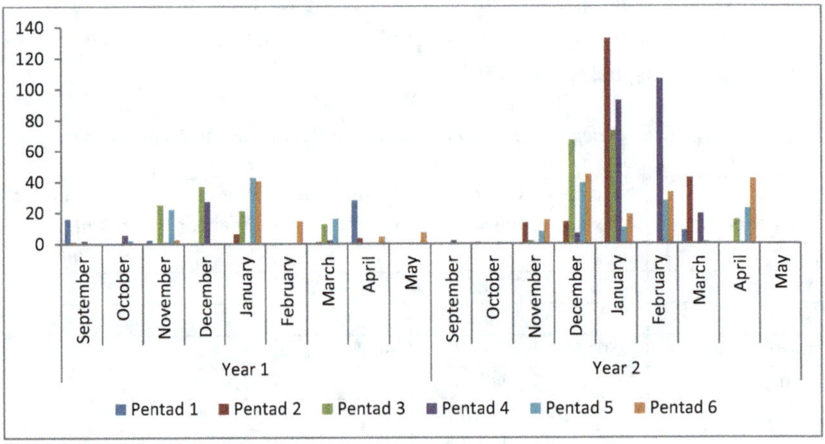

Fig. 1.2.1: Rainfall pentad distribution at Chiredzi Research Station for the two rainfall years

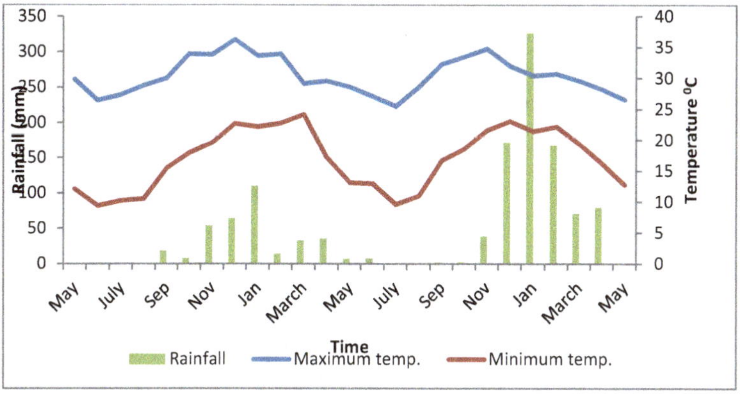

Fig. 1.2.2: Monthly mean rainfall and monthly mean maximum and minimum temperatures at Chiredzi Research Station

1.2.4.2 Leaf colour 2 weeks after topdressing and foliar application

Application of 75 kg N/ha or 150 kg N/ha with or without PK had significant leaf colour difference shown (fig. 1.2.3). Where soil nitrogen was not applied with or without PK leaf colour was significantly reduced. Application of soil nitrogen at 75 and 150 kg/ha with or without foliar increased leaf colour significantly compared to when nitrogen was not applied with or without foliar. Application of PK and N at 75 and 150 kg N/ha with or without foliar; application of N at both 75 and 150 kg/ha with or without foliar; and application of N without PK increased leaf colour score significantly. The results revealed that the effect of micronutrients leaf spraying was significant at the 5% level of probability on leaf colour chart. The increment of leaves greenness is explainable through foliar application of some essential micronutrients such as iron (Fe) that is involved with chlorophyll synthesis and in enzymes for electron transfer (Jones and Olson 2016). When soil nitrogen was applied at 150 kg N/ha, additional foliar application could not improve leaf colour significantly. However, significant leaf colour changes only appeared when foliar was applied with 75 Kg N/ha without PK. Application of foliar with or without P and K did not have significant effects on leaf colour. The summer crop showed significant high leaf colour differences compared to winter. This shows that summer has favourable conditions for maize growth and utilization of applied foliar fertilizers causing plants to be greener compared to winter. Sulphur is essential for many growth functions in plants including nitrogen metabolism, enzyme activity and protein and oil synthesis.

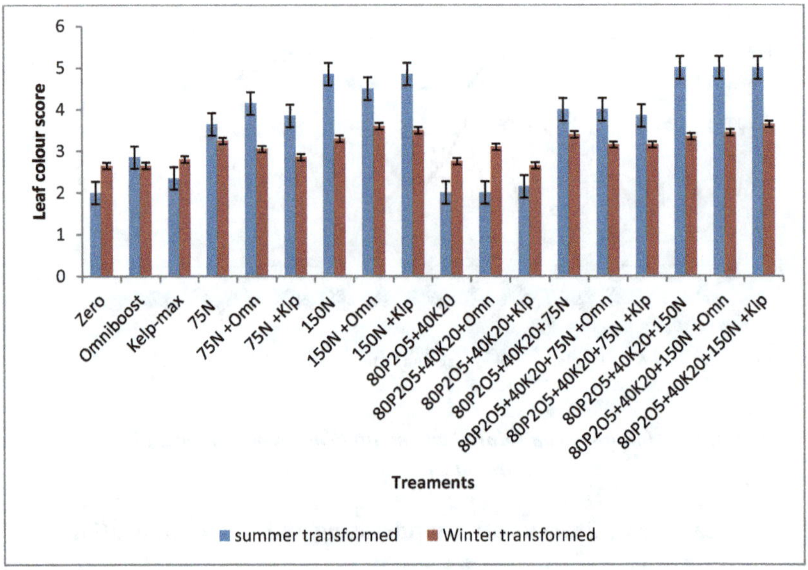

Fig. 1.2.2: Effects of applied soil and foliar fertilizers on Leaf colour

1.2.4.3 Plant heights

The height of plants is an important growth character directly related with the productivity potential of plants (Konuskan et al., 2019). Results showed that there were no significant differences on plant heights in year 1 and 2 for a summer crop Table 1.2.1. In winter significant plant height differences (p=0.033) were shown in the first year. Application of N with or without PK as basal showed significantly taller plants compared to where N was not applied in winter as top dressing table 1.2.2. However, significant plant height differences occurred across years in summer. Application of nitrogen (N) topdressing with or without Phosphate (P) and Potash (K) as basal exhibited significantly (p=0.04) taller plants compared to where N was not applied in summer. Across season analysis showed significantly taller plant height differences in winter. When N was applied as top dressing in conjunction with PK plant height increased significantly compared to application of N without PK in winter except when 150 kg/ha N was applied without PK. Application of folia did not show any significant differences on plant height in both years and seasons. However, on average summer plants were taller than winter plants. It has been reported by various researchers that foliar fertilizing has positive effects on plant heights (Brankov et al. 2020; Adhikari, 2019; Narimani, 2010)

1.2.4.4 1000 seed weight

Yield of maize is influenced by cob to grain yield ratio (Gabor et al 2017). The larger the weight of grain, compared to the weight of the cob, the higher the yield becomes. In this experiment there were no significant differences on 1000 seed weight for the summer crop shown in table 1.2.2. Winter crop showed significant differences on 1000 seed weight in year 1 (table 1.2.3) showing that there was better utilization of nutrient resources compared to summer. Application of foliar on its own did not show any increase in seed weight. Application of PK and folia did not show any increase in seed weight compared to application of NPK. However, in year 2 there were no significant weight shown on seed weight for a winter crop. Across season analysis for the summer and winter crop did not show significant differences on 1000 seed weight table 1.2.4. On average there were significant 1000 seed weight differences $p<0.001$ in both summer and winter. Application of N with or without PK significantly increased seed weight for the two years both summer and winter. This shows that there was more influence of soil applied nitrogen on seed filling. Application of folia on its own did not have a significant influence on seed filling.

Table 1.2.2: Response of summer maize to soil and foliar applied fertilizers on yield components

	Av. Plant	Av. 1000	Av. Cob	Av. Cob

1.2.4.5 Cob height

Year 2 in summer exhibited significant cob height differences p=0.016 table 1.2.1. Application of N with or without PK significantly influenced cob height compared to no N application. Year 1 did not have favourable conditions to have any significant differences on cob height. On average there were significant cob height differences p=0.016 with application of N with or without PK having significantly high cob height differences in the first-year p=0.023 table 1.2.3. Second year did not have any significant differences in winter. On average summer and winter seasons exhibited significant differences on cob placement height p<0.001 both summer and winter. In winter there were significant cob height differences in first year. Application of N with or without PK significantly p=0.023 increased cob height in first year table 1.2.3. In year 2 there were no significant differences in winter. On average there were significant cob height differences in winter with application of N with or without PK giving significant higher cob placement height p<0.00. Application of N with or without PK significantly influenced positively p<0.001 both summer and winter table 1.2.4.

1.2.4.6 Cob length

There were no significant differences on cob length in summer both first- and second-year table 1.2.2. However, significant differences occurred on average for the summer crop with application of N with or without PK significantly increasing cob length p<0.001. Application of N at 150 kg/ha and PK significantly produced the longest cob in summer. For the winter crop there were significant differences on cob length in first year p=0.001 table 1.2.3. Application of N with PK significantly increased cob length. However, in year 2 there were no significant differences p=0.76 on cob length table 1.2.3. Application of N with or without PK significantly influenced p<0.001 cob length for the winter crop. However, a combined analysis for both winter and summer showed no significant differences on cob length table 1.2.4. Application of PK + 150 kg N/ha + Kelp-P-Max increased cob length by 12% compared to 150 kg N/ha + Kelp-P-Max. Application of PK + 150 kg N/ha + Kelp-P-Max gave 7.8% increase in cob length compared to PK + 150 kg N/ha. Kelp-P-Max gave 10.7% increase in cob length compared to Omni-boost. The treatment results showed cob length increase compared to the control where no fertilizer was applied.

Table 1.2.3: Response of winter maize to soil and foliar applied fertilizers on yield components

Treatments	Plant height		Av.	1000 seed wt. (g)		Av. 1000 seed wt.	Cob height (cm)		Av. Cob height	Cob length (cm)		Av. Cob length
PK+150N +Kelp-P-	226^{abc}	282	254.0^{ab}	376^{abc}	405	390.5^{ab}	106^{abc}	135.3	120.7^{abc}	16.8^{ab}	18	17.40a

Table 1.2.4: Response of summer and winter maize to soil and foliar applied fertilizers on yield components

Treatments	Plant height (cm)		1000 seed weight (g)		Cob height (cm)		Cob length		Av. Grain Yield kg/ha	
	Summer	Winter	Summer	Winter	Summer	Winter	Summer	Summer	Summer	Winter
Zero (Control)	226.0^{ef}	191.0^{ghi}	336.5	304.0	91.40^{e}	69.85^{f}	13.10	12.70	1654	1120
Omni-boost	244.5^{bcde}	203.5^{fgh}	325.3	325.0	108.00^{e}	74.35^{f}	14.30	12.85	2257	1223
Kelp-P-max	240.0^{de}	194.0^{ghi}	337.3	324.5	99.35^{e}	67.35^{f}	14.00	13.10	2003	1478
75N	269.5^{abc}	250.0^{abcde}	364.5	354.0	122.30^{abcd}	110.15^{de}	16.45	16.55	6100	4291
75N +Omni-boost	257.0^{abcd}	212.5^{fg}	371.3^{j}	332.5	118.2^{abcd}	107.35^{e}	16.55	15.45	6179	4910
75N +Kelp-P-Max	257.5^{abcd}	212.5^{fg}	363.0	337.5	121.25^{abcd}	107.65^{e}	16.90	15.80	5534	4263
150N	261.5^{abcd}	244.5^{bcde}	407.1	386.5	117.65^{abcd}	120.65^{abcd}	17.00	18.15	7448	7768
150N +Omni-boost	264.5^{abcd}	258.0^{abcd}	411.3	381.0	116.3^{bcd}	123.65^{abc}	16.65	17.20	6668	7815
150N +Kelp-P-Max	267.5^{abcd}	255.5^{abcd}	404.5	390.5	125.3^{abcd}	122.5^{abc}	16.60	16.95	6818	7509
PK	252.5^{abcde}	178.0^{i}	341.0	321.5	102.6^{e}	65.00^{f}	13.35	12.40	2155	1013
PK+ Omni-boost	245.5^{bcde}	186.0^{hi}	344.1	305.5	92.2^{e}	71.50^{f}	13.05	12.30	2022	1451
PK+ Kelp-P-Max	251.0^{abcde}	204.5^{fgh}	337.1	309.5	103.2^{e}	79.35^{f}	14.90	14.05	2201	1654
PK+75N	264.0^{abcd}	242.0^{cde}	385.0	356.5	123.6^{abc}	119.50^{abc}	17.50	16.90	6033	6147
PK+75N +Omni-boost	276.0^{a}	266.0^{abcd}	394.6	367.5	130.45^{ab}	126.85^{abc}	17.10	16.10	6798	5404
PK+75N +Kelp-P-Max	263.0^{abcd}	252.0^{abcde}	371.5	345.0	118.75^{abc}	124.85^{abc}	16.75	16.15	5599	5551
PK+150N (Standard)	265.5^{abcd}	258.0^{abcd}	420.8	403.5	117.55^{abcd}	120.65^{abcd}	17.25	18.30	8400	8375
PK+150N +Omni-boost	272.0^{ab}	271.5^{ab}	414.3	404.0	125.45^{abc}	131.85^{a}	16.80	17.55	8418	8775
PK+150N +Kelp-P-Max	264.0^{abcd}	254.0^{abcd}	429.8	390.5	130.95^{a}	120.65^{abc}	18.60	17.40	9139	8269
Means	243.8		363.8		109.12		15.74		5068	
LSD	22.77		42.87		14.568		1.955		2239.4	
CV%	4.6		5.8		6.6		6.1		21.8	
SE	11.22		21.12		7.176		0.963		1103.1	
P-value	<0.001		0.997		<0.001		0.748		0.888	

1.2.4.7 Response of maize to soil and foliar applied fertilizers on grain yield

Mineral fertilizers incorporated into soil are subjected to leaching, impact of microorganisms, and other processes by which they become poorly available to plants (Araújo 2015; Lynch 2004). In this experiment results showed that there were significant differences p=0.032 on grain yield in first year whilst there were no significant differences in the second year. This can be ascribed to the favourable conditions that prevailed in the first year for better nutrient uptake when rainfall was not too much for leaching of nutrients (Gabor eta al 2017). In winter there were significant yield differences shown in year 1 whilst year 2 did not have significant differences. Application of PK+150+Kelp-P-Max gave 9139 kg/ha grain on average for the two years, giving 34% yield advantage to 150N+Kelp-P-Max in summer table 1.2.4. This shows that yield advantage emanated from the application of PK. Application of PK+150N+Kelp-P-Max gave 10% yield increase compared to PK+150N in winter. This yield advantage emanated from Kelp-P-Max which was applied. Foliar fertilizing as supplementary fertilization provides rapid absorption and utilization of nutrients, thus including absorbed elements directly in the metabolism (Bindraban et al., 2015). In this research positive effects of foliar application were observed with better results achieved by application of PK+150N+Kelp-P-Max in summer. Ryan 2002 concluded that foliar exhibited a secondary fertilizing role that determines a significant increase of productive consumption for soil elements and soil applied elements without substituting soil fertilization methods. Kelp-P-Max contains free amino acids which are the main product of organic N accumulation in plants (Masclauxlauxdaubrasse et al 2010). However, it was noted that without foliar fertilization the yield of maize in the experiment was relatively high 6100; 7448; 6033 and 8400 kg/ha in summer and 4291; 7768; 6147 and 8375 kg/ha in winter. This shows that this variety has good nutrient exploitation and utilization capacity. SC 727 is a high yielding variety and gave significant high yields in winter. This can be ascribed to moderate winter temperatures which are favourable for growth and development of this variety. In winter the crop had more grain filling period. In summer crop growth was rapid due to high heat units giving rapid crop development. Omni-boost gave 0.2% yield increase when applied in combination with PK+150N. However, there were no significant differences on year by season interaction p=0.888 (fig. 1.2.4). In winter second year there were significant yield increases when 150 kg/ha nitrogen was applied in combination with foliar and when PK and 150 kg/ha nitrogen was applied in combination with foliar fig 1.2.4. Without application of nitrogen PK and foliar did not have any significant increase in grain yield.

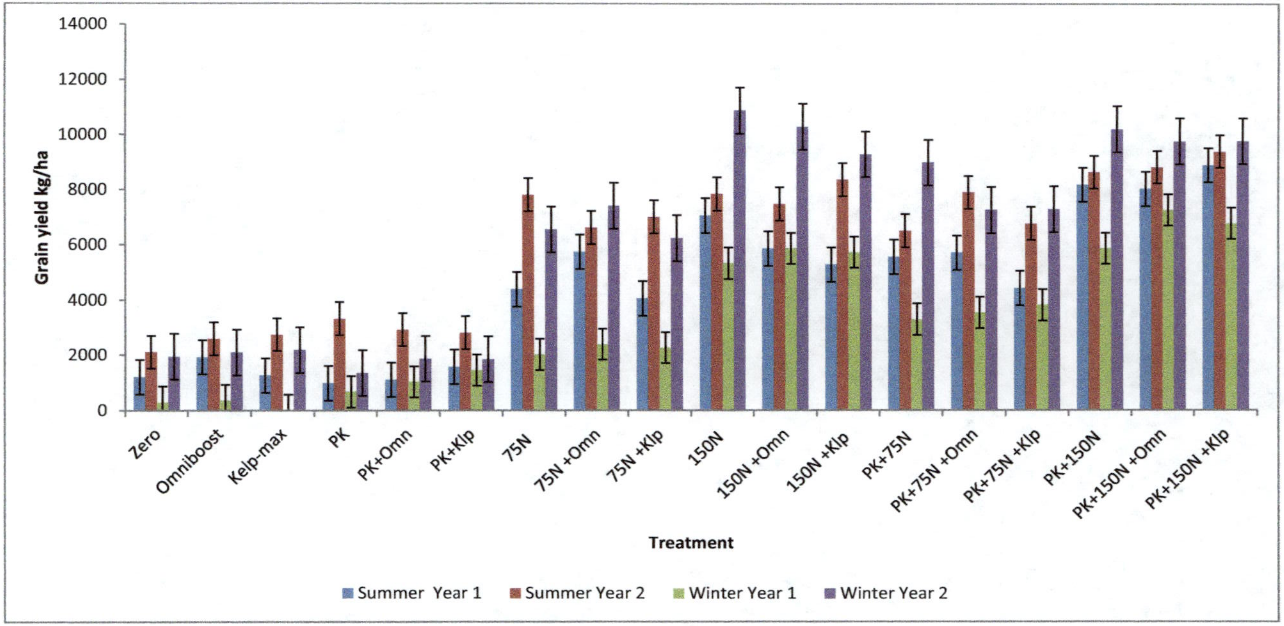

Fig. 1.2.4: Response of summer and winter maize grain yield to soil and foliar applied fertilizer

1.2.5 Conclusion

Based on obtained results, it can be concluded that foliar (Kelp-P-Max) fertilizer applied in combination with soil applied PK and N increased significantly yields of maize. Use of Kelp-P-Max was beneficial when applied in combination with PK and 150 kg N/ha. Winter maize can be as productive as summer maize in the South-East Lowveld of Zimbabwe. Application of soil P; K and N is very critical in both summer and winter in the South-East Lowveld of Zimbabwe.

1.2.6 Reference

1. Adhikari, B.; Dhungana, S.K.; Kim, I.D.; Shin, D.H. 2019. Effect of foliar application of potassium fertilizers on soybean plants under salinity stress. J. Saudi Soc. Agric. Sci. 2019.
2. Araújo, S.S.; Beebe, S.; Crespi, M.; Delbreil, B.; González, E.M.; Gruber, V.; Lejeune-Henaut, I.; Link, W.; Monteros, M.J.; Prats, E 2015. Abiotic stress responses in legumes: Strategies used to cope with environmental challenges. Crit. Rev. Plant Sci. 2015, 34, 237–280.
3. Baligar V. C., N. K. Fageria and Z. L. He, 2001. Nutrient use efficiency in plants. Commun. Soil Sci. Plant Anal., 32: 921–950.
4. Bindraban, P.S.; Dimkpa, C.; Nagarajan, L.; Roy, A.; Rabbinge, R. Revisiting fertilisers and fertilisation strategies for improved nutrient uptake by plants. Biol. Fertil. Soils 2015, 51, 897–911.
5. Brankov M., Simi M., Dolojanovic Z., Rajkkovic M., Mandi V. and Dragiˇcevi´c. 2020. The Response of Maize Lines to Foliar Fertilizing. Agriculture 2020, 10, 365; doi:10.3390/agriculture10090365.
6. Bruulsema, T., C. Witt and F. Garcia, 2008. A global framework for fertilizer BMPs. Better Crops, 92: 13-15.
7. Chien, S. H., L. I. Prochnow and H. Cantarella, 2009. Recent Developments of Fertilizer Production and Use to Improve Nutrient Efficiency and Minimize Environmental Impacts. Advances in Agronomy, 102:267- 321. DOI: https://doi.org/10.24925/turjaf.v7i10.1522-1525.2440
8. Gábor Péter Jakab, Zoltán-Dávid, Festő-Levente Komarek 2017. Investigation of Foliar Fertilization in Maize Production. Advanced Research Inlife Sciences 1(1), 2017, 1-6 .
9. Graham, R.D.; Welch, R.M.; Saunders, D.A.; Ortiz-Monasterio, I.; Bouis, H.E.; Bonierbale, M.; de Haan, S.; Burgos, G.; Thiele, G.; Liria,

R.;2007. Nutritious subsistence food systems. Adv. Agron. 2007, 92, 1–74.

10.Hirschi, K., 2009. Nutritional improvements in plants: time to bite on biofortified foods. Trends in Plant Science.

11.Jones C, Olson-Rutz K. 2016. Plant nutrition and soil fertility. Montana. St. Uni. Nutrient Management Module, 2016; 2: 4449–4452.

12.Konuşkan1 O., Yalçın M. and Gözübenli H. 2019. Effects of Foliar Applications of Boron at the Early Vegetative Stages on Plant Growth Parameters of Maize. Turkish Journal of Agriculture - Food Science and Technology, 7(10): 1522-1525,

13.Lynch, J.P.; St.Clair, S.B. 2004. Mineral stress: The missing link in understanding how global climate change will affect plants in real world soils. Field Crop. Res. 2004, 90, 101–115.

14.Marschner, P. Mineral Nutrition of Higher Plants, 3rd ed.; Elsevier Ltd.: Berlin/Heidelberg, Germany, 2012. https://www.sciencedirect.com/book/9780123849052/marschners-mineral-nutrition-ofhigher-plants.

15.Martinka, M.; Vaculík, M.; Lux, A., 2014. Plant Cell Responses to Cadmium and Zinc. In Applied Plant Cell Biology: Cellular Tools and Approaches for Plant Biotechnology, Plant Cell Monograph; pp. 209–246.

16.Maurawa K.L., Makaza K., Sakadzo N., Mubvuma M., Chidoko P. and Nyamushamba R. 2021. Bioactive leaf extracts of thorn apple (Datura stramonium l.) as an alternative to synthetic pesticides for fall armyworm (Spodoptera frugiperda (J.E Smith) management in maize (Zea mays L.). Octa Journal of Biosciences Octa J. Biosci. Vol. 9(2):99-105.

17.Narimani, H.; Rahimi, M.M.; Ahmadikhah, A.; Vaezi, B. 2010. Study on the effects of foliar spray of micronutrient on yield and yield components of durum wheat. Arch. Appl. Sci. Res. 2010, 2, 168–176.

18.Njinga R., L Moyo M.N., and Abdulmaliq S.Y. 2013. Analysis of Essential Elements for Plants Growth Using Instrumental Neutron Activation Analysis. International Journal of Agronomy. Volume 2013 Article ID 156520.

19.Shaaban, M., 2001. Effect of Trace-nutrient Foliar Fertilizer on Nutrient Balance. Pakistan Journal of Biological Sciences, 4: 770-774.

20.Tisdale S.L., Nelson W.L, Beaton J.D, and Havlin J.L.1993. Soil Fertility and Fertilizer, Prentice Hall, Upper Saddle River, NJ, USA, 5th edition, 1993.

21. Van der Ploeg, R.R., W. Böhm, and M.B. Kirkham. 1999. "On the origin of the theory of mineral nutrition of plants and the Law of the Minimum." Soil Sci. Soc. Am. J. 63-1055-1062.
22. Vasileidas, V.P.; Sattin, M.; Otto, S.; Veres, A.; Palinkas, Z.; Ban, R.; Pons, X.; Kudsk, P.; van der Weide, R.; Czembor, E.; et al. Crop protection in European maize-based cropping systems: Current practices and recommendations for innovative Integrated Pest Management. Agric. Syst. 2011, 104, 533–540.
23. Masclauxdaubresse, C.; Danielvedele, F.; Dechorgnat, J.; Chardon, F.; Gaufichon, L.; Suzuki, A. 2010. Nitrogen uptake, assimilation and remobilization in plants: Challenges for sustainable and productive agriculture. Ann. Bot. 105, 1141–1157.
24. Ryan, J. (2002), Fertilizers, applications, methods. Encyclopedia of Soil Sciences. Marcel Dekker Inc. New York. 553-556.

1.3 Evaluating 12 cassava varieties grown under irrigation for tuber yield and tolerance to cassava mosaic virus.

[1]*Godwin Mtetwa;* [2]*Tapiwa Nyakauru,* [2]*Fiona Robertson,* [1]*Terence Dube and* [1]*Rosemary Dube*

[1]*Chiredzi Research Station, P. O. Box 97, Chiredzi. Zimbabwe.*

[2]*University of Zimbabwe, P. O. Box MP 167, Harare, Zimbabwe.*

1.3.1 *Abstract:*

Cassava (Manihot esculenta Crantz) is becoming an important crop in the drier areas of Zimbabwe especially for household food security. Cassava varieties react differently to the changing environment when moved from one location to another. To address this challenge twelve newly introduced varieties namely Feyolami 1; Feyolami 2; I 30211; L9/304/4714 B2; 96/0113; TME 13; MM96/0016; I 96/1430; 91/0014; 96/0037; M 7 (local check) and XM6 (local check) were evaluated under irrigation at Chiredzi Research Station. Varieties were harvested at 12 months after planting. From the study Feyolami 1 and 2 showed stable tuber yields across years. Variety 96/0014 and 96/0037 gave higher yields above trial mean which were not significant but were susceptible to cassava mosaic virus disease. Mean yield of two best varieties Feyolami 1 and Feyolami 2 were 39.9 and 51.7 t/ha which was 32.6% and 70.8% above trial mean respectively. Seven varieties were tolerant to cassava mosaic virus disease. South African cassava mosaic virus and East African cassava mosaic virus-Uganda and Malawi strains were identified as the ones that infested the varieties. XM6 was the variety that had the highest score of 5 and infested with all the cassava virus strains.

Keywords: hydrogen cyanide, tuberous, cyanogenic glucosides, mottling, root yield

1.3.2 Introduction

Cassava (Manihot esculenta Crantz) is cultivated mainly in the tropic and sub-tropic regions of the world, over a wide range of environmental and soil conditions. It is very tolerant of drought and heat stress and produces well on marginal soils. As a subsistence crop, cassava is the third most important carbohydrate food source in the tropics after rice and maize, providing more than 60% of the daily calorific needs of the populations in tropical Africa and Central America (Richardson 2011). According to Sheela Immanuel, et al

(2024) cassava plays an important role in alleviating food problems, because it thrives and produces stable yields under conditions in which other crops fail. It is a crop that is generally grown on marginal lands with a minimum of agricultural inputs (Olabisi et al 2021). Once established, the cassava crop is given little attention, but still is able to tolerate weed competition, as well as insect pests and diseases. The potential exists for improving the productivity of cassava through better agronomic practices, superior varieties and pest and disease management.

Cassava varieties are generally distinguished from each other by their morphological characteristics which include leaf, stem and tuber colour, leaf shape, petiole colour and number of storage roots per plant. The plant produces all year round and can be harvested over an extended period of time. Although it is easily propagated by stem cuttings, the lack of quality planting material is a major constraint to the development of a viable cassava production industry. Eze and Ugwuoke (2010) reported that tuber yield of cassava is influenced by both the quality of planting material used and the agronomic practices employed. Access to high yielding cassava varieties and an improvement of the production system will result in increased economic benefits to local farmers. Cassava is a versatile crop and can be processed into a wide range of products such as starch, flour, tapioca, beverages and cassava chips for animal feed and pharmaceutical. Cassava is also gaining prominence as an important crop for the emerging biofuel industry and, as corroborated by Fathima et al. (2022) and Ziska et al (2009) is a potential carbohydrate source for ethanol production. A well-planned strategy for the development and utilization of cassava and cassava products can provide incentives for farmers, crop vendors and food processors to increase their incomes. It can also provide food security for households producing and consuming cassava products (Sheela Immanuel et al., 2024).

Cassava is as well gaining recognition in Zimbabwe mainly because of the change in climatic pattern and economic production of the crop. Movement of cassava from one location to another alters its cyanogenic potential (Kanaabi et al 2024), while different varieties react differently with changing environments (Ndungú et al 2012). This change in behaviour is attributed to climatic and soil factors Kanaabi et al. (2024). Cassava Mosaic virus Disease (CMD) is the most severe and widespread virus infection that affects cassava crops (Sadasivan et al 2016). It is caused by cassava mosaic Gemini-viruses (CMGs) (Geminiviridae: Begomovirus) and is the most important constraint to the production of cassava in Africa (Patil and Fauquet 2009). CMD which is considered as a major constraint for cassava cultivation is transmitted by white flies or by vegetative propagation (Oumar Doungouset al 2022). Typical symptoms of CMD consist of a yellow or pale green chlorotic mosaic of leaves, commonly accompanied by distortion and crumbling. CMD affected leaves appear reduced in size, twisted and distorted. CMD produces a variety of foliar symptoms that include mosaic, mottling, misshapen and twisted leaflets, and an overall reduction in size of leaves and plants. CMD-affected cassava plants produce few or no tubers depending on the severity of the disease and the age of the plant at the time of infection. However, it is not known for how long it will take for a virus free crop to reduce yield economically when infested with a virus.

A dearth of knowledge about the processing of cassava, and fears of cyanide (CN) poisoning, are some of the factors restraining the cultivation of cassava in Zimbabwe. To overcome that varietal constraint, new varieties with low cyanogenic potential were collected from Malawi for tasting their adaptability. In that vein the objectives of the trial were to select high yielding and low cyanogenic cassava varieties suitable for production in Zimbabwe and to select varieties that are tolerant to cassava mosaic virus disease.

1.3.3. Materials and Methods

A Randomized Complete Block Design with 3 replications was used with 10 clones from Malawi and 2 local checks namely Feyolami 1; Feyolami 2; I 30211; L9/304/4714 B2; 96/0113; TME 13; MM96/0016; I96/1430; 91/0014; 96/0037; M 7 (local check) and XM6 (local check). Planting was done on top of ridges 1m spaced and 1m intra-row spacing. Harvesting was done at 12 months after planting. Weeding was done using hand hoes and ridges were re-built from time to time whenever it was necessary. No fertilizers were used in the trials. Irrigation was given as supplementary. During dry season irrigation was given using 120 mm moisture depletion. Information of plant heights and number of tubers per plant were recorded. Control of pests was done using chemicals. Main problem pests were termites and mealy bug. Baboons and wild pigs were also problems. Data was analyzed using GenStat 14th Version and mean separation was done using Duncan multiple Range Test whenever necessary. The above ground mosaic symptoms were assessed by qualitative description of symptoms using a class scoring scale proposed by Hillocks et al. (1996) with class 1: no visible symptoms; class 2: slight foliar mosaic but no stem lesions; class 3: foliar mosaic with mild stem lesions but no die-back; class 4: foliar mosaic and pronounced stem lesions with beginning of die-back, and class 5: defoliation with pronounced die-back and stem lesions. Leaf samples were taken from plants that were showing and those that were not showing symptoms of Cassava mosaic virus for DNA extraction. DNA was extracted from cassava leaves using a modified method described by Doyle and Doyle, (1990). Briefly, leaves were sterilized using 70% ethanol and were ground by a mortar and pestle and were put in 1.5 ml microfuge tubes with 400 µl of CTAB extraction buffer (2% cetyl trimethylammonium bromide, 1% polyvinyl pyrrolidone, 100 mM Tris-HCl, 1.4 M NaCl, 20 mM EDTA, and 5% Beta mercaptoethanol) at 65 °C. Samples were mixed by flicking the tubes and were incubated for 20 min at 65 °C, with shaking every 5 min intervals. After incubation, 400 µl of chloroform isoamyl alcohol (24:1) was added and samples were shaken horizontally on a rotary shaker (200 rpm) at room temperature for 15 min. Samples were centrifuged at 10, 000 rpm for 5 min and the supernatant was transferred to a new tube and 400 µl of isopropanol was added and mixed thoroughly. Samples were incubated for 1 h at -20 °C and centrifugation was performed for 5 min at 10, 000 rpm. The supernatant was discarded and 400 µl of 100% cold ethanol was added and samples were centrifuged again for 5 min at 10, 000rpm. DNA was eluted in 100 µl of nuclease free deionized water. Extracted DNA was further purified using Genomic DNA Clean & Concentrator (D4010).

Degenerate primers TN.F (CKACGTCRTCAATGACGTTGTACC) and TN.R (ACCTTACATGGRCCTTCACAK) designed to amplify CMV strains were used in PCR. Each PCR contained 2.5 µL of 10X Standard Taq Reaction Buffer (Cat number: M0273), 1 mM magnesium chloride, 0.2 uM dNTP mix, 0.2 µM forward primer, 0.2 µM reverse primer, 0.125 µl Units Taq DNA Polymerase, gDNA template, and nuclease-free water to a final reaction volume of 25 µL. PCR conditions included an initial step of 95 °C for 10 min followed by 30 cycles of 95 °C for 30 s, 57 °C for 30 s and 72 °C for 30s, and 72 °C for 10 min. Five microlitres of PCR products were visualized using 1% agarose gel electrophoresis. DNA sequencing was performed using Oxford Nanopore sequencing according to manufacturer's instructions and bioinformatics analysis was performed to identify strains of CMV affecting cassava plants. Tuberous root samples were also taken for hydrogen cyanide (HCN) analysis (Usman-Macadaag, 2024) in the laboratory at the University of Zimbabwe.

1.3.4 Results and Discussion

1.3.4.1 Cassava mosaic virus tolerance (CMD) and hydrogen cyanide (HCN) content

The susceptibility or tolerance of cassava varieties to CMD depends on the genetic constitution of the varieties Stephen Kwame Torkpo and Emmanuel Amponsah (2023). Varieties Feyolami 1; I 30211; L9/304/4714B2; Feyolami 2; 91/0014; 96/0016 and TME 13 did not have virus symptoms as shown in table 1 below both scored in the field and analyzed in the laboratory. This shows that these varieties are tolerant to CMD because they have been growing in the same field for 12 years and being exposed to the virus by whatever means that the disease is known to spread. XM 6 was infested with all the viruses (East African Cassava Mosaic Virus (EACMV-Uganda), EACMV -Kenya and South African cassava mosaic virus (SACMV) strains. 96/0037 and 96/0113 were infested with EACMV-Uganda and SACMV. XM6 had the highest symptoms showing in the field when it was scored. I96/1430 had moderate score of 3. Variety M7 had score of 3 when it was scored in the field. However, it was negative when it was analyzed in the laboratory. Milcah Wambua (2024) classified cassava into three categories based on HCN content, namely non-bitter cassava (HCN <100 mg/kg of fresh cassava), bitter cassava (HCN between 100-450 mg/kg of fresh cassava), and very bitter cassava (HCN> 450 mg/kg of fresh cassava tubers). These varieties indicated that they are non-bitter varieties and are safe for consumption.

Table 1.3.1: Cassava variety Analysis for Cassava mosaic virus (CMV) using DNA sequencing and cyanogenic potential

Variety name	CMD visual assessment	CMD score	PCR (Degenerate primers)	Sequencing DNA-A	Sequencing DNA-B	HCN mg/kg fresh weight
96/0113	symptoms present	2	+ve	EACMV-Uganda	SACMV	22
96/0037	symptoms present	2	+ve	EACMV-Uganda	SACMV	4
Feyolami 1	No symptoms	1	-ve	-ve	-ve	10.9
XM6	Symptoms present	5	+ve	EACMV; EACMV-kenya; EACMV-	EACMV; EACMV-kenya; EACMV-	23

				Uganda; SACMV	Uganda; SACMV	
M7	symptoms present	3	-ve	-ve	-ve	5.3
I30211	No symptoms	1	-ve	-ve	-ve	8.2
I96/1430	Symptoms present	3	+ve	EACMV-Uganda	SACMV	5.6
L9/304/4714B2	No symptoms	1	-ve	-ve	-ve	20.5
Feyolami 2	No symptoms	1	-ve	-ve	-ve	10.8
91/0014	No symptoms	1	-ve	-ve	-ve	4.9
96/0016	No symptoms	1	-ve	-ve	-ve	11.9
TME 13	No symptoms	1	-ve	-ve	-ve	14.8

1.3.4.2 Plant height at 6; 9 and 12 months after planting

There was significant differences $p<0.001$ and at 0.025 in plant heights among varieties at 6 months after planting in second and third year respectively (table 1.3.2). Variety 91/0014 (152.3 cm), MM96/0016 (145 cm) were significantly the tallest in the first year, while 91/0014 was significantly the tallest in the second year at 6 months after planting compared to other varieties. In the first year there were no significant differences in plant height between variety 96/0014 and MM96/0016. In the second year there were no significant plant height differences among varieties Feyolami 1; TME 13; MM96/0016; Feyolami 2; 91/0014 and 96/0037. At 9 months after planting 91/0014 (206.7 cm) was significantly the tallest varieties in year 2. At 12 months after planting Feyolami 1 (221 cm), TME 13 (220 cm) and 91/0014 (213 cm) were significantly the tallest ($p<0.001$) varieties compared to the other varieties. This shows that 91/0014 is a vigorous growing variety. However, plant height was found to have no correlation with root yield. It had little or no direct linkage with root yield (Aina et. al. 2007).

1.3.4.3 Tuber diameter and number of tubers per plant

There were no significant differences on tuber diameter 12 months after planting (table 1.3.2) in year 2 and 3. However, in year 1 varieties with significantly the biggest tubers were Feyolami 2 (50 mm), 96/0037 (53 mm); 91/0014 (49 mm); 96/0113 (49 mm) and I30211 (50 mm) but with no significant differences among them. In year 2 and 3 there were no significant differences among varieties. There were significant differences in number of tubers per plant among varieties p=0.01; 0.011, 0.036 in year 1, 2 and 3 respectively (table 1.3.2). On average across years Feyolami 1 (7), I30211 (7); TME 13 (7); Feyolami 2 (7) and 96/0016 (7) had significantly the highest number of tubers per plant across the 3 years. Varieties 96/0037 and 96/0014 had tubers that were below average in two years out of three. However, variety Feyolami 1 had tubers above average across three years showing a variety that maintains the tuber sizes across different years. The number and size of tuber diameters were components of the main yield that contributed to the increase of cassava yield (Ntawuruhunga and Dixon 2010). Therefore, increasing root size and number of roots per plant would increase yield in cassava. It was reported by (Aina et al 2007) that increasing the number of tubers per plant without a corresponding increase in the tuber size will not increase tuber yield significantly.

1.3.4.4 Fresh tuber yield t/ha

Varieties Feyolami 2 (64.3t/ha), Feyolami 1 (51.7) and 96/0037 (49.5t/ha) in the first year had significantly (p=0.001) highest fresh tuber yield (table 1.3.3) showing no significant differences among the three varieties. In the second year, varieties Feyolami 2 (54.85 t/ha); 96/0037 (51.5 t/ha) and Feyolami 1 (38.5 t/ha) had significant the highest yields among other varieties. They also exhibited medium to large tuberous roots. Aina et al (2007) reported that medium sized roots showed the highest correlation coefficient with root yield, indicating that this character has a higher influence on root yield. In the third year, varieties Feyolami 2 (15.6 t/ha); Feyolami 1 (14.7 t/ha); I 30211 (13.67 t/ha); I96/1430 (12.8 t/ha); 96/0037 (12.3 t/ha) and L9/304/4714B2 (12.3 t/ha) had significantly the highest yields showing no significant differences among them. Variety 96/0037 performed well on tuber yield but was susceptible to CMD. Other varieties that were eliminated due to CMD susceptibility are 91/0113; 96/1430 and XM 6.

1.3.4.5 Dry tuber yield t/ha and Dry matter content

There were significant differences among varieties on p<0.001 with Feyolami 2 (23 t/ha; 20.5 t/ha and 5.695 t/ha) of dry matter in first, second and third year respectively giving the highest dry tuber yield. In the third year, there was no irrigation to an extent that the crop survived on rainfall. That was the reason why yields were low in the third year. Ntawuruhunga et al. 2001 and

Aina et al. 2007 Meneguccil et al 2024, reported that dry matter content of cassava roots was not correlated with storage roots yields. Meaning to say it is not an important indicator of storage roots in cassava. Amarullah (2020) also confirmed that dry matter was significantly negatively correlated with storage root weight.

Table 1.3.2: Plant heights, tuber diameter, number of tubers per plant and yield of 12 varieties of Cassava grown at Chiredzi Research Station

Variety	6	6	9 months	9 months	12	12	**Average tuber diameter**	**Average number of tubers**	**Av. Dry matter**
I.9/304/4714									

Table 1.3.3: Fresh and Dry tuber yield of 12 varieties across years at Chiredzi Research Station

	Fresh tuber yield t/ha			Dry tuber yield t/ha		
Variety	Year 1	Year 2	Year 3	Year 1	Year 2	Year 3
Feyolami 1	51.7^{ab}	38.46^{abcd}	14.47^{a}	17.1^{b}	12.02^{abc}	5.098^{abc}
I 30211	34.7^{bcde}	28.11^{cdef}	13.67^{ab}	13.4^{bc}	9.96^{bc}	4.749^{abc}
L9/304/4714 B2	40.0^{bcd}	9.35^{f}	12.29^{abc}	14.4^{bc}	2.764^{c}	4.442^{abcd}
96/0113	30.0^{de}	20.47^{def}	14.29^{ab}	9.2^{cd}	7.037^{c}	5.334^{ab}
TME 13	32.0^{cde}	21.08^{def}	7.18^{de}	12.6^{bcd}	8.347^{c}	2.547^{e}
MM96/0016	48.5^{abc}	32.25^{bcde}	6.23^{e}	14.7^{bc}	10.630^{abc}	2.215^{e}
I96/1430	28.5^{de}	26.25^{cdef}	12.80^{abc}	11.4^{bcd}	8.312^{c}	4.442^{abcd}
Feyolami 2	64.3^{a}	54.85^{a}	15.62^{a}	23.0a	20.5^{a}	5.695^{a}
91/0014	28.7^{de}	43.0^{abc}	9.65^{cde}	10.8^{bcd}	17.6^{ab}	3.357^{de}
96/0037	49.5^{ab}	51.5^{ab}	12.32^{abc}	15.8^{b}	16.44^{ab}	4.282^{bcd}
M 7	23.7^{de}	15.5^{ef}	10.61^{bcd}	8.7^{cd}	5.927^{c}	3.826^{cd}
XM 6	18.3^{e}	16.5^{ef}	9.14^{cde}	6.8^{d}	5.350^{c}	3.319^{de}
Trial Mean	**37.5**	**29.8**	**11.52**	**13.2**	**10.4**	**4.11**
CV	24.7	36.2	17.2	24.0	43.9	16.7
LSD	15.7	18.25	3.355	5.6	7.37	1.16
SE	9.25	10.78	1.98	3.2	4.35	0.686
P-value	<0.001	<0.001	<0.001	<0.001	0.007	<0.001

Table 1.3.4: Average tuber diameter, no. of tubers/plant, plant height and tuber yield across years

Variety	Tuber diameter	No of tuberous	6 months after	9 months after	12 months after	Fresh tuber yield t/ha	Dry tuber yield t ha

1.3.5. Conclusions

The study revealed that seven varieties namely Feyolami 1, 91/0014; 96/0016; TME 13; L9/304/4714B2; I30211; Feyolami 2 were tolerant to CMD (EACMV -Uganda and EACMV -Kenya and SACMV straits). Varieties Feyolami 2; I 30211 and Feyolami 1 were among the highest yielders and tolerant to CMD. Feyolami 1 yield was more stable across the years. Feyolami 1 and Feyolami 2 are the recommended varieties to grow under this environment.

1.3.6 Acknowledgments:

I would like to acknowledge Ms. Tapiwa Chitakunye for editing, Mr. Terence Dube, Mrs Rosemary Dube and Mr. Happyson Mashava for trial data collection and supervision of the trial. I would like to thank Mr. Charles Mhazo head of the Institute for the resources that enabled us to carry out the research work.

1.3.7 References

1. Aina, O.O., Dixon, A.G.O. and Akinrinde, E.A. 2007. Genetic variability in cassava as it influences storage root yield in Nigeria. J. Biol. Sci. 7: 765-770. https://doi.org/10.3923/jbs.2007.765.770

2. Alexandratos, N. ed. 1995. World Agriculture: Towards 2010: A FAO Study, Chichester, England. Food and Agriculture Organization of the United Nations; John Willey and Sons. pp. xxvi + 488.

3. Amarullah ID., 2020. Evaluation of quality and variety of Indonesian cassava (Manihot esculenta Crantz). Int. J. Agril. Res. Innov. Tech. 10(1): 108-116, June 2020. ISSN: 2224-0616.

4. Eze, S. C. and Ugwuoke, K. I. (2010). Evaluation of different stem portions of cassava (Manihot Esculentus) in the management of its establishment and yield. Research Journal of Agriculture and Biological Sciences, 6 (2): 181-185.

5. Fathima A.A., Sanitha M., Tripathi L., Muiruri S., 2022. Cassava (Manihot esculenta) dual use for food and bioenergy: A review. Food Energy Secur. 2022;00:e380. DOI: 10.1002/fes3.380

6. Hillocks R. J., Thresh J. M. and Bellotti A. C. (2002). Cassava; biology, production and utilization. New York: CABI Publishing.

7. Kanaabi Michael, Mukasa B. Settumba, Ephraim Nuwamanya, Nicholas Muhumuza, Paula Iragaba, Alfred Ozimati, Fatumah B. Namakula,Ismail S. Kayondo,Julius K. Baguma,Ann Ritah Nanyonjo, Williams Esuma and Robert S. Kawuki. 2024. Genetic Variation and Heritability for Hydrogen Cyanide in Fresh Cassava Roots: Implications for Low-Cyanide Cassava Breeding. Advanced Journals Plants Volume 13 Issue 9 10.3390/plants13091186.

8. Meneguccil N.C., Magali Leonel, Adalton Mazetti Fernandes, Jason Geter da Silva Nunes and Jesion Geibel da Silva Nunes 2024. Biomass yield and chemical composition of the cassava plant over two vegetative growth cycles. http://www.periodicos.uem.br/ojs/ISSN on-line: 1807-8621Doi: 10.4025/actasciagron.v46i1.63719

9. Milcah Wambua John Masani Nduko, Joseph Matofari Richard Mulwa, 2024. Nutritional, safety, and starch properties of flour from selected sweet cassava (Manihot esculenta Crantz) varieties grown in Nakuru County Kenya. Journal of Food Composition and Analysis Volume 135, November 2024, 106684

10. Ndungú J.N, Wachira F.N., Kinyua M.G., Lelgut D.K., Okwaro H., Njau P., Obiero H., 2012. Influence of Environment on Cassava Quality Triats in Central Rift Valleyof Kenta. American Journal of Plant Science Vol. 13 No. 10(2012) DOI:10.4236/ajps.2012.310181

11. Ntawuruhunga, P. and Dixon, A. 2010. Quantitative variation and interrelationship between factors influencing cassava yield. J. Appl. Biosci. 26: 1594-1602.

12. Ntawuruhunga, P., Rubaihaho, P.R., Whyte, J.B. A., Dixon, A.G.O. and Osiru, D.S.O. 2001. Inter-relationships among traits and path analysis for yield components of cassava: A search for storage root yield indicators. African Crop Sci. J. 9: 599-606. https://doi.org/10.4314/acsj.v9i4.27582

13. Olabisi Omolara Onasanya, Stefan Hauser, Magdalena Necpalova, Felix Kolawole Salako, Christine Kreye, Meklit Tariku, Johan Six, Pieter Pypers 2021. On-farm assessment of cassava root yield response to

tillage, plant density, weed control and fertilizer application in southwestern Nigeria. Field Crops Research Volume 262, 1 March 2021, https://doi.org/10.1016/j.fcr.2020.108038

14. Oumar Doungous, **B**outou Masky, Dopgima L Levai, Joseph AL Bahoya, Emile Minyaka, Jacques F Mavoungou, J Musembi Mutuku, Justin S Pita. 2022. Cassava mosaic disease and its whitefly vector in Cameroon: Incidence, severity and whitefly numbers from field surveys. Crop Prot. 2022 Aug;158:106017. doi: 10.1016/j.cropro.202 2.106017

15. Patil, B.L. and Fauquet, C.M. (2009) Cassava mosaic geminiviruses:actual knowledge and perspectives. Mol. Plant Pathol. 10, 685–701

16. Richardson K.V.A 2011. Evaluation of three cassava varieties for tuber quality and yield.

17. Sadasivan Nair Raji1 & Narayanan Subhash1,3 & Velumani Ravi2 & Raju Saravanan2 & Changatharayil N. Mohanan1 & Thangaraj MakeshKumar2 & Sukumar Nita1. 2016. Detection and Classification of Mosaic Virus Disease in Cassava Plants by Proximal Sensing of Photochemical Reflectance Index

18. Sheela Immanuel, D. Jaganathan, P. Prakash and Sethuraman Sivakumar 2024. Cassava for Food Security, Poverty Reduction and Climate Resilience: A Review. Indian Journal of Ecology (2024) 51(1): 21-31 DOI: https://doi.org/10.55362/IJE/2024/4191

19. Stephen Kwame Torkpo and Emmanuel Amponsah 2023. Challenges of Cassava Mosaic Begomoviruses, Cassava Brown Streak Ipomoviruses and Satellites to Cassava Production. DOI: 10.5772/intechopen.110103

20. Usman-Macadaag, O. 2024. Hydrocyanic acid (HCN) content of cassava in three different processing method. Global scientific journals. 12(2)

21. Ziska, L., Runion, G. B., Tomecek, M., Prior, S. A., Torbet, H. A. and Sicher. R. (2009). An evaluation of cassava, sweet potato and field corn as potential carbohydrate sources for bioethanol production in Alabama and Maryland. Biomass and Bioenergy 33:1503-1508.

1.4. Cassava in Sub-Saharan Africa: Origin, Agronomy, Disease Management and Future Priorities

Research Article: ISSN: 2643-671X

Journal of Agriculture and Horticulture Research

Tapiwa Nyakauru[1], Godwin Mtetwa[3], Ruvimbo Chimidzi[2] and Thomas Marangwana[2]*

[1]School of Agriculture and Food Science, University College Dublin, Belfield, Dublin 4, Ireland

[2]University of Zimbabwe, P.O. Box MP 167 Mt. Pleasant. Harare, Zimbabwe

[3]Chiredzi Research Station, P. O. Box 97, Chiredzi, Zimbabwe

**Corresponding Author*

Tapiwa Nyakauru, School of Agriculture and Food Science, University College Dublin, Belfield, Dublin 4, Ireland.

1.4.1 Abstract

Cassava (Manihot esculenta Crantz) is a foundational staple for millions of people in sub-Saharan Africa. Cassava is valued for its caloric density, drought tolerance, and flexibility across marginal environments. African production of cassava has expanded markedly since the year 2000. Nigeria, the Democratic Republic of Congo (DRC), Ghana, and Tanzania have been among leading producers. However, the production of cassava is heavily affected by various diseases such as cassava mosaic disease (CMD) and cassava brown streak disease (CBSD), and by structural challenges in seed systems, markets, and processing. This review focuses on current knowledge on the origin of cassava and domestication, ecophysiology and agronomy, climatic tolerances, the epidemiology and management of diseases, progress in genetic improvement, food safety, issues around cyanogenic glycosides, and priorities for the next decade. We highlight advances from IITA/CGIAR breeding programs, expanding surveillance including the use of AI-enabled tools, and integrated seed systems. We conclude with a forward-looking agenda spanning durable resistance pyramids, whitefly-vector ecology, climate-smart agronomy, and consumer-preferred quality traits.

Keywords: Cassava, Cassava Diseases, Climate-Smart Agronomy, Cassava Seed Systems, Genetic Resistance

1.4.2 Introduction

1.4.2.1. Cassava Domestication and Growth

Cassava is the third largest source of dietary carbohydrates in the tropics and a daily staple for >500 million people. Africa accounts for roughly half of global output, with Nigeria consistently the top producer of cassava, ≈60–63 Mt, followed by DRC and Tanzania among others [1-3]. Cassava was first domesticated in lowland South America from Manihot esculenta ssp. *flabellifolia*, likely in south-western Amazonia, within the last 10,000 years, placing it among the younger Neotropical domesticates [4-6]. Domestication of cassava involved careful selection for enlarged starchy storage roots with reduced cyanogenic toxicity that are palatable and suitable for human consumption [7]. Following domestication, cassava spread throughout the American continent, with evidence of cultivation in the Caribbean and Central America, prior to European continent. In the 16th century, Portuguese traders introduced cassava to Africa, where it rapidly became integrated into local food systems, particularly along the West African coast [4,8].

Cassava exhibits optimal growth and photosynthesis at 25-35 °C, with mean temperatures of 25-29 °C being most favourable, while physiological activity declines below this range [9]. Growth becomes inhibited at temperatures lower than 17 °C, primarily due to reduced photosynthetic efficiency and metabolic processes. Conversely, exposure to high temperatures above 37-40 °C leads to heat stress, causing impaired enzymatic activity, accelerated respiration, and potential leaf damage, which together reduce biomass accumulation and tuber formation [9,10]. Cassava is often cultivated in marginal environments, but the best performance of the crop is observed in well-drained, light-textured soils of moderate fertility [11-12]. Its root system is highly sensitive to waterlogging, making drainage essential for optimal root development. Furthermore, cassava demonstrates considerable tolerance to acidity, thriving across a pH spectrum of approximately 5.5-7.5 [11-13]. Growth is possible even at pH values as low as 4.5, although nutrient availability and uptake efficiency are enhanced within the slightly acidic to neutral range [11].

is generally drought tolerance, and this characteristic allows it to persist in environments receiving as little as 500 mm of annual rainfall. Cassava can also perform well under higher rainfall regions of up to 1500 mm per year, provided drainage is adequate [13-15]. Drought tolerance in cassava is due to its unique physiological mechanism traits, which is one of the primary reasons it has become a food security crop across much of sub-Saharan Africa [16,17]. These traits include rapid stomatal closure under both atmospheric and edaphic stress, maintenance of leaf water potential through osmotic adjustment, and slow depletion of soil moisture via an extensive but conservative root system [18, 21].

At the physiological level, cassava exhibits delayed leaf senescence during drought, followed by rapid recovery once rains resume, which helps sustain photosynthetic capacity over long growing cycles [22,23]. Proteomic and transcriptomic studies under long term water stress have revealed shifts in osmo-protective proteins such as dehydrins and Late Embryogenesis Abundant (LEA) proteins, enhanced reactive oxygen species (ROS) scavenging mechanisms, and reprogramming of carbohydrate metabolism, including increased mobilisation of stored starch to sustain root bulking [24-27]. These characteristics combine to confer high water-use efficiency and yield stability even in environments where annual rainfall falls below expected or where prolonged mid-season dry spells are common.

1.4.2.2. The Strategic Role of Cassava in Food Security

Global cassava production was reported at approximately 315–330 Mt in 2021–2022, with Africa's share increasing in response to growing demand for fresh roots, gari, fufu, flours, and industrial starch [28]. Cassava's importance in sub-Saharan Africa stems from multiple agronomic and socioeconomic advantages. It produces reliable yields on marginal, low-fertility, and acidic soils where other cereals often fail, making it a dependable crop for smallholder farmers [29, 30]. A study by which aimed to assess how climate change will affect cassava compared to other staple crops in Africa showed that cassava can experience neutral to positive changes in climate suitability across much of Africa (–3.7% to +17.5%), whereas other staples like beans, potato, banana, and sorghum can face significant declines [31]. Its flexible harvest window, ranging from 6 to 24 months, allows farmers to stagger harvests according to food needs, market demand, or labour availability, effectively acting as a food security buffer for many Africans [32].

In addition, cassava exhibits notable tolerance to drought and suboptimal rainfall, maintaining photosynthetic activity under water stress and recovering quickly after dry periods. In Thailand, a 60-day dry spell imposed early in the season reduced photosynthesis and starch accumulation, yet plants resumed growth and achieved substantial yields once rains returned, highlighting strong recovery capacity [33]. Similarly, multi-year trials in Kenya with 37 genotypes found that although drought reduced yields by up to 59% overall, certain lines maintained relatively high productivity (7.1 t/ha under drought) compared to susceptible ones (3.3 t/ha), confirming genetic variability in resilience and the cassava's notable tolerance to water stress [21]. Together, these findings reinforce cassava's potential as a drought-resilient crop capable of recovering quickly after dry periods.

Additionally, cassava's versatility extends across both food and industrial value chains. Cassava roots are consumed fresh, processed into traditional staples such as gari and fufu, or transformed into flours and starch [29, 34]. Cassava leaves are rich in protein, vitamins, and minerals contributing to household nutrition and providing a valuable feed source for livestock [35, 36]. On a dry matter basis, cassava leaves typically contain about 20–30% crude protein, with concentrates reaching nearly 49% [37, 38]. Mineral analyses across hundreds of genotypes report calcium levels of 3,600-17,600 mg/kg, potassium 3,100 27,000 mg/kg, and iron 43-660 mg/kg [39]. These attributes make cassava a valuable supplement for household nutrition and protein source for livestock. In Industry, cassava starch can be processed into ethanol, providing a renewable biofuel alternative to fossil fuels [40, 41]. Thailand operates 47 licensed bioethanol plants with a combined capacity of approximately 12.3 million litres per day, totalling about 3.7 billion litre per year. Cassava-based ethanol accounts for roughly one-third of this output, amounting to approximately 435 million litres annually [42]. Cassava starch is also increasingly used in the manufacture of biodegradable plastics, adhesives, paper, and packaging materials [44-46]. This offers environmentally friendly alternatives to petroleum-based products. These multifaceted uses enhance cassava's economic and nutritional value, making it an attractive crop.

1.4.2.3 Cyanogenic Glycosides in Cassava: Challenges and Solutions for Safe Consumption

Cassava tissues contain the cyanogenic glycosides (CGs) linamarin and *lotaustralin*, which, upon tissue disruption, undergo enzymatic hydrolysis by *linamarase* to release acetone cyanohydrin and ultimately hydrogen cyanide (HCN) [47,48]. Bitter varieties generally accumulate much higher levels of CGs than sweet types, with reported fresh root concentrations ranging from <20 mg CN^{-} kg^{-1} FW in sweet cassava to >300 mg CN^{-} kg^{-1} FW in bitter landraces [49]. Regulatory authorities, including Codex Alimentarius and

FAO/WHO, recommend a maximum limit of 10 mg CN^{-} kg^{-1} in edible cassava flour, while the acute reference dose is approximately 0.09 mg kg^{-1} body weight (cyanide equivalents) [50]. Exceeding these thresholds can pose serious health risks. Acute cyanide intoxication manifests as dizziness, vomiting, and, in severe cases, respiratory failure, while chronic low-level exposure has been linked to an irreversible spastic paraparesis in eastern and central Africa [51, 52]. Traditional African processing methods such as peeling, grating, pressing, fermenting, drying, and roasting are highly effective at reducing cyanogens by 80-95%, with some well-optimized systems achieving near-complete detoxification [53,54]. For example, gari production in Nigeria and Ghana involves fermentation and roasting that significantly lower residual HCN, while fufu production relies on wet fermentation that further enhances detoxification [55,56]. However, poorly executed or shortened processing steps such as inadequate fermentation time or insufficient drying can leave high cyanide residues, particularly in food sold in informal markets, and this will cause health implications to the consumers and may result in death [50,55,57]. Various concentrations of CN have been observed in processed cassava and are presented in Table 1.4.1.

Table 1.4.1: Cyanogenic Glycoside Content in Cassava Products Compared with Fresh Roots

Product	Residual HCN (mg/kg)	% reduction vs fresh root	Reference
Fresh root (bitter)	100–400	-	(JECFA, 2021.)
Gari	2–20	90–98%	(Padmaja & Steinkraus, 1995)
Fufu	5–25	85–95%	(Njankouo Ndam et al., 2019b)
Attiéké	3–10	90–97%	(Bouatenin et al., 2021)
Flour (well-processed)	<10	>95%	(Zhong et al., 2021)

It is important to note that routine monitoring of cyanide levels in cassava products is uncommon in most cassava producing countries because laboratory capacity is often limited, and smallholder farmers typically lack awareness of the health risks associated with inadequate processing [58-63]. Breeding programs aimed at developing low-cyanide cassava varieties have made significant progress in enhancing food safety [64, 65]. However, these efforts often encounter trade-offs, particularly concerning root bitterness and agronomic performance. Studies have shown that while low-cyanide cassava cultivars are essential for reducing health risks associated with cyanide toxicity, they may result in higher susceptibility to diseases, or reduced drought tolerance, further complicating breeding strategies. Therefore, breeding programs must carefully consider these trade-offs to develop cassava varieties that are both safe for consumption and acceptable to consumers [51, 66, 67].

1.4.2.4. Cassava Diseases: Mechanisms, Transmission, and Consequences

Cassava production in Africa is constrained by both viral and non-viral diseases. The diseases reduce yields, compromise root quality, and the generational ailments limit the availability of healthy planting materials like seeds and stalks for vegetative propagation [68]. The prevalence and severity of the diseases vary across regions, and their impact consistently affects yield potential, farmer income, and food security. The major diseases affecting cassava in Africa are summarized in Table 1.4.2 and their effects are shown in Figure 1.4.1.

Table 1.4.2: Major Cassava Diseases in Africa

Disease	Pathogen	Vecto	Epidemiology	Yield Loss	Reference
Cassava mosaic disease (CMD)	*Begomoviruses* (e.g., ACMV, EACMV)	*Bemisia tabaci*	West, East, Central Africa	70–100% in severe epidemics	(J. P. Legg et al., 2011) Combala et al. (2024)
Cassava brown streak disease (CBSD)	*Ipomoviruses* (CBSV, UCBSV)	*Bemisia tabaci*	East & Southern Africa	30–100% (root necrosis)	(Tomlinson et al., 2018)
Bacterial blight	*Xanthomonas axonopodis pv. manihotis*	Rain splash, contaminated cuttings	Central & West Africa	10–30%	(Fanou et al., 2018)
Anthracnose	*Colletotrichum gloeosporioides*	Wind, rain splash	Humid tropics	<20%	(Wydra & Verdier, 2002))

Figure 1.4.1: Symptoms of Major Cassava Diseases. (A) Cassava Mosaic Disease Symptoms [73]. (B) Cassava Anthracnose Disease Symptoms (C) Cassava Bacterial Blight Symptoms [74,75]. (D) Cassava Brown Streak Disease Symptoms [69-72,76]

1.4.2.5 Cassava Mosaic Disease

Cassava mosaic disease (CMD) is one of the most important viral diseases affecting cassava production in sub-Saharan Africa [68]. CMD is caused by several species of Begomoviruses of the family Geminiviridae, including African cassava mosaic virus (ACMV) and East African cassava mosaic virus (EACMV) [77]. The disease is primarily transmitted by the whitefly vector, Bemisia tabaci in a persistent, circulative manner. In addition, it can also spread efficiently through infected stem cuttings used for vegetative propagation [78].

The symptoms of CMV infection includes a mosaic or mottled yellow-green discoloration of leaves, leaf distortion, and overall stunted growth of the plant [79]. In susceptible cultivars, yield losses can range between 50% and 90% under conditions of high disease pressure. In Nigeria, yield loss assessments have shown that infection by CMV can cause reductions of up to 70 %, depending on cultivar susceptibility, infection timing, and viral strain [80,81]. In extreme cases, especially when infection occurs early in plant growth or when mixed infections of ACMV and EACMV are present, losses may approach 100% [69,82]. Similar trends have been reported across other cassava-producing countries in sub-Saharan Africa. In Uganda, severe CMD outbreaks involving mixed infections of ACMV and EACMV-UG have led to root yield losses of up to 82%, whereas single infections caused smaller but still substantial reductions (42–68 %) [83,84]. In Tanzania, field evaluations reported yield losses ranging between 72% and 100% for susceptible landraces under heavy disease pressure [85]. Furthermore, surveys in Kenya and Benin revealed that CMD, often in combination with cassava brown streak disease (CBSD), reduced yields by 25-95 %, depending on cultivar and agroecological zone [83,86]. More recently, coinfection of various CMV strains such as ACMV, SACMV, and EACMV have been observed in cassava plants in Zimbabwe [87].

The magnitude of yield loss due to CMV infection depends on multiple factors, including the strain involved, the host cultivar's genetic resistance, the time of infection, vector population dynamics, and environmental conditions. Early infections, mixed infections, and the cultivation of susceptible varieties exacerbate losses, whereas the use of resistant cultivars and virus-free planting materials remain the most effective management strategies [81,85]. The loss of yield due to CMV pose a serious threat to food security and livelihoods of smallholder farmers who rely heavily on cassava as a staple crop, and continuous monitoring and deployment of CMD-resistant genotypes are essential to sustain cassava productivity in sub-Saharan regions.

Management of CMD involves an integrated approach that includes the use of resistant or tolerant cassava varieties, phytosanitation practices such as removing and destroying diseased plants, and vector management [90,91]. Regional efforts, including breeding programs and community-based clean seed systems, are critical in reducing the impact of CMD in countries such as Kenya, Malawi, Mozambique, Tanzania and Uganda.

1.4.2.6 Cassava Brown Streak Disease

Cassava brown streak disease (CBSD) is caused by two closely related Ipomovirus species Cassava brown streak virus (CBSV) and Ugandan cassava brown streak virus (UCBSV) within the family Potyviridae. Both viruses are transmitted semi-persistently by the whitefly vector B. tabaci [72, 92]. In addition, cuttings can accelerate the spreading of the disease by planting diseased stalks, allowing CBSD to persist and expand rapidly through vegetative planting materials [93].

Historically, CBSD was first reported in Tanzania in the 1930s along the coastal belt near the Indian Ocean [94]. For several decades, it was largely confined to low-altitude coastal regions of Tanzania, Kenya, and Mozambique. However, since the early 2000s, the disease has undergone a major geographical expansion into inland East Africa, including the Lake Zone of Tanzania, Uganda, Rwanda, Burundi, Malawi, and parts of Democratic Republic of Congo [95,96,71]. This expansion coincided with the spread of B. tabaci super-abundant whitefly populations and the widespread exchange of infected cassava planting materials. In Uganda, CBSD was first detected around 2004 in coastal districts bordering Lake Victoria, but by 2010 it had spread to nearly all major cassava-growing regions [7,79]. Similarly, in Tanzania, severe epidemics have been recorded across the Lake Victoria basin, where the disease co-occurs with CMD, leading to devastating yield losses [98]. In Kenya, CBSD spread from coastal areas to the central and western highlands during the 2000s, with incidences exceeding 60% in some districts [99]. The disease has also been confirmed in Rwanda, Burundi, and Malawi, where surveys reported incidences between 20% and 70% in newly affected zones [100,101]. More recently, CBSD has been reported in Mozambique and Zambia, marking its southernmost distribution [98,102-104].

While foliar symptoms of CBSD are often variable and may include chlorotic mottling, vein yellowing, and leaf distortion, the most economically devastating symptom occurs in the storage roots. Infected roots develop necrotic, corky brown lesions that render them unmarketable and unsuitable for consumption or industrial use [93,72]. This root necrosis often occurs even in plants showing mild or no leaf symptoms, making the disease particularly insidious. Yield loss estimates vary depending on cultivar susceptibility, viral strain, and infection timing but can be extremely high up to 100% [95,98]. Both CBSV and UCBSV contribute to the disease, although CBSV is typically associated with more severe symptoms and greater yield losses [95]. Mixed infections of CBSV and UCBSV are frequently observed, compounding disease severity. Genetic diversity studies have shown that CBSV tends to dominate in lowland and coastal regions, while UCBSV is more prevalent in higher-altitude inland areas [105]. Management of CBSD remains a major challenge because resistant varieties are limited and the disease can be transmitted through symptomless cuttings. The International Institute of Tropical Agriculture (IITA) and national breeding programs in Tanzania, Uganda, and Kenya have developed moderately resistant cultivars (e.g., NAROCASS 1-4 and Kiroba), but resistance is often incomplete and can break down under high inoculum pressure [91]. Integrated disease management strategies emphasizing clean seed systems, vector control, and community Phyto sanitation are therefore critical to curbing CBSD spread [71,98].

CBSD now poses a continental threat, with its expansion into southern and central Africa raising major concerns for regional food security. Unless effective management and resistant cultivars are widely deployed, the combined effects of CBSD and CMD could severely undermine cassava production across sub-Saharan Africa in the coming decades [85].

1.4.2.7 Cassava Bacterial Blight

Cassava bacterial blight (CBB) is caused primarily by the bacterium Xanthomonas axonopodis pv. manihotis (Xam), a gram-negative, rod-shaped pathogen that infects the vascular tissues of cassava. CBB has been reported throughout cassava-growing regions of Africa, South America, and parts of Asia, with some of the most severe epidemics recorded in West and Central Africa. The disease can cause catastrophic yield losses, reaching up to total crop failure under conducive environmental conditions, particularly in areas with high humidity and rainfall [73,106].

The disease typically begins with angular, water-soaked lesions on young cassava leaves. These lesions are restricted by the veins and gradually enlarge, becoming necrotic and dark brown, usually bordered by yellow or chlorotic halos [106]. As infection progresses, the lesions coalesce, forming large, blighted areas that lead to premature leaf drop. A characteristic feature of the disease is the exudation of a sticky, amber or creamy-white bacterial gum from infected leaf veins, petioles, and stems, especially during humid weather [73]. In later stages, the infection spreads through the petioles and into the stem, causing vascular browning, wilting, and dieback. In advanced cases, systemic infection reaches the vascular tissues of roots and cuttings, leading to complete wilting and plant death. When infected cuttings are used for propagation, the disease spreads rapidly within and between fields [73,107].

In Nigeria, CBB has long been recognized as a major constraint to cassava production, particularly in humid agroecological zones of the southwest and southeast. Historical reports indicate that epidemics in the 1970s and 1980s led to yield losses of 75-100% in susceptible cultivars, wiping out entire fields in wet years [74,108]. More recent field surveys have shown that disease incidence remains high, ranging from 35-70% in states such as Oyo,

Ondo, and Enugu, with severity influenced by variety, environmental conditions, and poor planting material hygiene [109]. The continued reliance on infected cuttings and the lack of widespread adoption of resistant varieties have contributed to the persistence of CBB in Nigeria.

Similarly, in Uganda and Rwanda, epidemics have caused significant yield losses exceeding 70% in traditional landraces. The disease tends to intensify during rainy seasons when high humidity promotes bacterial multiplication and spread [110]. In the DRC, CBB is also widespread, particularly in regions such as Kisangani and Bas-Congo, where conditions favour disease development. Reports from these regions have documented yield losses of 60-100%, with severe epidemics causing total field destruction [106].

The epidemiology of CBB is closely linked to environmental and agronomic factors. The bacterium spreads mainly through rain splash, insects, wind-driven rain, and most importantly, through the use of infected stem cuttings as planting material. Warm and humid conditions, dense crop canopies, and poor drainage significantly enhance disease development. Once introduced into a field, the pathogen can survive for extended periods in plant debris and on alternative host species, serving as reservoirs for subsequent infections [74,106]

Management of CBB relies primarily on integrated control strategies rather than chemical treatments, which are largely ineffective against systemic bacterial infections in cassava. The use of resistant varieties is considered the most sustainable approach. In Nigeria, improved cultivars such as TMS 30572, TMS 92/0326, and NR 8082 have shown moderate resistance to CBB [111,112]. Other critical practices include Phyto sanitation, where farmers are encouraged to rogue infected plants and use certified disease-free planting materials. Crop rotation and fallowing are also recommended to reduce inoculum levels in the soil. Recent advances in molecular genetics and pathogenomics have enhanced understanding of the pathogen's virulence mechanisms and have facilitated the breeding of cassava varieties with broad-spectrum and durable resistance [82,113].

Despite ongoing research and management efforts, cassava bacterial blight continues to pose a serious threat to cassava production, particularly in Nigeria, Uganda, and the DRC, where cassava is a key food and industrial crop. The disease's ability to spread through vegetative propagation, its wide host range, and the emergence of new virulent strains complicate control efforts. Strengthening farmer awareness, implementing clean seed systems, and expanding resistant breeding programs remain essential strategies for mitigating the impact of CBB on cassava production and food security across Africa and beyond.

1.4.2.8 Cassava Anthracnose Disease (CAD)

Cassava anthracnose disease (CAD) is caused primarily by Colletotrichum gloeosporioides (Penz.) Penz. & Sacc., although other Colletotrichum species such as C. truncatum and C. dematium have occasionally been reported in association with the disease in some regions [114,115]. The disease affects all above-ground parts of the cassava plant, including stems, petioles, and leaves, and is recognized as one of the most important fungal diseases limiting cassava production in tropical Africa. The disease reduces root yield and starch content, directly impacting food and industrial starch supply chains. CAD is favoured by warm and humid conditions, making it particularly destructive in areas of high rainfall and dense canopy cover, such as those found in West and Central Africa [116].

Symptoms of anthracnose usually begin as small, dark-brown necrotic lesions on tender stems, petioles, and young leaves, often developing after mechanical injury or insect damage. As infection progresses, the lesions expand and may girdle the stem, leading to dieback, shoot wilting, and defoliation. In severe cases, the affected stems break at the lesion site resulting in the death of the entire shoot. Under moist conditions, pink to orange masses of fungal spores appear on the lesion surfaces, which serve as inoculum for further disease spread [117,118]. The pathogen can also survive on infected planting material, crop residues, and wild cassava relatives, serving as sources of secondary infection.

In Nigeria, studies in southwestern and southeastern Nigeria have shown disease incidences ranging from 25% to 70%, depending on environmental conditions and cultivar susceptibility [117,119]. Yield reductions of at least 30% have been reported in highly susceptible varieties such as TMS 30572 under severe outbreaks [120]. In Ghana, cassava anthracnose is also widespread, especially in humid regions such as the Ashanti and Eastern Regions, where rainfall exceeds 1,500 mm annually. Yield losses in these regions range between 20% and 50%, depending on the severity of infection [121,122]. In Benin and Cameroon, anthracnose epidemics have been reported during periods of heavy rainfall, where poor field sanitation and the use of infected cuttings exacerbate disease spread [73,121]. Similarly, outbreaks in Sierra Leone and Côte d'Ivoire have been linked to the introduction of highly susceptible cassava varieties and the lack of awareness among smallholder farmers about the disease's management [123].

The impact of cassava anthracnose extends beyond yield reduction to the quality and availability of planting material. Since the disease primarily attacks stems and petioles, it compromises the health and viability of stem cuttings used for propagation. This results in a shortage of clean planting materials, slowing down the re-establishment of crops after harvest. In areas where cassava leaves are consumed as leafy vegetables or used as livestock feed, anthracnose infection also reduces the nutritional value and palatability of foliage, indirectly affecting household nutrition and animal production [124,125]. Repeated severe epidemics can thus lead to food shortages and income decline, especially among subsistence farmers who rely on cassava as their main staple and source of livelihood.

The economic and social consequences of anthracnose are especially severe in smallholder systems where cassava serves multiple purposes. In regions where cassava is central to rural economies, recurring anthracnose outbreaks have contributed to economic vulnerability and household poverty. Losses in root yield and quality translate into reduced marketable output, while damaged stems limit access to quality planting material for the next production cycle. Consequently, the disease undermines food security and agricultural resilience across cassava-growing regions of Africa [111].

The persistence of anthracnose in Africa is largely due to poor management practices, limited farmer awareness, and inadequate adoption of resistant varieties. Most smallholder farmers continue to plant infected cuttings and lack access to extension services that promote integrated pest and disease management. Effective control strategies include the use of resistant or tolerant varieties, such as TMS 92/0326 and TMS 92/0057 in Nigeria, combined with field sanitation, crop rotation, and timely removal of infected debris to minimize inoculum build-up [111,112]. Research has also shown that fungal antagonists such as *Trichoderma harzianum* and biological control agents can help suppress Colletotrichum populations in the soil and on plant surfaces [126,127]. However, large-scale adoption of these practices remains limited, highlighting the need for increased farmer training, improved access to resistant planting materials, and regional surveillance programs.

1.4.2.9 Integrated Disease Management

Integrated Disease Management (IDM) represents the most sustainable framework for controlling cassava diseases because it reduces reliance on any single method and targets multiple mechanisms involved in cassava infection [128]. Effective IDM for cassava combines the use of genetically resistant or tolerant varieties, pathogen-free seeds, agronomic and phytosanitary practices, targeted vector management, and community-level coordination to achieve area-wide suppression of disease [129-131]. By integrating these complementary strategies, IDM improves efficacy, delays resistance breakdown, and enhances resilience of smallholder production systems.

1.4.2.10 Host Resistance and Varietal Improvement

Conventional breeding has delivered cultivars with partial to strong resistance to a range of cassava diseases, while genomic tools such as MAS, genomic prediction, and gene editing are accelerating trait discovery and trait deployment, including editing susceptibility genes and stacking resistance loci to broaden durability [132,133]. Introgression of resistance genes from Manihot glaziovii into African germplasm led to the release of CMD-resistant cultivars widely adopted in Nigeria, Uganda, and Tanzania [134]. These gene-editing advances offer opportunities to reduce viral replication, lower vector acquisition, and improve tolerance to bacterial pathogens. However, MAS is most effective for traits controlled by one or a few genes with a major effect, like resistance to certain cassava mosaic disease (CMD) strains. For complex, quantitative traits like yield and drought tolerance, MAS is less effective because these traits are governed by many genes with small, cumulative effects [133,135,136].

1.4.2.11 Clean Seed Systems

Clean seed systems for cassava involve producing and distributing disease-free, high-quality planting material, often using improved varieties and structured methods like rapid multiplication techniques and isolation fields to control pests and diseases [137,138]. However, because cassava is clonally propagated, dissemination of disease-free planting material is critical. Vegetative cuttings can easily harbour latent infections of CMD, CBSD, and bacterial blight, facilitating long-distance spread [139-140]. Clean seed initiatives, such as the IITA's community Phyto sanitation program and national clean seed systems in Uganda and Nigeria, combine tissue culture, rapid multiplication, and farmer training to ensure access to certified disease-free planting material. However, scaling remains limited by infrastructure and farmer awareness, highlighting the need for public-private partnerships [182,138].

1.4.2.12 Cultural Practices and Crop Management

Cultural practices are the activities involved in the cultivation of cassava from the decision to plant it, site selection, and right up to the harvesting and post harvesting operations. Traditional agronomic practices, when integrated with modern innovations, can significantly reduce disease pressure [142,143]. Timely rouging of infected plants, rotation with non-host crops, optimal planting densities, and synchronization of planting dates reduce inoculum build-up and vector colonization. Intercropping cassava with legumes has been shown to decrease whitefly abundance, while improving soil fertility and yields. Nonetheless, these strategies require collective adoption at the community level to be effective, as isolated implementation at single farms offers limited protection [128,142-146].

1.4.2.13 Vector Management

The whitefly vector poses a persistent challenge, not only transmitting CMD and CBSD but also damaging crops directly through sap feeding and honeydew deposition [147]. Chemical insecticides have limited utility due to environmental and cost constraints, coupled with the rapid increase in resistance that has been observed recently [148]. Integrated vector management emphasizes the deployment of biological control agents (e.g., parasitoid wasps), resistant crop varieties, and ecological approaches such as habitat diversification that reduce vector colonization [149]. Advances in whitefly population genomics are also informing targeted strategies for managing superabundant haplotypes associated with disease outbreaks [150]. A combination of tissue culture, chemotherapy, and thermotherapy can be used to clean cassava plants from viral infections that cause mosaic and brown streak disease [151].

1.4.2.14 Community-Based Surveillance and Digital Tools

Early detection of disease symptoms, coupled with information sharing across farmer networks, allows for coordinated responses and reduces the risk of epidemic spread. There are different types of digital tools for cassava surveillance which include AI powered mobile apps, satellite and drone imagery, Geospatial data platform, Web-based and SMS services, and molecular diagnostics and sequencing [152-154]. These will assist in early detection of diseases, and reduce cost and time, improve response and coordination, enhanced data collection, and empowering farmers in disease detection.

1.4.2.15 Policy and Institutional Support

Technical interventions alone cannot achieve sustainable disease management without an enabling policy and institutional environment. Effective implementation of IDM strategies requires coherent national policies, strong institutional coordination, and long-term investment. National cassava development programs should prioritise disease surveillance, variety dissemination, and capacity building through well-structured extension systems [155]. Regional initiatives such as the West African Virus Epidemiology (WAVE) project demonstrate the importance of coordinated disease monitoring, early-warning systems, and data sharing across borders. Likewise, international collaborations involving organizations such as the International Institute of Tropical Agriculture (IITA), CGIAR, and the Food and Agriculture Organization (FAO) play a pivotal role in integrating research outputs, harmonizing phytosanitary standards, and supporting the development of resilient seed systems [155,156]. Sustained policy commitment, adequate funding, and inclusive stakeholder engagement from researchers and extension agents to farmers and policymakers are essential for scaling IDM innovations and ensuring long-term resilience of cassava production, particularly within resource-constrained smallholder systems [157,158].

1.4.2.16 Critical Reflections

While IDM has proven conceptually effective, practical implementation remains uneven. Many smallholders lack consistent access to clean seed, improved varieties, or extension services. Reliance on community compliance for practices like rouging or synchronized planting can falter in heterogeneous farming landscapes [159]. Furthermore, emerging threats such as climate change, new virus strains, and insecticide-resistant whiteflies threaten to undermine progress. Future IDM strategies must therefore integrate biotechnological innovations including CRISPR-based virus resistance, RNA interference for vector suppression, and genomic prediction with strong social and institutional frameworks [160]. Without simultaneous attention to both biological and socioeconomic dimensions, IDM risks becoming a fragmented rather than transformative approach.

1.4.2.17 Breeding Programs in Africa

The International Institute of Tropical Agriculture (IITA) and national partners working within the Consultative Group on International Agricultural Research (CGIAR) Roots, Tubers and Bananas (RTB) program have delivered large numbers of improved cassava clones such as TME419 which is widely adopted for its CMD resistance, high dry-matter content, and good pounding quality, and TMS 30572 and TMS-980505 that are used extensively as sources of CMD resistance and high yield (Table iii) [136,161]. Modern cassava breeding technologies in Africa have evolved into highly integrated systems that combine several complementary approaches to accelerate genetic gain and deliver improved varieties to farmers. Marker-assisted selection is widely used to introgress major disease resistance loci, particularly those conferring durable resistance to CMD and CBSD, ensuring that new varieties retain robust protection against viral threats [162,163]. Genomic selection technology predicts the breeding value of cassava varieties based on genome-wide marker data and this information will be used to improve complex quantitative traits such as fresh root yield, dry matter content, and provitamin A carotenoids [164-166]. This allows cassava breeders to make informed decisions earlier, based on genomic estimated breeding values rather than waiting for full field evaluations [165,166]. Additionally, to ensure that selected clones perform reliably across diverse agroecological zones, breeders implement multi-environment trials and apply genotype by-environment analyses to identify varieties that combine high productivity with tolerance to diseases or abiotic stresses [167,168]. The efficiency of varietal delivery is enhanced through tissue culture approaches, enabling large-scale production of improved clones to supply to the farmers [127,169]. Together, these strategies represent a modernized cassava breeding framework that balances molecular tools, field-based evaluation, and seed system innovations to shorten breeding cycles and ensure that improved varieties reach smallholder farmers more rapidly. CRISPR/Cas approaches are being actively developed in cassava to broaden disease resistance and to improve key quality traits that impact both food safety and industrial utility [168]. With regard to viral infections, several research groups have applied CRISPR/Cas-based strategies aimed at either knocking out host susceptibility genes, directly disrupting viral genomes, or engineering immune signalling pathways. In CMD, CRISPR mediated targeting of ACMV has demonstrated a reduction in viral accumulation, highlighting the potential of direct antiviral editing [170-172]. However, challenges remain, including the risk of viral escape mutations, which necessitate multiplexed targeting strategies and careful evaluation of field-level durability. In CBSD, recent genome-wide association and molecular studies are beginning to identify host loci associated with root necrosis and viral load. These discoveries are now informing candidate targets for editing, although the development of stable, broad-spectrum resistance to CBSD continues to be complicated by viral diversity and the complexity of host–pathogen interactions [173,174]

A major constraint to cassava utilisation is its accumulation of cyanogenic glycosides, which must be carefully removed during processing to prevent cyanide poisoning. CRISPR/Cas knockouts of the CYP79D1 and CYP79D2 genes, which catalyse the first step in linamarin biosynthesis, have been shown to significantly reduce cyanide content in both leaves and storage roots [175,176]. Multiple independent studies confirm that disruption of these paralogs attenuates cyanogen production, providing a clear and practical path toward developing low-cyanide cassava varieties. These innovations could directly enhance food safety, particularly in regions where cassava is a dietary staple.

The composition and structural properties of starch is essential in cassava. Traits such as the amylose-to-amylopectin ratio and chain-length distribution strongly influence its performance in food, feed, and industrial applications [177,178]. Genome editing has successfully targeted key starch biosynthetic genes, including GBSS (Waxy), PTST1, and soluble starch synthases, to produce amylose-free or modified starch types [179-181]. Such modifications have the potential to improve food texture, expand industrial processing options, and enhance cassava's role as a biofuel feedstock. By tailoring starch properties, genome editing is opening new opportunities for both nutritional improvement and commercial diversification of cassava products.

Overall, the application of CRISPR/Cas technologies in cassava demonstrates remarkable progress in disease resistance, food safety, and starch quality. While technical challenges such as transformation efficiency and field-level validation remain, the advances made to date shows the transformative potential of genome editing for cassava improvement. By integrating these approaches with conventional breeding and ensuring robust biosafety evaluation, genome editing can play a pivotal role in delivering cassava varieties that are more resilient, safer, and better suited for diverse uses, ultimately benefiting both farmers and consumers worldwide.

Table 1.4.3: CMD- and CBSD-Resistant Cassava Varieties Released in Africa

Variety	Country	CMD Resistance	CBSD Tolerance	Dry Matter (%)	Carotenoid content	Reference
TMS 30572	Nigeria	Resistant	Susceptible	34	White	(Ossai et al., 2025)
Narocass 1	Uganda	Resistant	Moderate tolerance	33	White	(Mukiibi et al., 2019)
Kiroba	Tanzania	Resistant	Tolerance	32	White	(Shirima et al., 2019)
TMS 01/1368	Nigeria	Resistant	Susceptible	35	Yellow (provit. A)	(Ossai et al., 2025)

1.4.3 Future Research and Development in Cassava

Research should focus on integrating durable disease resistance, climate resilience, and improved quality traits into farmer preferred varieties [182-184]. Achieving durable resistance will require pyramiding CMD genes with multi-locus CBSD resistance or tolerance, alongside whitefly-resilient plant architecture to minimize resistance erosion under climate extremes. Understanding whitefly ecology and deploying landscape-level integrated pest management, including host mosaics, synchronized planting, and biological control can reduce dependence on insecticides. Scaling genomic selection across African breeding programs will accelerate progress on key traits such as cooking quality, dry matter content, cyanogenic potential, and virus resistance while shortening breeding cycles. Overcoming transformation and gene-editing delivery bottlenecks in elite cultivars, coupled with evaluation of field durability and regulatory pathways, will expand the practical impact of precision breeding.

Climate-smart agronomic practices, including optimized planting windows, soil cover, nutrient management, and crop model informed advisories, will help farmers adapt to drier conditions. Food safety can be mainstreamed by embedding Codex HCN limits, using simple test kits in small-scale processing, and standardizing fermentation and pressing protocols with appropriate training. Digital surveillance and extension, leveraging AI-enabled diagnostics and participatory monitoring, can shorten response times to CMD and CBSD outbreaks. Finally, aligning breeding programs with gendered preferences and urban market traits such as peelability, colour, and texture while linking production to starch and ethanol industries, will ensure inclusive markets and stable demand for improved cassava varieties.

1.4.4 Conclusions

Cassava's centrality to African food systems is critical, and its productivity, quality and safety must improve in tandem. The past two decades have shown an improvement in CMD-resistant varieties, SOPs for clean seed, and a maturing genomics toolkit. The next decade will be defined by whether Africa can (i) deploy disease-resilient, high-quality varieties at scale; (ii) manage whitefly and viral pressures amid climate volatility; (iii) mainstream safe, standardized processing; and (iv) connect producers to higher value markets. Strategic investment in breeding (including GS and gene editing), seed enterprises, climate-smart agronomy, and digital surveillance offer a credible pathway to durable impact.

Authors' contribution: Tapiwa Nyakauru: Conceptualization, investigation, writing original draft, review and editing. Ruvimbo Chimidzi: reviewing draft and editing. Godwin Mtetwa: Obtained images from farms, reviewing draft and editing. Thomas Marangwana: Reviewing draft and editing.

Statements and Declarations

Competing Interests: There are non-financial interests that are directly or indirectly related to this work.

Funding: Not applicable

Conflict of interest: The authors declare that they have no known conflict of interest

Ethics declaration: Not applicable.

Consent to Publish declaration: Not applicable

1.4.5 References

1. Adebayo, W. G. (2023). Cassava production in africa: A panel analysis of the drivers and trends. Heliyon, 9(9), e19939.

2. Borku, A. W. (2025). Cassava (Manihot esculenta Crantz): its nutritional composition insights for future research and development in Ethiopia. Discover Sustainability, 6(1), 404.

3. Ikuemonisan, E. S., Mafimisebi, T. E., Ajibefun, I., & Adenegan, K. (2020). Cassava production in Nigeria: trends, instability and decomposition analysis (1970–2018). Heliyon, 6(10), e05089.

4. Allem, A. C. (2002). The Origins and Taxonomy of Cassava.

5. Olsen, K. M., & Schaal, B. A. (1999). Evidence on the origin of cassava: Phylogeography of Manihot esculenta. Proceedings of the National Academy of Sciences, 96(10), 5586–5591.

6. Simon, M. F., Mendoza Flores, J. M., Liu, H., Martins, M. L. L., Drovetski, S. V., Przelomska, N. A. S., Loiselle, H., Cavalcanti, T. B., Inglis, P. W., Mueller, N. G., Allaby, R. G., Freitas, F. de O., & Kistler, L. (2022). Phylogenomic analysis points to a South American origin of Manihot and illuminates the primary gene pool of cassava. New Phytologist, 233(1), 534–545.

7. Carvalho, L. J. C. B., Anderson, J. V., Chen, S., Mba, C., & Doğramaci, M. (2018). Domestication Syndrome in Cassava (Manihot esculenta Crantz): Assessing Morphological Traits and Differentially Expressed Genes Associated with Genetic Diversity of Storage Root. In Cassava. InTech.

8. Mohidin, S. R. N. S. P., Moshawih, S., Hermansyah, A., Asmuni, M. I., Shafqat, N., & Ming, L. C. (2023). Cassava (Manihot esculenta Crantz): A Systematic Review for the Pharmacological Activities, Traditional Uses, Nutritional Values, and Phytochemistry. Journal of Evidence-Based Integrative Medicine, 28.

9. Santanoo, S., Vongcharoen, K., Banterng, P., Vorasoot, N., Jogloy, S., Roytrakul, S., & Theerakulpisut, P. (2022). Physiological and Proteomic Responses of Cassava to Short Term Extreme Cool and Hot Temperature. Plants, 11(17), 2307.

10. Phanthanong, P., Promnikorn, K., Kongsil, P., Kraichak, E., Jenweerawat, S., Vuttipongchaikij, S., & Kittipadakul, P. (2025). Variety-specific

responses to climatic and edaphic factors in cassava productivity. Frontiers in Agronomy, 7.

11. Adjei, E. O., Ayamba, B. E., Buri, M. M., Biney, N., & Appiah, K. (2023). Soil quality and fertility dynamics under a continuous cassava-maize rotation in the semi-deciduous forest agro-ecological zone of Ghana. Frontiers in Sustainable Food Systems, 7.

12. Howeler, R. H. (2017). Cassava cultivation and soil productivity (Clair. Hershey, Ed.; Vol. 2).

13. CARDI. (2021). Growing Cassava Soil Testing.

14. Antwi, B. O., Asante, S. K., & Yeboah, J. (2016). Drought Assessment for Reduced Climate Impact on Cassava Production. Journal of Applied Sciences, 17(1), 12–21.

15. Okogbenin, E., Setter, T. L., Ferguson, M., Mutegi, R., Ceballos, H., Olasanmi, B., & Fregene, M. (2013). Phenotypic approaches to drought in cassava: review. Frontiers in Physiology, 4.

16. Koundinya, A. V. V., Nisha, A., & Ajeesh, B. R. (2024). Early vigour: A key to drought tolerance in cassava based on physiological and biochemical traits including inherent non enzymatic antioxidant activity. Scientia Horticulturae, 331, 113110.

17. Sakadzo, N., Kugedera, A. T., Ranganai, N., & Kokerai, L. K. (2025). Cassava: practices and technologies to improve food security in sub-Saharan Africa. Cogent Food & Agriculture, 11(1).

18. Lopes, M. S., Araus, J. L., van Heerden, P. D. R., & Foyer, C. H. (2011). Enhancing drought tolerance in C4 crops. Journal of Experimental Botany, 62(9), 3135–3153.

19. Tardieu, F. (2012). Any trait or trait-related allele can confer drought tolerance: just design the right drought scenario. Journal of Experimental Botany, 63(1), 25–31.

20. Turyagyenda, L. F., Kizito, E. B., Ferguson, M., Baguma, Y., Agaba, M., Harvey, J. J. W., & Osiru, D. S. O. (2013). Physiological and molecular

characterization of drought responses and identification of candidate tolerance genes in cassava. AoB Plants, 5(0), plt007–plt007.

21. Orek, C., Gruissem, W., Ferguson, M., & Vanderschuren, H. (2020). Morpho-physiological and molecular evaluation of drought tolerance in cassava (Manihot esculenta Crantz). Field Crops Research, 255, 107861.

22. Pereira, L. F. M., Santos, H. L., Zanetti, S., Brito, I. A. de O., Tozin, L. R. dos S., Rodrigues, T. M., & Silva, M. de A. (2022). Morphology, biochemistry, and yield of cassava as functions of growth stage and water regime. South African Journal of Botany, 149, 222–239.

23. Charles, O. (2024). A review of drought-stress responsive genes and their applications for drought stress tolerance in cassava (Manihot esculenta Crantz). Discover Biotechnology, 1(1), 5.

24. Ding, Z., Fu, L., Tie, W., Yan, Y., Wu, C., Hu, W., & Zhang, J. (2019). Extensive Post-Transcriptional Regulation Revealed by Transcriptomic and Proteomic Integrative Analysis in Cassava under Drought. Journal of Agricultural and Food Chemistry, 67(12), 3521–3534.

25. Shan, Z., Luo, X., Wei, M., Huang, T., Khan, A., & Zhu, Y. (2018). Physiological and proteomic analysis on long-term drought resistance of cassava (Manihot esculenta Crantz). Scientific Reports, 8(1), 17982.

26. Wu, C., Hu, W., Yan, Y., Tie, W., Ding, Z., Guo, J., & He, G. (2018). The Late Embryogenesis Abundant Protein Family in Cassava (Manihot esculenta Crantz): Genome Wide Characterization and Expression during Abiotic Stress. Molecules, 23(5), 1196.

27. Sheela Immanuel, D., Jaganathan, P., Prakash P, & Sethuraman Sivakumar. (2024). Cassava for Food Security, Poverty Reduction and Climate Resilience: A Review. Indian Journal of Ecology.

28. Borku, A. W., Tora, T. T., & Masha, M. (2025). Cassava in focus: A comprehensive literature review, its production, processing landscape, and multi-dimensional benefits to society. Food Chemistry Advances, 7, 100945.

29. El-Sharkawy, M. A. (2012). Stress-Tolerant Cassava: The Role of Integrative Ecophysiology-Breeding Research in Crop Improvement. Open Journal of Soil Science, 02(02), 162–186.

30. Jarvis, A., Ramirez-Villegas, J., Herrera Campo, B. V., & Navarro-Racines, C. (2012). Is Cassava the Answer to African Climate Change Adaptation? Tropical Plant Biology, 5(1), 9–29.

31. Mtunguja, M., Beckles, D., Laswai, H., Ndunguru, J., & Sinha, N. (2019). Opportunities to Commercialize Cassava Production for Poverty Alleviation and Improved Food Security in Tanzania. African Journal of Food, Agriculture, Nutrition and Development, 19(01), 13928–13946.

32. Santanoo, S., Ittipong, P., Banterng, P., Vorasoot, N., Jogloy, S., Vongcharoen, K., & Theerakulpisut, P. (2024). Photosynthetic Performance, Carbohydrate Partitioning, Growth, and Yield among Cassava Genotypes under Full Irrigation and Early Drought Treatment in a Tropical Savanna Climate. Plants, 13(15), 2049.

33. Chijioke, U., Madu, T., Okoye, B., Ogunka, A. P., Ejechi, M., Ofoeze, M., Ogbete, C., Njoku, D., Ewuziem, J., Kalu, C., Onyemauwa, N., Ukeje, B., Achonwa, O., Forsythe, L., Fliedel, G., & Egesi, C. (2021). Quality attributes of fufu in South-East Nigeria: guide for cassava breeders. International Journal of Food Science & Technology, 56(3), 1247–1257.

34. Hossain, M. D., Yan, Q., Zhou, Z., Zhang, X., Wittayakun, S., Napasirth, V., Napasirth, P., Lukuyu, B. A., & Tan, Z. (2025). Cassava as a feedstuff for ruminant feeding system in Belt and Road countries: innovations, benefits and challenges. Journal of Agriculture and Food Research, 21, 101874.

35. Makieu, P., Kanu, M. S., Sillah, A., & Sheriff, A. (2025). Nutritional values of cassava leaves in three districts, Kenema, Kailahun, and Bo, Sierra Leone. Food and Humanity, 4, 100592.

36. Amare, T. A., Storebakken, T., Mørkøre, T., Nurfeta, A., & Ahlstrøm, Ø. (2024). Potency of cassava leaf as protein source for Nile tilapia (Oreochromis niloticus). Aquaculture International, 32(7), 10197–10214.

37. Oresegun, A., Fagbenro, O. A., Ilona, P., & Bernard, E. (2016). Nutritional and anti-nutritional composition of cassava leaf protein concentrate from six cassava varieties for use in aqua feed. Cogent Food & Agriculture, 2(1).

38. Alamu, E. O., Dixon, A., Eyinla, T. E., & Maziya-Dixon, B. (2022). Characterization of macro and micro-minerals in cassava leaves from genotypes planted in three different agroecological locations in Nigeria. Heliyon, 8(11), e11618.

39. Agu, R., Amadi, O., Nwagu, T., Moneke, A., & Okolo, B. (2024). Grades of cassava starch (flours) as sustainable substrates for use in neutral spirit and biofuel production—Industry approach. Food Chemistry Advances, 4, 100726.

40. Do, T. X., & Vu, T. A. (2024). Enhancing sustainable ethanol fuel production from cassava in Vietnam. Energy and Climate Change, 5, 100136.

41. Ponnarong, P., & Mariya, R. (2024). THIS REPORT CONTAINS ASSESSMENTS OF COMMODITY AND TRADE ISSUES MADE BY USDA STAFF AND NOT NECESSARILY STATEMENTS OF OFFICIAL U.S. GOVERNMENT POLICY.

42. Ezeoha, S. L. (2013). Production of Biodegradable Plastic Packaging Film from Cassava Starch. IOSR Journal of Engineering, 3(10), 14–20.

43. Shittu, T. A., Alimi, B. A., Wahab, B., Sanni, L. O., & Abass, A. B. (2016). Cassava Flour and Starch: Processing Technology and Utilization. In Tropical Roots and Tubers (pp. 415–450). Wiley.

44. Silveira, Y. D. O., Franca, A. S., & Oliveira, L. S. (2025). Cassava Waste Starch as a Source of Bioplastics: Development of a Polymeric Film with Antimicrobial Properties. Foods, 14(1), 113.

45. Jørgensen, K., Bak, S., Busk, P. K., Sørensen, C., Olsen, C. E., Puonti-Kaerlas, J., & Møller, B. L. (2005). Cassava Plants with a Depleted Cyanogenic Glucoside Content in Leaves and Tubers. Distribution of Cyanogenic Glucosides, Their Site of Synthesis and Transport, and Blockage of the Biosynthesis by RNA Interference Technology. Plant Physiology, 139(1), 363–374.

46. Ndubuisi, N. D., & Chidiebere, A. C. U. (2018). Cyanide in Cassava: A Review. International Journal of Genomics and Data Mining, 3(1).

47. Chiwona-Karltun, L., Brimer, L., Kalenga Saka, J. D., Mhone, A. R., Mkumbira, J., Johansson, L., Bokanga, M., Mahungu, N. M., & Rosling, H. (2004). Bitter taste in cassava roots correlates with cyanogenic glucoside levels. Journal of the Science of Food and Agriculture, 84(6), 581–590.

48. Forkum, A. T., Wung, A. E., Kelese, M. T., Ndum, C. M., Lontum, A., Kamga, E. B., Nsaikila, M. N., & Okwen, P. M. (2025). Safety of cassava and cassava-based products: a systematic review. Frontiers in Sustainable Food Systems, 9.

49. Osman, G., Maleta, W., Masamba, K., Ng'ong'ola-Manani, T., & Kalimbira, A. A. (2025). Cyanide in cassava: unveiling health risks in the lens of unsustainable food systems in Sub Saharan Africa – a systematic review. Frontiers in Sustainable Food Systems, 9.

50. Tshala-Katumbay, D. D., Ngombe, N. N., Okitundu, D., David, L., Westaway, S. K., Boivin, M. J., Mumba, N. D., & Banea, J. (2016). Cyanide and the human brain: perspectives from a model of food (cassava) poisoning. Annals of the New York Academy of Sciences, 1378(1), 50–57.

51. Amsalu Nebiyu. (2011). Soaking and drying of cassava roots reduced cyanogenic potential of three cassava varieties at Jimma, Southwest Ethiopia. African Journal of Biotechnology, 10(62).

52. Scaria, S. S., Balasubramanian, B., Meyyazhagan, A., Gangwar, J., Jaison, J. P., Kurian, J. T., Pushparaj, K., Pappuswamy, M., Park, S., & Joseph, K. S. (2024). Cassava (Manihot esculenta Crantz)—A potential source of phytochemicals, food, and nutrition—An updated review. EFood, 5(1).

53. Chikezie, P., & Ojiako, O. (2013). Cyanide and aflatoxin loads of processed cassava (Manihot esculenta) tubers (Garri) in Njaba, Imo state, Nigeria. Toxicology International, 20(3), 261.

54. Njankouo Ndam, Y., Mounjouenpou, P., Kansci, G., Kenfack, M. J., Fotso Meguia, M. P., Natacha Ngono Eyenga, N. S., Mikhaïl Akhobakoh, M.,

& Nyegue, A. (2019a). Influence of cultivars and processing methods on the cyanide contents of cassava (Manihot esculenta Crantz) and its traditional food products. Scientific African, 5, e00119.

55. Orhe OG, Ebereghwa EM, Emuoghenerue EO, & Anyanwu EB. (2022). The challenges of cassava fermentation for Garri production: A case report of a woman with muscle twitching possibly due to cyanide poisoning from poor cassava fermentation. World Journal of Advanced Pharmaceutical and Medical Research, 3(2), 038–042.

56. JECFA. (n.d.). Agenda Item 14 CX/CF 21/14/12 March 2021 JOINT FAO/WHO FOOD STANDARDS PROGRAMME CODEX COMMITTEE ON CONTAMINANTS IN FOODS 14 th Session (virtual) 3-7 and 13 May 2021 DISCUSSION PAPER ON HYDROCYANIC ACID AND MYCOTOXIN CONTAMINATION IN CASSAVA AND CASSAVA-BASED PRODUCTS.

57. Padmaja, G., & Steinkraus, K. H. (1995). Cyanide detoxification in cassava for food and feed uses. Critical Reviews in Food Science and Nutrition, 35(4), 299–339.

58. Njankouo Ndam, Y., Mounjouenpou, P., Kansci, G., Kenfack, M. J., Fotso Meguia, M. P., Natacha Ngono Eyenga, N. S., Mikhaïl Akhobakoh, M., & Nyegue, A. (2019b). Influence of cultivars and processing methods on the cyanide contents of cassava (Manihot esculenta Crantz) and its traditional food products. Scientific African, 5, e00119.

59. Bouatenin, K. M. J.-P., Kouame, K. A., Djeni, N. T., Koffi, N. G., & Dje, K. M. (2021). Production of Attieke by the Technique of Drying of Cassava Ferment. Journal of Food Quality, 2021, 1–7.

60. Zhong, Y., Xu, T., Wu, X., Li, K., Zhang, P., Ji, S., Li, S., Zheng, L., & Lu, B. (2021). Dietary exposure and risk assessment of cyanide via cassava consumption in Chinese population. Food Chemistry, 354, 129405.

61. Dhas, P., Chitra, P., Jayakumar, S., & Mary, A. (2011). Study of the effects of hydrogen cyanide exposure in Cassava workers. Indian Journal of Occupational and Environmental Medicine, 15(3), 133.

62. Alamu, E. O., Dixon, G. A., Adesokan, M., & Maziya Dixon, B. (2023). Correlate the cyanogenic potential and dry matter content of cassava roots and leaves grown in different environments. Scientific Reports, 13(1), 15382.

63. Kanaabi, M., Settumba, M. B., Nuwamanya, E., Muhumuza, N., Iragaba, P., Ozimati, A., Namakula, F. B., Kayondo, I. S., Baguma, J. K., Nanyonjo, A. R., Esuma, W., & Kawuki, R. S. (2024). Genetic Variation and Heritability for Hydrogen Cyanide in Fresh Cassava Roots: Implications for Low Cyanide Cassava Breeding. Plants, 13(9), 1186.

64. Apil, J., Mulugo, L., Atekyereza, P., & Obaa, B. (2025). Cassava cyanide knowledge, risk perception, and adoption of preventive measures in Northwestern Uganda: a moderated mediation perspective. Frontiers in Sustainable Food Systems, 9.

65. Burns, A., Gleadow, R., Cliff, J., Zacarias, A., & Cavagnaro, T. (2010). Cassava: The Drought, War and Famine Crop in a Changing World. Sustainability, 2(11), 3572–3607.

66. Gmakouba, T., Dzidzienyo, D. K., Koussao, S., Tongoona, P., & Asante, I. K. (2025). Identifying cassava production constraints, farmers preferences, and cassava mosaic disease perceptions in Togo: insights for a participatory breeding approach. Agriculture & Food Security, 13(1), 63.

67. Legg, J. P., Jeremiah, S. C., Obiero, H. M., Maruthi, M. N., Ndyetabula, I., Okao-Okuja, G., Bouwmeester, H., Bigirimana, S., Tata-Hangy, W., Gashaka, G., Mkamilo, G., Alicai, T., & Lava Kumar, P. (2011). Comparing the regional epidemiology of the cassava mosaic and cassava brown streak virus pandemics in Africa. Virus Research, 159(2), 161–170.

68. Tomlinson, K. R., Bailey, A. M., Alicai, T., Seal, S., & Foster, G. D. (2018). Cassava brown streak disease: historical timeline, current knowledge and future prospects. Molecular Plant Pathology, 19(5), 1282–1294.

69. Fanou, A. A., Zinsou, V. A., & Wydra, K. (2018). Cassava Bacterial Blight: A Devastating Disease of Cassava. In Cassava. InTech.

70. Wydra, K., & Verdier, V. (2002). Occurrence of cassava diseases in relation to environmental, agronomic and plant characteristics. Agriculture, Ecosystems & Environment, 93(1–3), 211–226.

71. Hormes, H. (2007). https://www.flickr.com/photos/129099219@N03/32638164343/in/photolist-RJ88arTAZ1CH-nfsGJs-88VdGX-RJ88o2-cqiaZ9-81L8yBTpxydv-2dPfti-adke4r-SjpWxC-TpxxFZ-UbgMz3-TAZ1jrdWUT2E-SjpWqo-fuhYHs-dWUK3j-TAYZG4-T1ZHcCTpxy1g-TpxxP4-fuigZA-fu49Gp-fu16Ug-SjpVUo-fu47wiTxmsw9-c15D8s-dWPbfc-dWPcz2-fuhvz3-fuhD7G-fuivAL dWUFNq-fu3g3V-fu1Te8-fu2kjP-fugGrf-fu2aZ2-ftZXiF-fu19Kt-fuhFNu-dWPffp-fuhPiq-fuhJLh-fu3vXD-fufErAftZVLD-fu1fA6.

72. Aumentado, H. D. R., & Balendres, M. A. O. (2025). Three Colletotrichum species associated with cassava anthracnose in the Philippines. Journal of Plant Pathology, 107(3), 1537-1547.

73. APS. (2019). https://www.plantdiseases.org/bacterial-leafspot-bacterial-necrosis-cassava.

74. Ndunguru, J., Sseruwagi, P., Tairo, F., Stomeo, F., Maina, S., Djinkeng, A., Kehoe, M., & Boykin, L. M. (2015). Analyses of Twelve New Whole Genome Sequences of Cassava Brown Streak Viruses and Ugandan Cassava Brown Streak Viruses from East Africa: Diversity, Supercomputing and Evidence for Further Speciation. PLOS ONE, 10(10), e0139321.

75. Patil, B. L., & Fauquet, C. M. (2009). Cassava mosaic geminiviruses: actual knowledge and perspectives. Molecular Plant Pathology, 10(5), 685–701.

76. Du, H., Wang, Y.-M., & Wang, X.-W. (2025). Begomovirus Transmission by the Insect Vector, the Whitefly Bemisia tabaci (pp. 35–47).

77. Chikoti, P. C., Mulenga, R. M., Tembo, M., & Sseruwagi, P. (2019). Cassava mosaic disease: a review of a threat to cassava production in Zambia. Journal of Plant Pathology, 101(3), 467–477.

78. Adjéi, E. A., Traoré, K., Sawadogo-Compaore, E. M. F. W., Kouakou, B. S. M., Séka, J. S. S., Ahoya, D. K. D., Kouassi, K. M., Kouassi, N. K., & Pita, J. S. (2025). Enhancing Farmers' Capacity for Sustainable

Management of Cassava Mosaic Disease in Côte d'Ivoire. Agriculture, 15(12), 1277.

79. Eni, A. O., Efekemo, O. P., Onile-ere, O. A., & Pita, J. S. (2021). South West and North Central Nigeria: Assessment of cassava mosaic disease and field status of African cassava mosaic virus and East African cassava mosaic virus. Annals of Applied Biology, 178(3), 466–479.

80. Bisimwa, E., Walangululu, J., & Bragard, & C. (2015). Cassava Mosaic Disease Yield Loss Assessment under Various Altitude Agro¬ ecosystems in the Sud¬Kivu Region, Democratic Republic of Congo. In TROPICULTURA (Vol. 33).

81. Legg, J. P., Lava Kumar, P., Makeshkumar, T., Tripathi, L., Ferguson, M., Kanju, E., Ntawuruhunga, P., & Cuellar, W. (2015). Cassava Virus Diseases: Biology, epidemiology, and management (pp. 85–142).

82. Saffa, M. D., Samura, A. E., Bah, M. A., Eni, A. O., Tibiri, E. B., Sagnon, A., Tiendrébéogo, F., Pita, J. S., Norman, P. E., & Johnson, R. A. B. (2025). Detection of Cassava Mosaic Disease and Assessment of Selected Agronomic Traits of Cassava (Manihot esculenta). Horticulturae, 11(6), 618.

83. Shirima, R. R., Legg, J. P., Maeda, D. G., Tumwegamire, S., Mkamilo, G., Mtunda, K., Kulembeka, H., Ndyetabula, I., Kimata, B. P., Matondo, D. G., Ceasar, G., Mushi, E., Sichalwe, K., & Kanju, E. (2020). Genotype by environment cultivar evaluation for cassava brown streak disease resistance in Tanzania. Virus Research, 286, 198017.

84. Musopole, H. T., Mtonga, A. P., Pankomera, P., & Maruthi, M. N. (2024). Survey of cassava mosaic and brown streak diseases and their vector whiteflies in Malawi. Discover Agriculture, 2(1), 113.

85. Nyakauru, T., & Robertson, F. (2025). Unveiling Cassava Mosaic Virus Diversity in Zimbabwe: A Nanopore Sequencing Approach.

86. Alonso Chavez, V., Milne, A. E., van den Bosch, F., Pita, J., & McQuaid, C. F. (2022a). Modelling cassava production and pest management under biotic and abiotic constraints. Plant Molecular Biology, 109(3), 325–349.

87. Maruthi, M. N., Whitfield, E. C., Otti, G., Tumwegamire, S., Kanju, E., Legg, J. P., Mkamilo, G., Kawuki, R., Benesi, I., Zacarias, A., Munga, T., Mwatuni, F., & Mbugua, E. (2019a). A method for generating virus-free cassava plants to combat viral disease epidemics in Africa. Physiological and Molecular Plant Pathology, 105, 77–87.

88. Mulenga, R. M., Boykin, L. M., Chikoti, P. C., Sichilima, S., Ng'uni, D., & Alabi, O. J. (2018). Cassava Brown Streak Disease and Ugandan cassava brown streak virus Reported for the First Time in Zambia. Plant Disease, 102(7), 1410–1418.

89. Hillocks, R. J., & Thresh, J. M. (2000). Cassava Mosaic and Cassava Brown Streak Virus Diseases in Africa: A Comparative Guide to Symptoms and Etiologies. 7(December), 1–8.

90. Storey, H., (1939). Report of the Plant Pathologist. Annual Report East African Agricultural Research.

91. Alicai, T., Omongo, C. A., Maruthi, M. N., Hillocks, R. J., Baguma, Y., Kawuki, R., Bua, A., Otim-Nape, G. W., & Colvin, J. (2007). Re-emergence of Cassava Brown Streak Disease in Uganda. Plant Disease, 91(1), 24–29.

92. Munguti, F. M., Nyaboga, E. N., Kilalo, D. C., Yegon, H. K., Macharia, I., & Mwango'mbe, A. W. (2023). Survey of cassava brown streak disease and association of factors influencing its epidemics in smallholder cassava cropping systems of coastal Kenya. Frontiers in Sustainable Food Systems, 6.

93. Kaweesi, T., Kawuki, R., Kyaligonza, V., Baguma, Y., Tusiime, G., & Ferguson, M. E. (2014). Field evaluation of selected cassava genotypes for cassava brown streak disease based on symptom expression and virus load. Virology Journal, 11(1), 216.

94. Shirima, R. R., Wosula, E. N., Hamza, A. A., Mohammed, N. A., Mouigni, H., Nouhou, S., Mchinda, N. M., Ceasar, G., Amour, M., Njukwe, E., & Legg, J. P. (2022a). Epidemiological Analysis of Cassava Mosaic and Brown Streak Diseases, and Bemisia tabaci in the Comoros Islands. Viruses, 14(10), 2165.

95. Mware, Benard., Ateka, Elijah., Songa, Josephine., Narla, Rama., Olubayo, Florence., & Amata, R. U. (2009). Transmission and distribution of cassava brown streak virus disease in cassava growing areas of Kenya.

96. Nyirakanani, C. (2023). Analysis of Cassava Brown Streak Disease in Rwanda: Incidence, Dissemination, Genetic Diversity, and Innovative Mitigation Strategies.

97. Robson, F., Hird, D. L., & Boa, E. (2024). Cassava brown streak: A deadly virus on the move. Plant Pathology, 73(2), 221–241.

98. Ano, C. U., Ochwo-Ssemakula, M., Ibanda, A., Ozimati, A., Gibson, P., Onyeka, J., Njoku, D., Egesi, C., & S. Kawuki, R. (2021). Cassava Brown Streak Disease Response and Association with Agronomic Traits in Elite Nigerian Cassava Cultivars. Frontiers in Plant Science, 12.

99. Chikoti, P. C., & Tembo, M. (2022a). Expansion and impact of cassava brown streak and cassava mosaic diseases in Africa: A review. Frontiers in Sustainable Food Systems, 6.

100. Toko, M., Hanna, R., Legg, J. P., Andrade, M. I., Jone, A., Agboton, B., Otema, M. A., Okao-Okuja, G., Obonyo, R., & Mambo, E. (2007). Distribution, incidence and severity of cassava diseases and pests in Mozambique. In Proceedings of the 9th ISTRC-AB Symposium.

101. Nandudu, L., Sheat, S., Winter, S., Ogbonna, A., Kawuki, R., & Jannink, J.-L. (2024). Genetic complexity of cassava brown streak disease: insights from qPCR-based viral titter analysis and genome-wide association studies. Frontiers in Plant Science, 15.

102. Zárate-Chaves, C. A., Gómez de la Cruz, D., Verdier, V., López, C. E., Bernal, A., & Szurek, B. (2021). Cassava diseases caused by Xanthomonas phaseoli manihotis and Xanthomonas cassava. Molecular Plant Pathology, 22(12), 1520–1537.

103. Popoola, A. R., & Olanloye, O. A. (2012). Rapid screenhouse assessment of bacterial blight in cassava genotypes in Nigeria. Archives Of Phytopathology and Plant Protection, 45(5), 574–584

104. Okechukwu, R. U. (2004). Distribution and current status of bacterial blight and fungal diseases of cassava in Nigeria. 105. Chukwuka, K.,

Okechukwu, R., & Azorji, J. (2013). Farmer perceptions of cassava bacterial blight disease in Oyo State, southwest, Nigeria. African Journal of Root and Tuber Crops, 10(1).

106. Night, G., Asiimwe, P., Gashaka, G., Nkezabahizi, D., Legg, J. P., Okao-Okuja, G., Obonyo, R., Nyirahorana, C., Mukakanyana, C., Mukase, F., Munyabarenzi, I., & Mutumwinka, M. (2011). Occurrence and distribution of cassava pests and diseases in Rwanda. Agriculture, Ecosystems & Environment, 140(3–4), 492–497.

107. López, C. E., & Bernal, A. J. (2012). Cassava Bacterial Blight: Using Genomics for the Elucidation and Management of an Old Problem. Tropical Plant Biology, 5(1), 117–126.

108. Ntui, V. O., Tripathi, J. N., Kariuki, S. M., & Tripathi, L. (2024a). Cassava molecular genetics and genomics for enhanced resistance to diseases and pests. Molecular Plant Pathology, 25(1).

109. Cucu, M. A., Choudhary, R., Trkulja, V., Garg, S., & Matić, S. (2025). Utilizing Environmentally Friendly Techniques for the Sustainable Control of Plant Pathogens: A Review. Agronomy, 15(7), 1551.

110. Dahniya, M. T. (1994). An overview of cassava in africa. African Crop Science Journal.

111. de Oliveira, S. A. S., da Silva, L. L., Nascimento, D. de S., Diamantino, M. S. A. S., Ferreira, C. F., & de Oliveira, T. A. S. (2020). Colletotrichum species causing cassava (Manihot esculenta Crantz) anthracnose in different eco-zones within the Recôncavo Region of Bahia, Brazil. Journal of Plant Diseases and Protection, 127(3), 411–416.

112. Sangpueak, R., Phansak, P., & Buensanteai, N. (2018). Morphological and molecular identification of Colletotrichum species associated with cassava anthracnose in Thailand. Journal of Phytopathology, 166(2), 129–142.

113. Fokunang, C. N., Ikotun, T., Dixon, A. G. O., & Akem, C. N. (1997). First Report of Colletotrichum gloeosporioides manihotis, Cause of Cassava Anthracnose Disease, Being Seed-borne and Seed-Transmitted in Cassava. Plant Disease, 81(6), 695–695.

114. Moses, E., & Woode, R. (2014). Anthracnose disease of cassava-Ghana. In PlantwisePlus Knowledge Bank.

115. Fokunang, C. N., Dixon, A. G. O., Ikotun, T., Asiedu, R. T.F. E. A., & Akem, C. N. (2002). In vitro, greenhouse, and field assessment of cassava lines for resistance to anthracnose disease caused by colletotrichum gloeosporiodes f. sp. Manihotis. Mycopathologia, 154.

116. Dimkpa, S. O. N., Lawson, T. S., & Ukoima, H. N. (2021). Growth Performance of Eleven Improved Cassava Varieties and their susceptibility to Some Insect Pests and Diseases in Humid Tropics, Rivers State,. SSRN Electronic Journal.

117. Appiah-Kubi, Z., Apetorgbor, A. K., Moses, E., & Appiah Kubi, D. (2015). Farmers Knowledge of Anthracnose Disease of Cassava and Yam in Four Ecological Zones in Ghana. Greener Journal of Agricultural Sciences, 5(6), 204–209.

118. Moses, E., Asafu-Agyei, J. N., Adubofour, K., & Adusei, A. (2007). Guide to identification and control of cassava diseases.

119. Samura, A., Kebbie, J., Amara, V., Quee, D., & Musa, J. (2025). Incidence, Severity, and Distribution of Cassava Bacteria Blight Disease in Sierra Leone. American Journal of Plant Biology, 10(1), 1–9.

120. Legg, J., & Alvarez, E. (2017). Diseases affecting cassava (pp. 213–244).

121. Obilo, O. P., Ikotun, B., Ihejirika, G. O., Ibeawuchi, I. I., & Oben, T. T. (2010). The effect of the incidence of cassava anthracnose disease (CAD) on the performance and yield of cassava cultivars. Crop Protection, 29(5), 482–486.

122. Khuong, N. Q., Duy, L. B., Thuan, V. M., Ngan, N. T., Hiep, P. C., Quang, L. T., Trong, N. D., Thu, H. N., Xuan, D. ., Thu, L. T. M., Nguyen, T. T. K., Xuan, L. N. T., & Phong, N. T. (2025). Application of Trichoderma spp. to Control Colletotrichum sp. and Pseudopestalotiopsis spp., Causing Agents of Fruit Rot in Pomelo (Citrus maxima (Burm.) Merr.). Applied Microbiology, 5(3), 66.

123. Yao, X., Guo, H., Zhang, K., Zhao, M., Ruan, J., & Chen, J. (2023). Trichoderma and its role in biological control of plant fungal and nematode disease. Frontiers in Microbiology, 14.

124. Alonso Chavez, V., Milne, A. E., van den Bosch, F., Pita, J., & McQuaid, C. F. (2022b). Modelling cassava production and pest management under biotic and abiotic constraints. Plant Molecular Biology, 109(3), 325–349.

125. Ntui, V. O., Tripathi, J. N., Kariuki, S. M., & Tripathi, L. (2024b). Cassava molecular genetics and genomics for enhanced resistance to diseases and pests. Molecular Plant Pathology, 25(1).

126. Otun, S., Escrich, A., Achilonu, I., Rauwane, M., Lerma Escalera, J. A., Morones-Ramírez, J. R., & Rios-Solis, L. (2023). The future of cassava in the era of biotechnology in Southern Africa. Critical Reviews in Biotechnology, 43(4), 594–612.

127. Tankam Chedjou, I., Donnelly, R., & Gilligan, C. A. (2025). Optimizing crop varietal mixtures for viral disease management: A case study on cassava virus epidemics. PLOS Computational Biology, 21(9), e1012842.

128. Bizimana, J. P., Ngapout, Y., Nyirakanani, C., Shakir, S., Kanju, E., Legg, J. P., Rukundo, P., Shimelis, H., Nduwumuremyi, A., & Vanderschuren, H. (2024). Breeding strategies for mitigating cassava brown streak disease in Africa. Tropical Plants, 3(1), 0–0.

129. Ceballos, H., Kawuki, R. S., Gracen, V. E., Yencho, G. C., & Hershey, C. H. (2015). Conventional breeding, marker assisted selection, genomic selection and inbreeding in clonally propagated crops: a case study for cassava. Theoretical and Applied Genetics, 128(9), 1647–1667.

130. Fondong, V. N. (2017). The Search for Resistance to Cassava Mosaic Geminiviruses: How Much We Have Accomplished, and What Lies Ahead. Frontiers in Plant Science, 8.

131. Blair, M. W., Fregene, M. a, Beebe, S. E., & Ceballos, H. (2007). Marker-assisted selection in common beans and cassava. Marker-Assisted

Selection: Current Status and Future Perspectives in Crops, Livestock, Forestry and Fish, 81–115.

132. Hasan, N., Choudhary, S., Naaz, N., Sharma, N., & Laskar, R. A. (2021). Recent advancements in molecular marker-assisted selection and applications in plant breeding programmes. Journal of Genetic Engineering and Biotechnology, 19(1), 128.

133. Kwibuka, Y., Nyirakanani, C., Bizimana, J. P., Bisimwa, E., Brostaux, Y., Lassois, L., Vanderschuren, H., & Massart, S. (2022). Risk factors associated with cassava brown streak disease dissemination through seed pathways in Eastern D.R. Congo. Frontiers in Plant Science, 13.

134. Legg, J. P., Diebiru-Ojo, E., Eagle, D., Friedmann, M., Kanju, E., Kapinga, R., Kumar, P. L., Lateef, S., Magige, S., Mtunda, K., Thiele, G., Yabeja, J., & Nitturkar, H. (2022). Commercially Sustainable Cassava Seed Systems in Africa. In Root, Tuber and Banana Food System Innovations (pp. 453–482). Springer International Publishing.

135. Chikoti, P. C., & Tembo, M. (2022b). Expansion and impact of cassava brown streak and cassava mosaic diseases in Africa: A review. Frontiers in Sustainable Food Systems, 6.

136. Shirima, R. R., Wosula, E. N., Hamza, A. A., Mohammed, N. A., Mouigni, H., Nouhou, S., Mchinda, N. M., Ceasar, G., Amour, M., Njukwe, E., & Legg, J. P. (2022b). Epidemiological Analysis of Cassava Mosaic and Brown Streak Diseases, and Bemisia tabaci in the Comoros Islands. Viruses, 14(10), 2165.

137. de Boef, W. S., Agbara, C., Bello, Z., Ali, H., Ezekiel Aladele, S., Chibuzor Iloh, A., Ignatius Angarawai, I., Osabuohien Omoigui, L., Osho-Lagunju, B., & Thijssen, M. H. (2025). Farmer- and community-managed seed systems: inputs for a Nigerian agenda.

138. Kidasi, P. C., Chao, D. K., Obudho, E. O., & Mwang'ombe, A. W. (2021). Farmers' Sources and Varieties of Cassava Planting Materials in Coastal Kenya. Frontiers in Sustainable Food Systems, 5.

139. Olusola Sanusi, R., Lordbanjou, D.-T., Olalekan Ibrahim, A., Babakatcha Abubakar, M., & Olalekan Oke, O. (2023). Cassava Production

Enterprise in the Tropics. In Tropical Plant Species and Technological Interventions for Improvement. IntechOpen.

140. Chen, H., Ruan, L., Cao, S., He, W., Yang, H., Liang, Z., Li, H., Wei, W., Huang, Z., & Lan, X. (2025). Cassava–soybean intercropping alleviates continuous cassava cropping obstacles by improving its rhizosphere microecology. Frontiers in Microbiology, 16.

141. Zou, Y., Liu, Z., Chen, Y., Wang, Y., & Feng, S. (2024). Crop Rotation and Diversification in China: Enhancing Sustainable Agriculture and Resilience. Agriculture, 14(9), 1465.

142. Onasanya, O. O., Hauser, S., Necpalova, M., Salako, F. K., Kreye, C., Tariku, M., Six, J., & Pypers, P. (2021). On-farm assessment of cassava root yield response to tillage, plant density, weed control and fertilizer application in southwestern Nigeria. Field Crops Research, 262, 108038.

143. Silva, R. B., Santos Neto, A. L. dos, Santos, W. M. dos, Teodoro, I., Barros, A. C., Santos, V. R. dos, Souza, A. A. de, Martins, G. M. C., Costa, B. R. dos S., Souza, J. W. G. de, & Silva, D. M. R. (2025). Cassava growth and productive performance at different planting times in the Agreste of Alagoas, Northeastern Brazil. Revista Brasileira de Engenharia Agrícola e Ambiental, 29(6).

144. Uzokwe, V. N. E., Mlay, D. P., Masunga, H. R., Kanju, E., Odeh, I. O. A., & Onyeka, J. (2016). Combating viral mosaic disease of cassava in the Lake Zone of Tanzania by intercropping with legumes. Crop Protection, 84, 69–80.

145. Naveed, H., Islam, W., Jafir, M., Andoh, V., Chen, L., & Chen, K. (2023). A Review of Interactions between Plants and Whitefly-Transmitted Begomoviruses. Plants, 12(21), 3677.

146. Ntui, V. O., Tripathi, J. N., Kariuki, S. M., & Tripathi, L. (2024c). Cassava molecular genetics and genomics for enhanced resistance to diseases and pests. Molecular Plant Pathology, 25(1).

147. Lizzi, K. M., Qualls, W. A., Brown, S. C., & Beier, J. C. (2014). Expanding Integrated Vector Management to promote healthy environments. Trends in Parasitology, 30(8), 394–400.

148. Yang, K., Li, H., Guo, D., Sun, Z., Li, F., & Chu, D. (2025). Spatiotemporal Population Genomics of the Invasive Whitefly Bemisia tabaci MED in China: Implications for Surveillance and Sustainable Control. Insects, 16(9), 975.

149. Maruthi, M. N., Whitfield, E. C., Otti, G., Tumwegamire, S., Kanju, E., Legg, J. P., Mkamilo, G., Kawuki, R., Benesi, I., Zacarias, A., Munga, T., Mwatuni, F., & Mbugua, E. (2019b). A method for generating virus-free cassava plants to combat viral disease epidemics in Africa. Physiological and Molecular Plant Pathology, 105, 77–87.

150. Ferris, A. C., Stutt, R. O. J. H., Godding, D., & Gilligan, C. A. (2020). Computational models to improve surveillance for cassava brown streak disease and minimize yield loss. PLOS Computational Biology, 16(7), e1007823.

151. Mrisho, L. M., Mbilinyi, N. A., Ndalahwa, M., Ramcharan, A. M., Kehs, A. K., McCloskey, P. C., Murithi, H., Hughes, D. P., & Legg, J. P. (2020). Accuracy of a Smartphone-Based Object Detection Model, PlantVillage Nuru, in Identifying the Foliar Symptoms of the Viral Diseases of Cassava–CMD and CBSD. Frontiers in Plant Science, 11.

152. Sawadogo, S., Tiendrebeogo, F., Tibiri, E. B., Name, P. E., Djigma, F., Traoré, L., Pita, J. S., & Eni, A. O. (2025). Artificial intelligence tool for cassava viral diseases diagnosis using participatory surveillance in Burkina Faso. Frontiers in Sustainable Food Systems, 9.

153. Zhou, W., Arcot, Y., Medina, R. F., Bernal, J., Cisneros Zevallos, L., & Akbulut, M. E. S. (2024). Integrated Pest Management: An Update on the Sustainability Approach to Crop Protection. ACS Omega, 9(40), 41130–41147.

154. Krisdiana, R., Noerwiyati, K., Mejaya, I. M. J., Ismu Wisnubroto, E., & Ngoc Huu, N. (2023). Policies and Models of Agribusiness Development of Cassava in North Sumatera. E3S Web of Conferences, 432, 00033.

155. King, N. R. (2011). Royal Government of Cambodia National Strategic Plan. March.

156. Zhang, X., & Whitney, E. (2004). Policy Options to Strengthen the Cassava and Rice Supply Chains. 32.

157. Beach, R. H., Milliken, C., Franzen, K., & Lapidus, D. (2025). Meta-analysis of the impacts of digital information interventions on agricultural development. Global Food Security, 45, 100866.

158. Yuan, J., Chen, D., Ren, Y., Zhang, X., & Zhao, J. (2008). Characteristic and Expression Analysis of a Metallothionein Gene, OsMT2b, Down-Regulated by Cytokinin Suggests Functions in Root Development and Seed Embryo Germination of Rice. Plant Physiology, 146(4), 1637.

159. Wossen, T., Abdoulaye, T., Alene, A., Feleke, S., Rabbi, I., Asumugha, G., Kulakow, P., & Manyong, V. (2017). Impact of improved cassava varieties in Nigeria. November, 1–45.

160. Olasanmi, B., Kyallo, M., & Yao, N. (2021). Marker-assisted selection complements phenotypic screening at seedling stage to identify cassava mosaic disease-resistant genotypes in African cassava populations. Scientific Reports, 11(1), 2850.

161. Tokunaga, H., Nhan, P. T., Huong, P. T., Anh, N. H., Huong, L. T. M., Hoa, T. M., Trang, N. T. H., Tung, N. B., Thuy, C. T. Le, Zhang, X., Seki, M., & Ham, L. H. (2025). Development of DNA markers for assisted selection of cassava resistant to cassava mosaic disease (CMD). Breeding Science, 75(2), 24046.

162. de Andrade, L. R. B., Sousa, M. B. e, Wolfe, M., Jannink, J.-L., de Resende, M. D. V., Azevedo, C. F., & de Oliveira, E. J. (2022). Increasing cassava root yield: Additive-dominant genetic models for selection of parents and clones. Frontiers in Plant Science, 13.

163. Esuma, W., Ozimati, A., Kulakow, P., Gore, M. A., Wolfe, M. D., Nuwamanya, E., Egesi, C., & Kawuki, R. S. (2021). Effectiveness of genomic selection for improving provitamin A carotenoid content and associated traits in cassava. G3 Genes|Genomes|Genetics, 11(9).

164. Wolfe, M. D., Del Carpio, D. P., Alabi, O., Ezenwaka, L. C., Ikeogu, U. N., Kayondo, I. S., Lozano, R., Okeke, U. G., Ozimati, A. A., Williams, E., Egesi, C., Kawuki, R. S., Kulakow, P., Rabbi, I. Y., & Jannink, J.

(2017). Prospects for Genomic Selection in Cassava Breeding. The Plant Genome, 10(3).

165. Argaw, T., Fenta, B. A., Zegeye, H., Azmach, G., & Funga, A. (2025). Multi-environment trials data analysis: linear mixed model-based approaches using spatial and factor analytic models. Frontiers in Research Metrics and Analytics, 10.

166. Sampaio Filho, J. S., Olivoto, T., Campos, M. de S., & Oliveira, E. J. de. (2023). Multi-trait selection in multi-environments for performance and stability in cassava genotypes. Frontiers in Plant Science, 14.

167. Chavarriaga-Aguirre, P., Brand, A., Medina, A., Prías, M., Escobar, R., Martinez, J., Díaz, P., López, C., Roca, W. M., & Tohme, J. (2016). The potential of using biotechnology to improve cassava: a review. In Vitro Cellular & Developmental Biology - Plant, 52(5), 461–478.

168. Divya, K., Thangaraj, M., & Krishna Radhika, N. (2024). CRISPR/Cas9: an advanced platform for root and tuber crops improvement. Frontiers in Genome Editing, 5.

169. Ma, Q., Tong, W., Cheng, L., Zhang, Y., Wang, Y., Wang, X., Feng, Y., Li, R., Lu, X., Yan, W., & Zhang, P. (2025). Advancing cassava molecular breeding through genome editing: a promising pathway. Tropical Plants, 4(1), 0–0.

170. Talakayala, A., Ankanagari, S., & Garladinne, M. (2022). CRISPR-Cas genome editing system: A versatile tool for developing disease resistant crops. Plant Stress, 3, 100056.

171. Gomez, M. A., Lin, Z. D., Moll, T., Chauhan, R. D., Hayden, L., Renninger, K., Beyene, G., Taylor, N. J., Carrington, J. C., Staskawicz, B. J., & Bart, R. S. (2019). Simultaneous CRISPR/Cas9-mediated editing of cassava eIF4E isoforms nCBP-1 and nCBP-2 reduces cassava brown streak disease symptom severity and incidence. Plant Biotechnology Journal, 17(2), 421–434.

172. Tripathi, J. N., Muiruri, S., & Tripathi, L. (2024). Advancements and challenges in gene editing for improvementof vegetatively propagated crops. Current Opinion in Plant Biology, 82, 102653.

173. Gomez, M. A., Berkoff, K. C., Gill, B. K., Iavarone, A. T., Lieberman, S. E., Ma, J. M., Schultink, A., Karavolias, N. G., Wyman, S. K., Chauhan, R. D., Taylor, N. J., Staskawicz, B. J., Cho, M.-J., Rokhsar, D. S., & Lyons, J. B. (2023). CRISPR Cas9-mediated knockout of CYP79D1 and CYP79D2 in cassava attenuates toxic cyanogen production. Frontiers in J Agri Horti Res, 2025 Plant Science, 13.

174. Juma, B. S., Mukami, A., Mweu, C., Ngugi, M. P., & Mbinda, W. (2022). Targeted mutagenesis of the CYP79D1 gene via CRISPR/Cas9-mediated genome editing results in lower levels of cyanide in cassava. Frontiers in Plant Science, 13.

175. Chamorro, A. F., Palencia, M., & Lerma, T. A. (2025). Physicochemical Characterization and Properties of Cassava Starch: A Review. Polymers, 17(12), 1663.

176. Wang, Y., Ou, X., Al-Maqtari, Q. A., He, H.-J., & Othman, N. (2024). Evaluation of amylose content: Structural and functional properties, analytical techniques, and future prospects. Food Chemistry: X, 24, 101830.

177. Bull, S. E., Seung, D., Chanez, C., Mehta, D., Kuon, J.-E., Truernit, E., Hochmuth, A., Zurkirchen, I., Zeeman, S. C., Gruissem, W., & Vanderschuren, H. (2018). Accelerated ex situ breeding of GBSS - and PTST1 -edited cassava for modified starch. Science Advances, 4(9).

178. Chen, J., Hawkins, E., & Seung, D. (2021). Towards targeted starch modification in plants. Current Opinion in Plant Biology, 60, 102013.

179. Zhou, W., Zhao, S., He, S., Ma, Q., Lu, X., Hao, X., Wang, H., Yang, J., & Zhang, P. (2020). Production of very-high-amylose cassava by post-transcriptional silencing of branching enzyme genes. Journal of Integrative Plant Biology, 62(6), 832–846.

180. Ossai, C. O., Ojobor, S. A., Ogwuche, T. O., Akpeji, S. C., Alama, I. S., & Diebiru-Ojo, E. M. (2025). Performance of four newly released cassava varieties in a fallowed land over two farming seasons.

181. Mukiibi, D. R., Alicai, T., Kawuki, R., Okao-Okuja, G., Tairo, F., Sseruwagi, P., Ndunguru, J., & Ateka, E. M. (2019). Resistance of

advanced cassava breeding clones to infection by major viruses in Uganda. Crop Protection, 115, 104–112.

182. Shirima, R. R., Maeda, D. G., Kanju, E. E., Tumwegamire, S., Ceasar, G., Mushi, E., Sichalwe, C., Mtunda, K., Mkamilo, G., & Legg, J. P. (2019). Assessing the Degeneration of Cassava Under High-Virus Inoculum Conditions in Coastal Tanzania. Plant Disease, 103(10), 2652–2664.

1.5 Enhancing Onion (Allium cepa) Yields: Integrating Precision Irrigation with Real-Time Soil Moisture and Mulching Strategies for Optimum Productivity

Research on World Agricultural Economy

https://journals.nasspublishing.com/index.php/rwae

Jean Nzuma[1], Abigirl Matsaure[2], Godwin Mtetwa[2], Liana-LisaSakwa[1], Linda Munyaradzi[1], Vimbai Samukange[1], Leonard Madzingaidzo[1]*

[1]Biotechnology Research Institute, Scientific and Industrial Research and Development Centre (SIRDC), Harare P. O. Box 6640, Zimbabwe

[2]Chiredzi Research Station, Chiredzi P.O. Box 97, Zimbabwe

1.5.1 Abstract

This study investigated the effects of various mulching conditions and irrigation techniques on the yield of Texas Grano onions (Allium cepa), a key crop for smallholder farmers in Zimbabwe. The primary objective was to identify optimal practices that enhance productivity, water productivity (WP), and water use efficiency (WUE) in onion cultivation. Four mulching treatments were evaluated: control (no mulch), maize residue, soybean trash, and grass. Concurrently, three irrigation methods were assessed: drip irrigation with Chameleon sensors for real-time moisture monitoring, drip irrigation without sensors, and traditional furrow irrigation. A 3 × 4 factorial split-plot design was employed within a randomized complete block design (RCBD) framework, with three replications. The Main plots were allocated for irrigation techniques, while the sub-plots comprised the mulching treatments, including a control. A comprehensive cost-benefit analysis evaluated the economic viability of each treatment. Results indicated that combining drip irrigation with Chameleon sensors and maize residue mulch significantly increased both yields and economic returns. These findings highlight the advantages of adopting precision agricultural practices, particularly real-time moisture monitoring, to improve WP and WUE while enhancing resource utilization, thereby increasing system resilience. To fully harness these benefits farmers should be capacitated and empowered through training on precision irrigation and mulching techniques via demonstration plots and farmer field schools (FFS). Establishing innovation hubs within irrigation schemes is recommended to foster resource sharing and technology adoption, improving productivity and market access for high-value crops. Policymakers must prioritize support for these initiatives to promote sustainable agricultural practices.

Keywords: Irrigation Techniques; Mulching; Moisture Monitoring; Productivity; Economic Viability; Smallholder Farmers; Sustainability; Resource Use

1.5.2 Introduction

Precision irrigation technologies have been extensively researched and widely adopted in many developed countries to improve water-use efficiency and crop yields. The use of drip irrigation systems equipped with soil moisture sensors has significantly increased yields while reducing water consumption by as much as 40% [1–3]. Precision agriculture technologies, such as real-time irrigation scheduling and soil moisture monitoring with Chameleon sensors, are revolutionizing water management in farming by optimizing water use. These sensors ensure water is delivered only when and where it is needed, effectively preventing losses due to evaporation and runoff. By providing real-time information on soil moisture levels, Chameleon sensors enable farmers to make informed irrigation decisions, leading to improved water productivity. This targeted approach not only conserves water but also enhances crop health and yields, making it a sustainable solution for modern

Agriculture [4–9]. In Africa, the adoption of precision irrigation technologies has been slower due to financial and knowledge constraints. Despite these barriers, studies conducted in various countries show promising results when precision irrigation is applied. For instance, drip irrigation combined with soil moisture monitoring has shown significantly increases in crop yields compared to traditional methods [1,3]. Similarly, farmers using advanced irrigation systems experienced substantial water savings while maintaining high yields [9–11]. These findings underscore the potential for widespread bene it if precision irrigation technologies are adopted more broadly across the continent.

Mulching is another practice that has shown significant improvements in crop yields. It functions by conserving soil moisture, moderating soil temperature, and inhibiting weed growth. Studies indicate that organic mulches like maize straw and soybean residue improve water use efficiency and boost crop productivity. However, relatively few studies have examined the combined effects of mulching and precision irrigation for onion production, particularly in Africa. Maize residue has been highlighted as particularly effective mulch due to its high carbon-to-nitrogen ratio, which provides long lasting bene it in terms of moisture retention and soil fertility [8, 12–14]. The interaction between irrigation methods and mulching types can also affect nutrient cycling and soil structure. For example, studies have shown that organic mulches, when used in combination with drip irrigation, improve nutrient availability in the soil, leading to enhanced crop growth and yield [9,14]. Similarly, the use of soybean trash and maize residue has been shown to improve soil aeration and reduce soil compaction, factors critical for optimal root development in onions [4,9]. Another key area of research is the impact of mulching on water-use efficiency. Studies have found that maize residue mulch reduces the amount of water needed for irrigation by improving soil moisture retention. This result is particularly important for regions facing water scarcity. In a similar study, it was observed that mulching reduced water evaporation by up to 30%, leading to higher yields with less water. These results are consistent with finding from studies conducted in India and sub-Saharan Africa, where mulching helped to mitigate the effects of drought on crop production [5, 12, 14].

Research conducted in Zimbabwe, while limited, suggests that smallholder farmers can benefited from the adoption of mulching and improved irrigation practices. One study found that the application of maize residue mulch on maize crops improved yields by 20% in semi-arid regions 12]. In light of these considerations, enhancing agricultural water productivity through precision irrigation is essential for addressing global food security challenges. Water productivity refers to the amount of crop yield produced per unit of water used. This makes it an important indicator that reflects both efficiency and sustainability in agricultural practices [5,8]. By integrating advanced irrigation technologies with effective mulching strategies, farmers can optimize their resource use while maximizing crop outputs [14]. Nonetheless, there remains a gap in thorough research specifically examining the synergy between drip irrigation and mulching for high-value crops such as onions.

Overall, the literature suggests that integrating precision irrigation with mulching could lead to significant improvements in water use efficiency and crop yields. While most studies have focused on irrigation or mulching independently, this study aims to il a critical gap by exploring the interaction between these two factors in the production of Texas Grano onions in

Zimbabwe. The findings of this research will contribute not only to the academic body of knowledge but also provide practical recommendations for smallholder farmers seeking to enhance productivity in areas with limited water resources [1–3,7–9,14,15].

1.5.3 Materials and Methods

1.5.3.1. Study Site

The study was conducted over the winter seasons of 2022 and 2023 at Chiredzi Research Station in Zimbabwe's Natural Region IV, an area known for low rainfall and high temperatures, making it ideal for water efficiency agriculture trials. During the winter seasons, rainfall was scarce, with 2023 receiving slightly lower amounts in May, June, and July compared to 2022, while average minimum and maximum temperatures slightly increased in 2023 (13°C to 31°C) relative to 2022 (12°C to 30°C) (Table1.5.1). Soils at the site were classified as clay loam with a pH ($CaCl_2$) of 6.5, organic matter content of 2.2%, nitrogen (N) at 0.18%, phosphorus (P) at 15 mg kg–1, and potassium (K) at 0.48 meq 100g−1[16,17].

1.5.3.2 Experimental Design

A3×4factorial split-plot design within a randomized complete block design (RCBD) was employed to examine the effects of three irrigation methods and four mulching treatments on Texas Grano onion yield. Each treatment combination was replicated three times on 6 m^2 plots, with 7 cm spacing between plants and 20 cm between rows, resulting in a planting density of approximately 71,429 plants per hectare. This factorial approach allowed for detailed examination of both individual and interactive effects of irrigation and mulching on onion yield (Table 1.5.2).

1.5.3.3. Irrigation and Soil Moisture Management

Texas Grano onion seedlings were transplanted at a density of approximately 71,429 plants per hectare, with in-row and inter-row spacing of 7 cm and 20cm, respectively. To maintain consistent moisture levels, furrow irrigation was applied manually. For the drip irrigation system, two distinct approaches were implemented:

1. Fixed Irrigation Schedule: This method was used for the system without sensors, where irrigation occurred on a predetermined schedule.

2. Real-Time Adjustments: The system equipped with Chameleon soil moisture sensors allowed for real-time adjustments based on feedback from the sensors.

These Chameleon sensors monitor moisture levels in the root zone and utilize color indicators (blue, green, and red) to convey soil suction thresholds:

• Green: Signifies optimal moisture levels, corresponding to suction readings between 20 and 50 kPa.

• Blue: Indicates sufficient moisture, with suction readings below 20 kPa.

• Red: Signals potential water stress, with suction readings above 50 kPa, prompting timely irrigation to prevent yield losses.

Table 1.5.1. 2022 and 2023 rainfall and temperature data (April to July) at the study site.

Month	Total Rainfall (mm)	Min. Temperature (°C)	Max. Temperature (°C)
April	50.0 (29.1)	15.0 (19.0)	32.0 (31.0)
May	40.0 (23.6)	16.0 (20.0)	30.0 (30.0)
June	20.0 (5.8)	17.0 (21.0)	28.0 (29.0)
July	10.0 (0.0)	18.0 (22.0)	27.0 (28.0)

Note: Values in parentheses represent data for the winter season of 2023, while those without parentheses are from 2022. Source:[16]

Table 1.5.2. Design and treatments of the study.

Factor	Treatment	Description
Irrigation systems	Furrow	Manual irrigation using standard buckets. Drip irrigation system equipped with Chameleon sensors for real-time soil moisture data
	Drip irrigation with soil moisture monitoring	Conventional drip irrigation system delivering water

		directly to the base of the plants
Mulch conditions	Control (no mulch) Maize straw Soya bean trash Grass	No mulch applied. Dried maize straw was applied as mulch. Dried soya bean trash was applied as mulch. Dried cut grass

Note: Grass species used as mulch was Bermudagrass (Cynodon dactylon).

By providing real-time feedback, these color-coded sensors ensure efficiency irrigation tailored to the crop's soil moisture needs, which is crucial for optimizing water use in arid climates [7,10,11,15].

1.5.3.4. Water Productivity and Water Use Efficiency Calculations

To assess the effectiveness of various irrigation methods in onion cultivation, calculations for Water Productivity (WP) and Water Use Efficiency (WUE) were conducted [10]. To calculate the metrics, the following formulas were employed:

• Water Productivity (WP) = Yield (t ha−1) / Total Water Applied (m^3 ha−1)

• Water Use Efficiency (WUE) = Yield (t ha−1) / Total Water Used (m^3 ha−1)

In this context, Total Water Applied included both irrigation and effective rainfall received during the growing season. The calculations were performed for each treatment to provide insights into how different irrigation practices impacted water utilization [10].

1.5.3.5. Mulch Application and Management

The mulch materials were uniformly applied to each subplot at the rate of 5 tons per hectare, initially covering the soil surface completely around the plants to a thickness of approximately 5 cm. No additional mulch was applied after the initial application, allowing for the natural decomposition process to take place and simulate the real-life practices of smallholder farmers. The effects of mulching on soil moisture content in the mulched and non-mulched plots were monitored using Chameleon sensors to compare the ability of each mulch type or retain soil moisture and to guide irrigation scheduling.

1.5.3.6. Soil and Plant Management

Texas Grano onion seedlings were transplanted at the 4-leaf stage, and standard fertilizer practices were applied uniformly across all plots to prevent fertility differences from influencing the results. A compound fertilizer (NPK5:15:12) was applied during planting at a rate of 600 kg per hectare. This was followed by a top dressing of ammonium nitrate at a rate of 100 kg per hectare four weeks after transplanting. Regular pest and disease control measures were implemented throughout the season, including the use of registered pesticides and fungicides, to ensure that pest pressure did not confound the experimental results. Weeds were manually removed from all plots.

1.5.3.7. Data Collection

During the study period, irrigation frequency was monitored across all treatments. After harvesting, a comprehensive evaluation of both fresh and dry biomass was conducted to assess crop productivity. This evaluation was essential for determining the effectiveness of the various treatments implemented throughout the growing season. To accurately quantify onion production, key yield components were measured, including average bulb weight, total yield, and marketable yield. A total of twenty pre-tagged plants from each plot were sampled to ensure that the data collected was representative of the entire plot. This method allowed for a more precise analysis of how different irrigation and mulching treatments affected onion growth. The marketable bulb yield was defined as the total weight of disease-free, undamaged bulbs weighing more than 21grams. This criterion ensured that only high-quality produce was included in yield calculations, which is crucial for assessing the economic viability of the onion crop from all plots.

1.5.3.8. Statistical Analysis

The gathered data were analysed using analysis of variance (ANOVA) to assess the effects of various irrigation methods, mulching types, and their interactions on key factors like yield, water-use efficiency, and soil moisture retention. The statistical software SPSS 12.0 facilitated robust interpretations of the results. Mean comparisons among treatment groups were conducted using least significant differences (LSD) at the 5% significance level to identify statistically significant treatment combinations. Additionally, correlation analysis was performed to explore relationships between critical variables, such as bulb weight and total yield, aiming to reveal significant associations that could inform future agricultural practices and improve onion production understanding.

1.5.4. Results

1.5.4.1. Irrigation Frequency across Various Irrigation Techniques

The results of the study revealed distinct irrigation frequencies across three irrigation techniques, as sum marizedinTable3. Each technique exhibited the following irrigation intervals:

• Drip Irrigation with Chameleon Sensors: Operated on an irrigation frequency of every 2 to 3 days. The integration of Chameleon sensors allowed for real-time soil moisture monitoring, enabling precise and frequent watering. This approach ensured consistent soil moisture levels, optimizing conditions for onion growth.

• Drip Irrigation without Sensors: Followed a fixed schedule, with irrigation occurring every 4 to 5 days. While this method is more efficient than traditional practices, it lacks the adaptability to respond to changing soil moisture conditions.

• Traditional Furrow Irrigation: Required irrigation every 5 to 7 days, depending on rainfall and evaporation rates. Although it involved larger volumes of water applied less frequently, this method was less efficient, leading to higher evaporation losses and less consistent soil moisture levels.

Table 1.5.3. Irrigation frequency among various irrigation technique

Irrigation Technique	**Irrigation Frequency (Days)**
Drip irrigation with chameleon sensors	Every 2 to 3days
Drip irrigation without sensors	Every 4 to 5days
Traditional furrow irrigation	Every 5 to 7days

1.5.4.2. Mean Bulb Weight of Onions under Varying Irrigation and Mulch Treatments

The mean bulb weight of onions was significantly influence by the irrigation methods and mulch types employed in the study. The results for mean bulb weight (in grams) for both the 2022 and 2023growing seasons are summarized in Table 1.5.4.

In the Furrow Irrigation treatment, the mean bulb weight was observed to be 58.4 g in 2022 and increased to 60.6 g in 2023, indicating a consistent performance across the years. However, these values were significantly lower than those observed with the other irrigation methods. For Drip Irrigation with Sensors, a substantial increase in mean bulb weight was recorded, with values of 99.4 g in 2022 and 102.0 g in 2023.This method yielded the highest bulb weights across both seasons, demonstrating its effectiveness in enhancing onion growth.

In the Drip Irrigation without Sensors treatment, mean bulb weights were also notable, with averages of 84.9g in 2022 and 86.0 g in 2023. Although these figures were lower than those from the sensor-assisted drip irrigation, they still exceeded the weights obtained under the furrow irrigation method.

The mean bulb weights for onions grown with various mulch types were also evaluated. Under maize residue, mean bulb weights were 96.7 g in 2022 and 99.7 g in 2023, showing as light improvement in the second year. The Soya Bean Trash treatment yielded average bulb weights of 85.6 g in 2022 and 87.6 g in 2023, while the Grass mulch resulted in mean bulb weights of 74.9g in 2022 and 76.4g in 2023

Table 1.5.4: Mean bulb weight (g) for different irrigation methods and mulch types (2022 and 2023).

Irrigation Method	Control		Maize Residue		Soybean trash residue		Grass		Mean Bulb weight		CV%	SE	LSD
	2022	2023	2022	2023	2022	2023	2022	2023	2022	2023			
Furrow irrigation	50.5^{a}	52.3^{a}	70.2^{b}	73.8^{b}	58.3^{c}	60.1^{c}	54.5^{c}	56.2^{c}	58.4^{c}	60.6^{c}	10%	1.82	3.65
Dip irrigation with sensors	78.5^{d}	80.1^{d}	120.3^{e}	123.5^{e}	108.2^{d}	110.4^{d}	90.6^{d}	93.8^{d}	99.4^{d}	102^{d}	14%	2.56	5.12
Drip without sensors	70.2^{c}	72.1^{c}	99.5^{d}	102.0^{d}	90.3^{c}	92.5^{c}	79.5^{c}	81.2^{c}	84.9^{c}	86.0^{c}	18%	3.28	6.56
Mean bulb weight	66.4	68.1	96.7	99.7	85.6	87.6	74.9	76.4	81.6	83.0	20%	3.65	7.30

Note: Values in parentheses represent the average bulb weight for the 2023 winter season. The mean separation letters (a,b, c,d,e) indicate groups that are statistically similar based on the analysis of variance (ANOVA) conducted on the data from both years (2022 and 2023). For instance, treatments sharing the same letter are not significantly different from each other, highlighting comparable performance in bulb weight across the two seasons. This ensures that any observed differences in average bulb weight can be attributed to their irrigation methods and mulch types rather than seasonal variability.

1.5.4.3. Interaction Effects on Yield and Marketable Yield

Results suggest significant interaction effects between irrigation methods and mulch types on yield and marketable yield, with both factors showing highly significant results ($P<0.001$) in 2022, with a similar trend in 2023 (Table 1.5.5). The interaction also showed a trend towards significance ($P=0.05$), suggesting a potential combined influence of irrigation methods and mulch types on yield outcomes.

1.5.4.4. Furrow Irrigation

Marketable yields ranged from 8.91 to 12.92 tha^{-1}in 2022 and 9.45 to 13.30 tha^{-1} in 2023, while total yields ranged from 9.91 to 13.92 tha^{-1}. The addition of maize residue mulch significantly increased yield under furrow irrigation, with a mean boost of approximately 4.01tha^{-1} over the control ($p<0.05$), highlighting maize residue's potential in optimizing yield despite furrow irrigation's lower efficiency.

1.5.4.5. Drip Irrigation with Chameleon Sensors

This method produced the highest marketable yields, with values from14.47 to 22.80 tha^{-1} in 2022 and 15.00 to 23.10 tha^{-1} in 2023, and total yields from 15.47 to 23.80 tha^{-1}. Utilizing Chameleon sensors significantly enhanced yield, with up to an 8.33 tha^{-1} increase over the control ($p<0.05$). This emphasizes the benefit of sensor technology in maximizing irrigation efficiency and yield outcomes.

1.5.4.6. Drip Irrigation without Sensors

Marketable yields for this method were between 12.92 to 18.70 tha^{-1} in 2022 and 13.40 to 18.95 tha^{-1} in 2023, while total yields ranged from 13.92 to 19.70 tha^{-1}. Although effective, it delivered yields approximately 3.88 tha^{-1} lower than sensor-assisted drip irrigation ($p<0.05$), under scoring the added value of sensors in further improving efficiency.

Table 1.5.5: Marketable yield (tha^{-1}) for different irrigation methods and mulch types (2022 and 2023)

Irrigation Method	Control		Maize Residue		Soya Bean Trash		Grass		Mean Marketable Yield		CV %	SE	LSD
	2022	2023	2022	2023	2022	2023	2022	2023	2022	2023			
Furrow irrigation	8.91^{a}	9.45	12.9^{b}	13.3	10.7^{c}	11.1	9.90c	10.2	10.60^{c}	11.02	12%	1.76	3.50
Drip with sensors	14.5^{d}	15.0	22.8^{e}	23.1	20.5^{d}	20.8	16.9^{d}	17.3	18.68^{d}	19.05	16%	2.47	4.95
Drip without sensors	12.9^{c}	13.4	18.7^{d}	18.9	16.9^{c}	17.2	14.70^{c}	15.1	15.80^{c}	16.16	19%	3.15	6.35
Mean marketable yield	12.10	12.62	18.14	18.45	16.02	16.4	13.84	14.2	15.0	12.2	21%	3.52	7.20

Note: Values in parentheses represent the average bulb weight for the 2023 winter season. The mean separation letters (a,b, c,d,e) indicate groups that are Statistically similar based on the analysis of variance (ANOVA) conducted on the data from both years (2022 and 2023). For instance, treatments sharing the same letter are not significantly different from each other, highlighting comparable performance in bulb weight across the two seasons. This ensures that any observed differences in average bulb weight can be attributed to the irrigation methods and mulch types rather than seasonal variability.

1.5.4.7 Impact of Mulching Conditions on Yield

1. Control (No Mulch): Marketable yields for the control were the lowest, ranging from8.91 to 12.92 tha^{-1} in 2022 and 9.45 to 13.30 tha^{-1} in 2023, with total yields spanning 9.91to 13.92t ha^{-1}. These yields were significantly lower than all other mulch treatments ($p<0.05$), confirming mulching's positive role in boosting yield.

2. Grass Mulch: Grass mulch marketable yields ranged from 9.90 to 16.92tha$-^{1}$ in 2022 and 10.20 to 17.30 tha^{-1} in 2023, showing improvement over the control but falling short of maize Residue and soybean trash by 1.92 to 5.88tha^{-1} ($p<0.05$). Total yields were between 10.90 to 17.92 tha^{-1}.

3. Maize Residue: Maize residue application led to the highest marketable yields, ranging from12.92 to 22.80 tha^{-1}in 2022 and 13.30 to 23.10 tha^{-1} in 2023, with a notable 4.01 to 9.8 8 tha^{-1} increase over the control ($p<0.05$). Total yields ranged from 13.92 to 23.80 tha^{-1}, making maize residue the most effective mulch type.

4. Soya Bean Trash: Marketable yields with soybean trash ranged from 10.66 to 20.53 tha^{-1} in 2022 and 11.12 to 20.80 tha^{-1} in 2023, exceeding the control and grass mulch by 1.75 to 8.61 tha^{-1} ($p<0.05$). Total yields ranged from11.66 to 21.53 tha^{-1}, indicating soybean trash as a viable mulching option, although maize residue out per formed it.

1.5.5. Water Productivity and Water Use Efficiency Analysis Gross Irrigation Applied

Table 1.5.6 summarizes the gross irrigation applied for each treatment, taking into account the irrigation methods used and the rainfall received during the growing season. The table provides a comparison of gross irrigation applied, total water applied, yield, and water related efficiency metrics for various irrigation treatments:

1. Drip Irrigation with Sensors: This method achieved the highest yield of 25tha^{-1}, resulting in a water productivity (WP) of 0.0093tm^{-3} and a water use efficiency (WUE) of 0.0093 tm^{-3}.

2. Drip Irrigation without Sensors: This approach yielded 23tha^{-1}, maintaining a competitive WP of 0.0072 tm^{-3} and WUE of 0.0072 tm^{-3}.

3. Furrow Irrigation: This method resulted in a lower yield of 20tha^{-1}, with a WP of 0.0042t m^{-3} and WUE of 0.0042 tm^{-3}, indicating less efficiency water use.

4. Control Treatment (No Mulch): This treatment yielded the least at 18tha−1, with a WP of 0.0035 tm^{-3} and WUE of 0.0035tm^{-3}.

1.5.5.1 Interactions between Irrigation and Mulching

The correlation analysis performed in this study highlights significant relationships among the variables yield, marketable yield, and average bulb weight (Table1.5.7). Relationships among Yield, Marketable Yield, and Average Bulb Weight •Yield and marketable yield: A robust positive corelation of 0.95 indicates that increases in total production are closely linked to a rise in the quantity of bulbs that meet market quality standards.

Table 1.5.6. Gross irrigation, yield, water productivity (WP), and water use efficiency (WUE) in onion cultivation (2022 and 2023)

Treatment	Gross Irrigation Applied (m3 ha−1)	Rainfall (m3 ha−1)	Total Water Applied (m3 ha−1)	Yield (t ha−1)	WP (tm−3)	WUE (tm−3)
Drip irrigation with sensors	2550	130(2022)/60 (2023)	2680/2610	25	0.0093	0.0093
Drip irrigation without sensors	3050	130(2022)/60 (2023)	3180/3110	23	0.0072	0.0072
Furrow irrigation	4600	130(2022)/60 (2023)	4730/4660	20	0.0042	0.0042
Control (no mulch)	5050	130(2022)/60 (2023)	5180/5110	18	0.0035	0.0035

• Yield and average bulb weight: The correlation coefficient of 0.89 implies that higher yields are associated with larger bulb sizes, likely resulting from favourable growing conditions. • Marketable yield and average bulb weight: A positive correlation of 0.87 indicates that as marketable yield increases, the average weight of the bulbs also tends to increase.

1.5.6 Economic Implications of Mulching Treatments in Onion Production

The cost-benefit analysis (CBA) was conducted using a partial budget approach to assess the economic viability of four mulching treatments (maize residue, soybean trash, grass, and a control group with no mulch). This analysis examined parameters such as mulch costs, yield data, revenue generation, cost savings, and net benefits which are summarized in Table 1.5.8 below.

1.5.6.1. Yield and Revenue Analysis of Mulching Treatments in Onion Production

The evaluation of mulching treatments in onion production reveals significant differences in yield and revenue generation among the various options. As presented in Table 1.5.8, maize residue emerged as the most productive treatment, achieving a yield increase of 3.85 tons per hectare. This substantial yield translated into a remarkable revenue increase of $1,925 in 2022 and $1,585 in 2023, highlighting the economic advantages of using maize residue as mulch. Following maize residue, soybean trash demonstrated a yield increase of 2.30 tons per hectare, resulting in a revenue increase of $1,150 in 2022 and $805 in 2023. The grass mulch treatment offered a more modest yield increase of 1.50 tons per hectare, leading to a revenue increase of $780. In stark contrast, the control group, which did not utilize any mulch, resulted in no yield increase or revenue generation, underscoring the critical role of mulching in enhancing onion productivity.

1.5.6.2. Net Benefit Analysis

The net benefit analysis reveals the economic viability of the mulching treatments. Maize residue achieved a net benefit of $1,615 per hectare in 2022 and $1,585 in 2023, establishing it as the most profitable option among the treatments evaluated. Conversely, soybean trash provided a net benefit of $835 in 2022 and $805 in 2023. Grass mulch, while contributing positively to yield, yielded a stable net benefit of $430 across both years. The control group, however, displayed a negative net benefit of –$300, highlighting the detrimental economic impact of not employing any mulching strategy.

1.5.7 Discussion

The findings from the 2022 and 2023 studies on onion cultivation reveal critical insights into the impact of various irrigation methods on Gross Irrigation, Yield, Water Productivity (WP), and Water Use Efficiency (WUE). The comparative analysis of these parameters illustrates the significant advantages of advanced irrigation technologies, particularly sensor-equipped drip irrigation systems.

1.5.7.1. Implications of Irrigation Frequency on Crop Productivity

The study demonstrates significant variations in irrigation frequency across different onion cultivation techniques. Drip irrigation with Chameleon sensors, allowing irrigation every 2 to 3 days, is identified as the most effective method for maintaining optimal soil moisture.

Table 1.5.7. Correlation matrix for yield and marketable yield, and average bulb weight

Variable	Yield	Marketable Yield	Average Bulb Weight
Yield	1	0.95	0.89
Marketable yield	0.95	1	0.87
Average bulb weight	0.89	0.87	1

This technology promotes efficiency water use and prevents over-irrigation by relying on real-time soil moisture data [7,8,11,15,18]. In contrast, drip irrigation without sensors, which operates on a fixed schedule of every 4 to 5 days, may lead to sub optimal moisture conditions. Traditional furrow irrigation, requiring watering every 5 to 7 days, is the least efficient, resulting in greater evaporation losses and inconsistent soil moisture that can adversely affect onion growth.

These finding highlight the critical need for advanced irrigation technologies, like sensor-based systems, to enhance water use efficiency (WUE) and improve crop yields. This transition is vital for smallholder farmers facing water scarcity and climate variability challenges. Investing in such innovations not only supports sustainable agricultural practices but also promotes better economic outcomes and resilience within farming communities [3,10,11,15,18]. Advantages of Drip Irrigation with Chameleon Sensors Drip irrigation with Chameleon sensors offers significant benefit in terms of water use efficiency (WUE) due to their ability to provide precise, real-time soil moisture data. By measuring soil moisture tension, these sensors allow farmers to apply water only when and where it is needed, minimizing waste and preventing over-irrigation. The study demonstrates that the use of drip irrigation with Chameleon sensors leads to enhanced WUE by allowing for precise irrigation scheduling based on real-time soil moisture data. This targeted approach minimizes water waste and ensures that crops receive adequate moisture, thereby maximizing productivity while conserving water resources [2,7,10,15]. As discussed in Section 4.2.1, the analysis shows that using these technologies significantly improves yield performance compared to traditional methods, with drip irrigation with sensors achieving a yield of 25 t ha−1 and a corresponding WUE of 0.0093 t m^{-3}. This reinforces the study's aim to foster the application of these technologies in the cultivation of Texas Grano onions (Allium cepa).

1.5.7 Yield Performance across Irrigation Methods

1.5.7.1 Water Productivity and Water Use Efficiency Analysis

The analysis of water productivity (WP) and water use efficiency (WUE) further highlights the benefit of precision irrigation techniques. As summarized in Table 1.5.6, drip irrigation with sensors achieved the highest yield of 25 t ha−1, resulting in a WP of 0.0093 t ma^{-3} and a WUE of 0.0093 t m^{-3}. These metrics illustrate a clear advantage in resource use, demonstrating that sensor-based irrigation not only maximizes yields but also optimizes the efficiency of water use. This aligns with finding from [2,7,8,10,11] who demonstrated that integrating soil moisture sensors significantly reduces water usage while maintaining or increasing crop yields. The study clearly demonstrates that the implementation of drip irrigation with Chameleon sensors leads to enhanced WUE by allowing for precise irrigation scheduling based on real-time soil moisture data. This targeted approach minimizes water waste and ensures that crops receive adequate moisture, thereby maximizing productivity while conserving water resources [4,6,15,19]. Similarly, Tiruye et al. [20], demonstrated that efficient water management technologies significantly improve water productivity and nutrient balances for irrigated crops. Conversely, drip irrigation without sensors yielded 23tha^{-1}, maintaining a competitive WP of 0.0072 t m^{-3} and WUE of 0.0072 t m^{-3}. Although slightly less efficient than the sensor-equipped system, this method still demonstrates the benefit of drip irrigation over traditional methods. In contrast, furrow irrigation resulted in a lower yield of 20 t ha−1, with a WP and WUE of 0.0042 t m^{-3}, indicating less efficiency in water use.

Table 1.5.8: Cost-benefit analysis of mulching treatments in onion production

Parameter	Maize Residue	Soybean Trash	Grass	Control (No Mulch)
Cost of mulch	$50	$45	$40	$0
Application cost	$10	$10	$10	$0
Fertilizer cost (2022)	$200	$200	$200	$200
Fertilizer cost (2023)	$230	$230	$230	$230
Labor cost	$100	$100	$100	$100
Total cost (2022)	$360	$355	$350	$300
Total cost (2023)	$390	$385	$380	$300
Market price of onions (per ton)	$500	$500	$500	$500
Cost savings	$50	$40	$30	$0
Total benefit	$1,975	$1,190	$780	$0

Yield increase (tons)	3.85	2.30	1.50	0.00
Market price of onions (per ton)	$500	$500	$500	$500
Revenue increase (2022)	$1,925	$1,150	$750	$0
Revenue increase (2023)	$1,585	$805	$750	$0
Cost savings	$50	$40	$30	$0
Total benefit	$1,975	$1,190	$780	$0
Net benefit (2022)	$1,615	$835	$430	–$300
Net benefit (2023)	$1,585	$805	$430	–$300

Notes:

1. Cost of Mulch: This includes the expenses associated with the materials used for mulching treatments (maize residue, soybean trash, and grass), which contribute to moisture retention and weed suppression in onion production.

2. Application Cost: This refers to the consistent expenses incurred for applying the mulch materials across all treatments.

3. Fertilizer Cost: Fertilizer costs are indicated for both years to demonstrate changes in input expenses. The increase from 2022 to 2023 reflects market fluctuation and inflation impacting overall production costs.

4. Labor Cost: The labor cost of $100 per hectare remains constant for both years, reflecting standard expenses for the application and management of the mulching treatments.

5. Total Cost: Total costs represent the sum of all input costs (mulch, application, fertilizer, and labor) for each treatment in both years. The control treatment (no mulch) shows significantly lower total costs but results in no yield increase.

6. Yield Increase: The yield increase is expressed in tons per hectare and reflects the additional onion production achieved through the application of different mulching treatments compared to the control.

7. Market Price of Onions: The market price of onions is assumed to be constant at $500 per ton for this analysis, although actual prices may vary based on market conditions.

8. Net Benefit: Net benefits are calculated as the difference between revenue increases and total costs. Negative net benefit indicates losses associated with the control treatment, under scoring the economic advantages of mulching.

9. The term “per hectare” has been removed from the table headings to streamline presentation, as all values are inherently based on a per-hectare assessment. Key finding indicate that maize residue provides the most significant yield increase of 3.85 tons and a net benefit of $1,585 after accounting for increased fertilizer costs, highlighting its economic viability in onion production.

The control treatment (no mulch) yielded the least at 18 tha^{-1}, with a WP and WUE of 0.0035 tm^{-3}, reinforcing the notion that insufficient irrigation strategies lead to decreased productivity. The results highlight the critical role of irrigation technology in enhancing water productivity (WP) and water use efficiency (WUE). For instance, [1–5,7,8,10,11,15,19,20] found that precision irrigation systems, including those using Chameleon sensors, optimize water use efficiency through accurate soil moisture data, which helps in making informed irrigation decisions. Furthermore, Lakhiar et.al and Abebe et.al. [17,18], demonstrated that smallholder farms employing advanced irrigation systems show improved water efficiency and crop performance, supporting the notion that such technologies are essential for sustainable agricultural practices.

1.5.7.2 Economic Benefit and Sustainability Implications of Mulching Practices

The economic analysis of various mulching treatments highlights the substantial benefit of integrating precision irrigation and mulching practices in onion production. The cost-benefit analysis demonstrated that maize residue mulch achieved the highest net benefit of $1,615 per hectare in 2022 and $1,585 per hectare in 2023 (Table 1.5.8), representing the most favorable return on investment among the mulching options. This finding reinforces the notion that sustainable agricultural practices can lead to significant economic advantages for farmers, aligning with previous studies that emphasize the economic benefit of organic mulches for improving crop performance [9,11,18,21]. The notable increase in yield, with maize residue providing a yield increase of 3.85 tons per hectare, coupled with significant cost savings of $50 per hectare for irrigation and weed management, positions maize residue as a highly efficient choice for enhancing onion production while minimizing overall input expenses. This economic implication extends beyond immediate cost savings; it contributes to the long-term sustainability of farming operations by promoting soil health and resilience 12, 22]. In contrast, while soybean trash offered considerable economic advantages with a net benefit of $835per hectare in 2022 and $805 per hectare in 2023, it did not match the net benefit of maize residue. Nonetheless, it still provided a substantial yield increase of 2.30 tons per hectare and cost savings of $40 per hectare, making it a practical option for farmers aiming to boost their productivity and profitability. Grass mulch, resulting in a net benefit of $430 per hectare in both years, demonstrated lower yield increases compared to maize residue and soybean trash but still provided some economic advantages, particularly in maintaining soil moisture and temperature [14,23].

1.5.8 Implications, Considerations, and Directions for Smallholder Farmers

1.5.8.1. Implications for Smallholder Farmers

The finding from this study highlights the significant economic advantages associated with the application of mulch, particularly maize residue. This type of mulch has emerged as the most beneficial option for smallholder farmers, offering a cost-effective method to enhance crop yields while simultaneously minimizing overall input expenses. By utilizing maize residue, farmers can improve their onion production and reduce reliance on costly inputs such as fertilizers and pesticides, thereby increasing their profit margins. Additionally, the use of mulch serves a dual purpose: it boosts crop yields and leads to substantial cost reductions. By decreasing the need for additional irrigation and reducing the frequency of weed control measures, mulching becomes an essential agricultural practice for smallholder farmers. This cost efficiency is particularly critical for farmers operating with limited resources, as it allows them to allocate their finance more effectively while achieving high levels of productivity. The economic implications of adopting these practices extend beyond immediate cost savings; they contribute to the long-term sustainability of farming operations by promoting soil health and resilience. For smallholder farmers, the adoption of these practices can lead to improved food security and economic stability in their communities. Enhanced onion production can contribute to local markets, providing fresh produce and increasing farmers' incomes [22–25].

1.5.8.2. Environmental Considerations

The environmental benefit of adopting precision irrigation and mulching practices are significantly. By enhancing water-use efficiency, these practices help conserve water resources, which is becoming increasingly crucial due to climate change and rising water scarcity. Additionally, employing organic mulches, such as maize residue, improves soil health by boosting organic matter levels, enhancing soil structure, and fostering beneficial microbial activity.

These environmental benefits extend beyond individual farms, contributing to broader ecosystem health. Healthy soils are more resilient to erosion and degradation, supporting biodiversity and ecosystem services that are vital for sustainable agriculture. Furthermore, by decreasing reliance on chemical fertilizers and pesticides, mulching practices can result in reduced agricultural runoff, thereby safeguarding the water quality of nearby streams and rivers [6,8,9,11,12,14,18,19,23,25].

1.5.8.3. Socio-Economic Factors

The socio-economic factors influencing the adoption of precision irrigation and mulching practices among smallholder farmers warrant further investigation. Key barriers include access to credit, knowledge gaps, and resource limitations [11]. Understanding these barriers can inform targeted interventions that promote sustainable agricultural practices.

• Access to Financial Resources: Many smallholder farmers struggle with limited access to credit due to perceived risks by financial institutions. Targeted micro finance programs or subsidies could alleviate these financial barriers and encourage broader adoption of precision irrigation and mulching practices [10].

• Education and Knowledge Transfer: Low educational levels can hinder farmers' understanding of modern techniques. Enhancing extension services and farmer training programs can provide crucial knowledge about technology benefit and implementation strategies [4,10,12,19].

• Social Networks: Membership in farmer organizations can facilitate resource sharing and collective action, enhancing market participation [11,19].

• Gender Dynamics: Addressing gender disparities is vital for ensuring that both male and female farmers benefit from sustainable practices [10].

• Infrastructure Challenges: Improving infrastructure is necessary to support market access for Smallholders [24].

1.5.8.4. Policy Recommendations

To effectively address socio-economic barriers, policymakers should consider the following strategies:

1.5.8.4.1 Developing Targeted Microfinance Programs:

These programs should provide accessible credit options specifically for investments in precision irrigation and mulching. Enhancing Agricultural Extension Services: Focus on education and training related to sustainable practices to empower farmers with the knowledge they need.

Establishing Innovation Hubs: Creating hubs within irrigation schemes can promote farmer organization membership, facilitating resource sharing, knowledge exchange, and technology adoption among farmers, thereby enhancing productivity and market access for high-value crops [5,7,10,11,15,19]. To implement these policy recommendations effectively, specificity in execution is essential. Policymakers could introduce a subsidy scheme that covers a percentage of the costs associated with purchasing precision irrigation equipment. Additionally, establishing Farmer Field Schools (FFS) can facilitate hands-on training in precision irrigation and mulching practices. These training programs should include workshops and demonstration trials conducted in Lead Farmer field to show case successful case studies and the tangible benefit of adopting these technologies [11,19].

1.5.9. Study Limitations

Controlled Conditions vs. Field Variability: The study was conducted under controlled conditions, which may not fully reflect the variability encountered in actual settings.

1. Short Duration of Study: Data collection over two years may not capture the long-term sustainability and impacts of the implemented practices.

2. Limited Geographic Scope: The findings are based on a specific geographical location, potentially limiting their applicability to other regions with different climatic and soil conditions.

3. Technology Adoption Barriers: The study does not extensively explore farmer perceptions or resistance to adopting new precision irrigation and mulching practices.

1.5.10. Research Gaps and Future Directions

Although this study offers important insights into the advantages of precision irrigation and mulching, several research gaps remain:

1. Long-Term Effects: Future investigations should examine the long-term effects of precision irrigation and mulching on soil health, crop yield, and economic sustainability to assess their viability over extended periods [2].

2. Diverse Organic Mulches: Research should explore the effects of different types of organic mulches on various crops, as the impact of mulch can vary significantly depending on the material used [12,14].

3. Integration of Cover Crops: There is a need to investigate the potential for integrating cover crops into precision irrigation and mulching systems, which could enhance soil health and improve overall crop productivity [9].

4. Farmer Adoption Factors: Further studies should focus on understanding the socio-economic factors influencing farmer adoption of these practices, including access to resources and training, to facilitate broader implementation in diverse agricultural contexts [11,18].

1.5.11 Conclusions

This study highlights the significant benefit of precision irrigation and mulching in enhancing onion productivity in semi-arid regions. Key finding demonstrates that drip irrigation equipped with Chameleon sensors markedly improves water use efficiency, allowing for real-time moisture management and resulting in higher yields compared to traditional methods. Additionally, the use of maize residue mulching enhances moisture retention, soil health, and nutrient availability, contributing to greater resilience against climate stressors. Smallholder farmers can achieve economic gains by adopting these practices, as they lead to reduced input costs and increased profitability. The research underscores the need for supportive policies and resources to encourage the adoption of these innovative technologies.

Ultimately, the combination of advanced irrigation and effective mulching not only fosters sustainable agricultural practices but also enhances food security and promotes environmental stewardship, paving the way for a more resilient agricultural landscape.

Author Contributions

Conceptualization, J.N., L.-L.S., and V.S.; methodology, J.N., L.-L.S., L.M. (Linda Munyaradzi), and A.M.; software, G.M.; validation, J.N., A.M., and G.M.; formal analysis, G.M.; investigation, J.N., A.M., L.M. (Leonard

Madzingaidzo), and L.-L.S.; resources, V.S.; data curation, G.M., A.M., and J.K.N.; writing—original draft preparation, J.N.; writing—review and editing, J.N., G.M., A.M., and L.-L.S.; visualization, G.M.; supervision, project administration and funding acquisition, L.M.(Linda Munyaradzi). All authors have read and agreed to the published version of the manuscript. Authorship is limited to those who have contributed substantially to the work reported.

Funding

This research was supported by the Korea Africa Food and Agriculture Cooperation Initiative.

Institutional Review Board Statement

Not applicable.

Informed Consent Statement

Not applicable.

Data Availability Statement

The data that support the results presented in this article are available and can be accessed by the public at no cost. For further details or specific inquiries, please contact thecorrespondingauthoratjknzumayahoo.com.

1.5.12 Acknowledgments

We extend our sincere gratitude to our extension partners, Agricultural Rural Development and Advisory Services (ARDAS) in Zimbabwe, for their invaluable support and assistance throughout this study. Their knowledge and commitment have been crucial to the success of our research. We also wish to acknowledge the significant support provided by the stage gate teams at the Scientific and Industrial Research and Development Centre (SIRDC) and Chiredzi Research Institute. Their collaboration and resources have greatly facilitated our research efforts.

Conflict of Interest

The authors declare that they have no financial or personal relationships that might be viewed as potential conflict of interest in the execution and reporting of this research.

1.5.13 References

[1] Moursy, M.A.M., ElFetyany, M., Meleha, A.M.I., et al., 2023. Productivity and profitability of modern irrigation methods through the application of on-farm drip irrigation on some crops in the Northern Nile Delta of Egypt. Alexandria Engineering Journal. 62, 349–356. DOI: https://doi.org/10.1016/j.aej.2022.06.063

[2] Yu, L.M., Gao, W.L., Shamshiri, R.R., et al., 2021. Review of research progress on soil moisture sensor technology. International Journal of Agricultural and Biological Engineering. 14(4), 32–42. DOI: https://doi.org/10.25165/j.ijabe.20211404.6404

[3] Yang, P., Wu, L., Cheng, M., et al., 2023. Review on Drip Irrigation: Impact on Crop Yield, Quality, and Water Productivity in China. Water, 15(9), 1733. DOI: https://doi.org/10.3390/w15091733

[4] Bwambale, E., Abagale, F.K. Anornu, G.K., 2022. Smart irrigation monitoring and control strategies for improving water use efficiency in precision agriculture: A review. Agricultural Water Management. 260, 107324. DOI: https://doi.org/10.1016/j.agwat.2021.107324

[5] Levidow, L., Zaccaria, D., Maia, R., et al., 2014. Improving water-efficiency irrigation: Prospects and difficulties of innovative practices. Agricultural Water Management. 146, 84–94. DOI: https://doi.org/10.1016/j.agwat.2014.07.012

[6] Zhang, X., Feng, G., Sun, X., 2024. Advanced technologies of soil moisture monitoring in precision agriculture: A Review. Journal of Agriculture and Food Research. 18, 101473. DOI:https://doi.org/10. 1016/j.jafr. 2024.101473

[7] Stirzaker, R., Driver, M., 2024. Soil water sensors that display colours as thresholds for action. International Journal of Water Resources Development. 1–19. DOI:https://doi.org/10.1080/07900627.2024.23 22153

[8] Mdemu, M., Kissoly, L., Bjornlund, H., et al., 2020. The role of soil water monitoring tools and agricultural innovation platforms in improving food security and income of farmers in smallholder irrigation schemes in Tanzania. International Journal of Water Resources Development.

36(Suppl. 1), S148–S170. DOI:https://doi.org/10.1080/07900627.2020.1765746

[9] Vanella, D., Guarrera, S., Ferlito, F., et al., 2025. Effects of organic mulching and regulated deficit irrigation on crop water status, soil and yield features in an orange or chard under Mediterranean climate. Science of The Total Environment. 958, 177528. DOI:https://doi.org/10.1016/j.scitotenv.2024.177528.

[10] Moyo, M., Van Rooyen, A., Bjornlund, H., et al., 2020. The dynamics between irrigation frequency and soil nutrient management: transitioning smallholder irrigation towards more profitable and sustainable systems in Zimbabwe. International Journal of Water Resources Development. 36(1), 102–126. DOI:https://doi.org/10.1080/07900627.2020.1739513

[11] Chilundo, M., De Sousa, W., Christen, E.W., et al., 2020. Do agricultural innovation platforms and soil moisture and nutrient monitoring tools improve the production and livelihood of smallholder irrigators in Mozambique? International Journal of Water Resources Development. 36(1), 127–147. DOI:https://doi.org/10.1080/079006 27.2020.1760799

[12] Mupangwa, W., Nyagumbo, I., Mutsamba, E., 2016. Effect of various mulching materials on maize growth and yield in conservation agriculture systems of sub-humid Zimbabwe. AIMS Agriculture and Food. 1(2), 239–253. DOI:https://doi.org/10.3934/agrfood.2016.2.239

[13] Bjornlund, H., Van Rooyen, A., Stirzaker, R., 2017. Profitability and productivity barriers and opportunities in small-scale irrigation schemes. International Journal of Water Resources Development. 33(5), 690–704. DOI:https://doi.org/10.1080/07900627.2016.1263552

[14] Nouri, H., Stokvis, B., Galindo, A., et al., 2019. Water scarcity alleviation through water foot print reduction in agriculture: The effect of soil mulching and drip irrigation. Science of The Total Environment. 653, 241–252. DOI:https://doi.org/10.1016/j.scito tenv.2018.10.311

[15] Stirzaker, R., Mbakwe, I., Mziray, N.R., 2017. A soil water and solute learning system for small-scale irrigators in Africa. International Journal of Water Resources Development. 33(5), 788–803. DOI:https://doi.org/10.1080/07900627.2017.1320981

[16] Climate-Data.org, 2022–2023. Climate: Chiredzi. Climate-Data.org. Available from: https://en.climate-data.org/africa/zimbabwe/masvingo-province/chiredzi-55090/

[17] Lakhiar, I.A., Yan, H., Zhang, C., et al., 2024. A Review of Precision Irrigation Water-Saving Technology under Changing Climate for Enhancing Water Use Efficiency, Crop Yield, and Environmental Footprints. Agriculture. 14, 1141. DOI:https://doi.org/10.3390/agriculture14071141

[18] Abebe, F., Zuo, A., Wheeler, S.A., et al., 2020. Irrigators' Willingness to Pay for the Adoption of Soil Moisture Monitoring Tools in South-Eastern Africa. International Journal of Water Resources Development. 36(Suppl. 1), S246–S267. DOI:https://doi.org/10.1080/07900627.2020.1755956

[19] Van Rooyen, A.F., Ramshaw, P., Moyo, M., et al., 2017. Theory and application of Agricultural Innovation Platforms for improved irrigation scheme management in Southern Africa. International Journal of Water Resources Development. 33(5), 804–823. DOI:https://doi.org/10.1080/07900627.2017.1321530

[20] Tiruye, A.E., Asres Belay, S., Schmitter, P., et al., 2022. Yield, water productivity and nutrient balances under different water management technologies of irrigated wheat in Ethiopia. PLoS Water. 1(12), e0000060. DOI:https://doi.org/10.1371/journal.pwat.0000060

[21] Kumar, P., Lata, S., Swapnil, S.A., et al., 2022. Effect of Tillage and Mulching on Growth, Yield and Economics of Maize Crops. International Journal of Bioresource and Stress Management. 13(4), 332–338. DOI:https://doi.org/10.23910/1.2022.2362a

[22] Qin, W., Hu, C., Oenema, O., 2015. Soil mulching significantly enhances yields and water and nitrogen use efficiencies of maize and wheat: a meta-analysis. Scienti i Reports. 5, 16210. DOI:https://doi.org/10.1038/srep16210

[23] Ayala, G., Bulti, A., Muleta, B., 2023. Evaluation of Mulching and Tied Ridges on Soil Moisture and Yield of Maize at Daro Lebu District, Western Hararghe Zone, Oromia, Ethiopia. International Journal of

Agricultural Economics. 8(3), 116–121. DOI:https://doi.org/10.11648/j.ijae.20230803.16

[24] Adetoro, A.A., Abraham, S., Paraskevopoulos, A.L., et al., 2020. Alleviating water shortages by decreasing water footprint in sugarcane production: The impacts of different soil mulching and irrigation systems in South Africa. Groundwater for Sustainable Development. 11, 100464. DOI: https://doi.org/10.1016/j.gsd.2020.100464

[25] Telila, R.G., Naneso, D.A., Diriba, M.E., et al., 2024. Evaluation of different moisture conservation practices on maize production and productivity in Dallomanna Districts of Bale Lowland Southeastern Ethiopia. Journal of Scientific Agriculture. 8, 59–65. DOI: https://doi.org/10.25081/jsa.2024.v8.8995

1.6 Evaluation of the Interaction Between *Beauveria Bassiana* and an Aphicide for the Management of the Tobacco Aphid *Myzus Persicae Nicotianea*

BY: DESIRE MTETWA

A dissertation submitted in partial fulfilment of the requirements for Bachelor of Science Honours Degree in Biological Sciences Department of Biological Sciences Faculty of Science and Technology, May 2015.

Abstract

Aphids (Myzus persicae nicotianae) cause damage by sucking sap from the plants leading to distortion and wilting of the foliage, and eventually reducing economic yields. In order to minimise losses to viral diseases, it is important to use insecticides in association with fungi to control insect pests in an environmental and ecological friendly way without contaminating the environment. A study was carried out from October, 2013 at Kutsaga Research Station in the laboratory to evaluate the efficacy of Beauveria bassiana (contact insecticide) and imidacloprid (systemic insecticide) 200 SL for the control of tobacco aphid, (Myzus persicae nicotianae). Three doses of 200 SL Imidacloprid (Imida) namely: Imida low rate 0.55 ml per 500ml water (LR); Imida medium rate 1.1 ml per 500ml water (MR) and Imida high rate (2.2 ml per 500ml water (HR) were mixed with Broadband® Beauveria bassiana (B. Bass) at three different rates namely: B. bass 4 x 103 (LR); B. bass. 4 x 104 (MR) or B. bass 4 x 105 (HR) to evaluate the most effective application combination rate of Beauveria bassiana and imidacloprid for the control of the tobacco aphid. Treatments were: water (control); Imida MR; B. bass MR; B. bass LR + Imida LR; B. bass LR + Imida MR; B. bass LR + Imida HR; B. bass MR + Imida LR; B. Bass MR + Imida MR; B. Bass MR + Imida HR; B. Bass HR + Imida LR; B. Bass HR + Imida MR and B. Bass HR + Imida HR. Inoculation of aphids was done and data collected from day 1 to day 7 after inoculation. Mortality of aphids was recorded at 8 o'clock in the morning cumulatively as a percentage of the initial number infested (10 per petri dish). Results showed that there were significantly high p<0.001 cumulative mortality from B. Bass HR + Imida HR and B. Bass HR + Imida MR in comparison with the control and other treatments. The number of deaths on day one indicated clearly that the level of mortality was dose-dependent and the presence of the contact insecticide (Beauveria bassiana) compared with systemic insecticide (imidacloprid) which took three days to show mortality.

In Imida MR death rates started at day 3 and accumulated up to 98% at day 7 whilst B. Bass started at day 1. However, death in day 1 increased as the dosage of B. Bass increased. It was also evident that as the rates of B. Bass and Imida increased the cumulative mortality also increased significantly ($p<0.001$). For controlling aphids B. bass HR + Imida MR was the best because its efficacy was the same as B. bass HR + Imida HR showing the same mortality in day one and by day five, both mortalities had reached 100%.

Acknowledgements

Firstly, all praise goes to the Lord Almighty for the gift of life, blessings and his guidance during my university years. I wish to show appreciation to the Tobacco Research Board for giving me the opportunity to do my project at Kutsaga Research Station; special mention goes to the Plant Health Services Division staff. I wish to extend sincere gratitude to my academic supervisor Mr Bare, for his assistance during the research project. For the same assistance I also wish to acknowledge my co-supervisor Rhoda Masukwedza, for her assistance was greatly appreciated and not forgetting Godwin Mtetwa, Stanford Banana and Betty Kachembere for their assistance during data analysis. I am also grateful to the Entomology staff at Kutsaga Research Station and fellow students, for every bit of knowledge and skills they passed on to me and their support. Profound gratitude also goes to my family; special mention goes to my father Godwin and my mother Miriam for their financial, spiritual and moral support, not forgetting my siblings Answer, Evidence and Beloved for their team effort, motivation and continual support throughout my studies. I would do injustice to my fellow acquaintances Estias, Tendai, Farai, Claudius, Vimbiso, Nobert, Jaison and Paidamoyo for not thanking them for their encouragement and their team work throughout my studies.

Dedication

To my wonderful parents Miriam and Godwin Mtetwa, and my siblings Answer, Evidence and Beloved.

1.6.0 Introduction

1.6.1 Background

Tobacco originated as a natural hybrid in South and Central America and has been under cultivation for many centuries (Breen, 1985). Tobacco is a plant within the genus *Nicotianae* of the Solanaceae family. While there are more than 70 species of tobacco, the chief commercial crop is *Nicotiana tabacum.* The more potent species *Nicotiana rustica* is also widely used around the world. There are over 50 species of the botanical genus *Nicotianae*, with all being derived from one of the two species *N. tabacum* and *N. rustica* (Breen, 1985). By far, the most important is *N. tabacum* which is an allotetraploid that resulted from natural hybridization of the two wild species *Nicotiana sylvestris* and *Nicotiana tementosiformis* (Shew, 1991). Tobacco is grown for commercial purposes in at least 97 countries around the world. It is a cash crop and is grown in warm climates. Tobacco contains the alkaloid nicotine which is used as a stimulant. The leaf turns from green to lemon gold at harvest when nicotine, its major product is extracted (Tucker, 1982). Major tobacco producing countries include China, Brazil, United States of America (U.S.A) and Zimbabwe (Takada, 1981). Zimbabwe produces some of the finest flavoured tobacco in the world owing to the climate and physical conditions which favour farming of the crop in the country (The Herald, 2005). It is grown as an annual from unusually small seeds. It has very large leaves which are pointed at the tip. There are many variations of tobacco under cultivation but this classification is mainly based on the curing method used (e.g. fire, air, sun and flue curing) (Breen, 1985).

Three types of tobacco have been traditionally grown in Zimbabwe: flue-cured barley and oriental tobacco (www.fao.org). Flue-cured tobacco constitutes over 95% of Zimbabwe's tobacco yield while barley constitutes about 4%. Tobacco rose to become a major export in Zimbabwe in the 1990's. According to Masuka, Cole and Mguni (1998), Zimbabwean tobacco is grown as a summer crop and it requires a 7-9 month growing season in order to produce a full crop. Income from tobacco in 2013, accounted for at least 10.7% of the Zimbabwe's gross domestic product and 21.8% of all exports, compared to 9.2% for other agricultural commodities.

The tobacco market worldwide is under threat from pests like various Lepidoptera such as cutworms (Agrotis species) and budworm (Heliothis species) while aphids (Myzus species) Black and White flies (Bemisia tabaci) suck nutrients and transmit viral diseases to the tobacco plant hence causing yield loses (Blair, 1990). *Myzus persicae* is a major vector in the transmission of Tobacco Bushy-top virus (TBTV) and potato virus Y (PVY) (DiFonzo, Ragsdale, Radcliffe, Gudmestad and Secor, 1997). Because of rapid acquisition and transmission times for these non-persistent viruses, a few aphids, through probing, can transmit these viruses to many plants in a short time, and greatly reduce the effectiveness of chemical control in limiting the spread of viruses (Chamberlin, 1958).

Currently, management of these cosmopolitan pests is achieved by a combination of chemical, biological and cultural methods (Blair, 1990). And of these, chemical control has been the most frequently used. However, this intensive spraying of chemical pesticides has left a lot of negative impacts on the environment and human health (Tamaki and Weeks, 1968). Most importantly, aphids also develop general resistance to almost all classes of chemical insecticides (Ye, Dun and Feng, 2005). Consequently, there is increased interest in using alternative and integrated approaches to control aphids and therefore, biological controls have been initiated (Blair, 1990).

The incorporation of fungal biological control agents into management systems has been considered as a feasible alternative to frequent chemical use (Feng and Johnson, 1990). The most practical integrated approach is incorporation of fungal and chemical agents into combined formulations with enhanced efficacy on the control of target pests (Ye et al., 2005). Imidacloprid, a versatile neonicotinoid insecticide, has been widely used to combat the sucking insect pests such as aphids over the past decade (Weichel and Nauen, 2004) and is one of few insecticides recommended for insect control on residue sensitive crops. Previous studies have shown that imidacloprid is highly compatible with *Beauveria bassiana* and may enhance fungal action in laboratory bioassays or field trials (Ye et al., 2005).

Several entomopathogenic fungi including B. bassiana (Balsamo) Vuellemi and Metarhizium anisopliae (Metschnikoff) Sorokin, *Paecilomyces fumosoroseus* (Wize) Brown & Smith and Verticillium lecani (Zimmermann) are potential biocontrol agents of various aphids (Feng and Johnson, 1990; Feng, Johnson and Kish, 1990a; Dorschner, Feng and Baird, 1991; Stary, 1988; Chandler, 1997). Both *B. bassiana* and *P. fumosoroseus* have established certain aphid control in the field (Ye et al., 2005, Vandenberg, Sandvol, Jaronski, Jackson, Souza and Halbert, 2001; Hatting, Wraight and Miller, 2004). As true pathogens, the fungal biological agents usually have a latent period of several days to kill aphid hosts after infection (Milner, 1997). Due to this inherent feature, fungal formulations lack knockdown action, kill target pests slower than chemical insecticides, and often discourage commercial efforts (Ye et al., 2005, Feng, Poprawski and Khachatourians, 1994; Wraight, Jackson and de Kock, 2001). For successful control, there is a strong need to evaluate interaction between *B. bassiana* and an applicable commercially available pesticide such as imidacloprid for the improved management of the tobacco aphid *Myzus persicae nicotianae*.

1.6.1.2 Justification

Zimbabwe is the largest producer of tobacco leaf in Africa and the world's fourth producer (www.fao.org) after China, Brazil and U.S.A. This is because the country does not have a large tobacco manufacturing industry and so it exports 98% of the tobacco it produces (Banga, 2012). Agriculture emerged as one of the nation's leading economic performers in 2011 with tobacco being the highest contributor (England and Hawkins, 2011). Tobacco is the crop of economic importance in Zimbabwe since tobacco production makes an important contribution to gross domestic product and to export revenue, and plays a major role in the national economy (www.fao.org). The tobacco aphid *Myzus persicae nicotianae* is an economically important pest of tobacco in Zimbabwe and is the most damaging pest (Chamberlin, 1958). Thus, it is in every grower's interest to minimise yield losses due to the attack by aphids so as to maintain high yields.

Aphids cause damage by sucking sap from the plants leading to distortion and wilting of the foliage, and reduction in crop yields. Aphids excrete large amounts of honey-dew on which a sooty mould develops. In addition to interfering with photosynthesis, the sooty mould reduces the quality of the cured leaf. Winged aphids act as vectors of viral diseases. In Zimbabwe, viruses transmitted by aphids include TBTV, PVY and Tobacco Mosaic Virus (TMV). If plants are infected at an early stage of growth, 100% yield losses can occur. In order to minimise losses to viral diseases the use of chemical insecticides in association with fungi to control insect pests can be regarded as a solution to environmental and ecological threats posed by using chemical insecticides only.

The use of chemical pesticides without considering the complexities of the soil environment has been a major case of disruption. This is because the target pest species in the process becomes tolerant to the pesticide leading to pesticide resistance and can no longer be controlled economically with chemicals (Dixon, 1998). Aphids develop general resistance to all classes of chemical insecticides (Ye et al., 2005). This has led to the population of the target species quickly recovering from the pesticide action and for a variety of reasons, rising to new resurgence.

The continuous cultivation of tobacco in Zimbabwe has resulted in chemical control methods being heavily relied on for the control and prevention of aphid transmitted bushy top virus disease (FAO, 2010). Evidence suggests that the aphid in Zimbabwe is already developing some complex resistance due to the over-use of chemicals (FAO, 2010) and there is great need to evaluate other formulations to ensure successful control of the aphid. The new methods need to be effective, less expensive and environmentally friendly. There is therefore need to evaluate the most effective application rates to effectively control aphids without adversely affecting the overall quality and yield of the tobacco crop.

1.6.1.3 Objectives

1.6.1.3.1 Main objective:

To evaluate the efficacy of the combination of *B. bassiana* and imidacloprid for the control of the tobacco aphid

1.6.1.3.2 Specific objective:

To evaluate the most effective application combination rate of *B. bassiana* and imidacloprid for the control of the tobacco aphid.

1.6.2 Literature Review

The tobacco aphid (Myzus persicae nicotianae)

1.6.2.1 Taxonomic classification

Myzus persicae nicotianae (Blackman) is classified as follows:

Kingdom: Animalia

Phylum: Arthropoda

Class: Insecta

Order: Homoptera

Family: Aphididea

Genus: Myzus

Species: Myzus persicae

Subspecies: Myzus persicae nicotianae

1.6.2.2 Description

Aphids are small (1-10 mm), soft-bodied plant-sucking insects and they have an intricate life cycle. It is believed that all generations of aphids comprise parthenogenetic females which reproduce without egg fertilization and are viviparous (i.e. produce live young) (Dixon, 1998). Embryos developing in parthenogenetic females also have embryos developing within them. This parthenogenesis and telescoping of generations enable aphids to achieve very high rates of increase which increases the complexity of controlling the pest (Dixon, 1998).

Aphids have a proboscis through which they suck plant juices from the phloem. The skin of the aphid is more or less provided with glands that secrete either waxy or mealy matter. The form is generally oval or globose, sometimes flattened below, the head is more or less distinct in the wingless females than winged females (Theobald, 1926). Three forms of females occur in the majority of aphids, an alate viviparous female, an apterous viviparous female and an apterous oviparous female. *M. persicae* and its tobacco adapted subspecies M. *persicae nicotianae* are pests of economic importance in Zimbabwe and they are practically indistinguishable from each other morphologically (Clements, Sorenson, Wiegmana, Neese and Roe, 2000). Based on the distinctive morphometric and anholocyclic development, Blackman (1987) first identified the tobacco feeding form of M. *persicae as M. nicotianae.*

However, the general consensus among aphidologists now is to treat this as a subspecies of *M. persicae* hence the name in common usage *M. persicae nicotianae* (Capinera, 2005). The body structure of aphids is simplified to perform only the function of feeding and reproduction, while retaining the ability to walk is aided by their gressorial legs (Blackman, 1987). Aphids exist in two body forms, the wingless (apteral) and the winged (alate). The tobacco aphid is found in two distinct colour forms, green and red. Colour variation between aphids is common, and it appears to be interclonal and sympatric in occurrence (Takada, 1981; Miyazaki, 1987).

The success of *M. persicae* in colonizing different host plants has been related to the presence of aphid enzymatic mechanisms of detoxification, which are responsible for the metabolism of host plant allelochemicals (Francis, Gerkens, Harmel, Mazzucchelli, DePauw and Haubruge, 2006). The performance of *M. persicae nicotianae* on tobacco should therefore have a genetic/ biochemical base, which could be related to the ability of the subspecies to colonize a well-defended host plant (i.e. tobacco with glandular trichomes and cuticular sucrose esters) (Cabrera-Brandt, Fuentes-Contreras and Figueroa, 2010).

1.6.2.3 Life cycle of aphids

Aphids are often parthenogenetic (asexual) for part or all of their life and both viviparous and oviparous at different times of the year (www.insect-world.com). Parthenogenetic reproduction generally results in vivipority (live birth) (www.insect-world.com). In temperate countries, the typical life cycle of an aphid species is an alternation between a primary host, usually woody, on which the over wintering eggs are laid on an herbaceous secondary host (Dixon, 1977). Sexual reproduction is induced by low temperature and shortening day length in many aphids (Tamaki and Weeks, 1968). This is not known to occur among tobacco aphids in Zimbabwe. All aphids in Zimbabwe are females and their potential for rapid expansion is unique among insects (Tamaki and Weeks, 1968). The life cycle of aphids in Zimbabwe consists essentially of the infestation of a host plant by winged migrants which give birth to living young. Seven to ten days after its birth, the nymph becomes a mature female. A mature female can live for up to 30 days during which time she can produce from 30 to 80 nymphs. These nymphs develop to wingless adults by a series of four moults, and they in turn give birth to a number of nymphs (Moran, 1990). The embryonic development of the young begins before the mother's birth, in the body of the grandmother, so that the food intake of one individual sustains three generations of about 6 000 individuals (Blair, 1990).

It is this telescoping of generations combined with a rapid life cycle of 7 to 14 days that results in the extremely high reproductive potential of aphids (Blair, 1990). By this life cycle, colonies quickly develop but eventually a physiological change in the host plant, caused by the feeding aphids themselves or the natural senescence of the plant tissues, a less degenerate, darker winged form of the insect is produced. When mature these animals fly from the host plant and start fresh infestations on other suitable plants (Moran, 1990).

1.6.2.4 Feeding ecology and damage

Aphids feed from the phloem of plants which they tap into with the stylets of their proboscis (www.insect-world.com). They gain access to the phloem vessels from three main parts of the plants: stems, leaves and roots (www.insect-world.com). Their stylets which are contained within the proboscis when the aphid is not feeding, are very thin and could suffer damage while being pushed into the plant or bent in an unwanted direction (www.insect-world.com). When the stylets reach a phloem tube the aphid injects saliva into the plant cell to avoid the plant cell sealing the puncture (www.insect-world.com). Sucking the sap from plants leads to distortion and wilting of the foliage, which leads to reduction in crop yields. Aphids excrete large amounts of honeydew because of the amount of sap they need to suck up in order to get enough nitrogen in the diet which means they have far more sugar and liquid than they need (www.insect-world.com). This honeydew does not only clog the pores of the leaves interfering with its assimilation, but also encourages the growth of black, sooty mould, which can prevent light from reaching the photosynthetic tissues of the plant (Tamaki and Weeks, 1968).

1.6.2.5 Aphid-Transmitted Diseases

Potato virus Y belongs to the Potyvirus genus. The genus is currently known to be the largest of all the plant virus genera and is thought to constitute the most destructive families of plant viruses affecting potato crops (England and Hawkins, 2011) and many other economically important plant species. These plants include tobacco, tomato and pepper (Miyazaki, 1987). The level of damage to a crop is determined by the strain of PVY infecting the plants, the viral load, the time at which infection occurs, as well as the tolerance the host possesses toward the virus (Warren, Kruger and Schoeman, 2005). Resistance to PVY infection by hosts is low in many cases. Infection of a potato field with PVY may ultimately result in 10-100% loss in yield (Warren et al., 2005).

1.6.2.6 Tobacco Bushy-top disease (TBTV)

Tobacco Bushy-top virus was first reported in Zimbabwe in 1958 (Gates, 1962). The virus usually occurs as a complex with other viruses, for example PVY. The virus is readily transmitted by mechanical inoculation from plants that are infected with another virus (e.g. Tobacco Vein Distorting Virus), but not from plants infected with it alone (Gates, 1962). Tobacco Bushy-top disease is caused by a complex of the Tobacco Bushy-top virus, a member of the genus Umbravirus and Tobacco Vein Distorting Virus, a member of the genus Poleovirus, which acts as a vector encapsidating the Tobacco Bushy-top genomic RNA (Mo, Qin, Wu, Yang, Wu, Duan, Li and Chen, 2011). Tobacco Bushy-top virus causes stunted growth in plants, and leaves show symptoms of vein distortion, vein clearing and mottling, and rounding (Mo, Qin, Tan, Li, Wu and Chen, 2002). It also stimulates the sprouting of axillary shoots from the main stem (Gates, 1962). These early sprouts form lateral shoots on which other shoots are produced, resulting in a 'bushy' appearance (Mo et al., 2002).

1.6.2.7 Sooty mould and secondary infections

The occurrence of sooty mould on tobacco has always been linked to the presence of parasitic insects on plants (Gates, 1962). Among them, aphids and whiteflies are most commonly associated with black mould on the lamina. This is related to the feeding behaviour of the insects. The aphids must collect large amounts of sap for the protein they need. This makes them reject the excess sugar in the form of honeydew (Gates, 1962). It can be found in large amounts in places where aphid colonies grow and taint the surface of the invaded lamina. The honeydew is an ideal substrate for several phylloplane fungi that use it for food and thus, progressively, the dew is covered by black mould (Alternaria spp. *Cladosporium spp* or *Capnodium spp*).

1.6.2.8 Tobacco Mosaic Virus (TMV)

TMV was the first virus to be discovered (nist.rcsb.org). Late in the 19th century, researchers found that a tiny infectious agent, too small to be a bacterium, was the cause of a disease of tobacco plants (nist.rcsb.org). TMV is very stable, so stable that it can survive for years in cigars and cigarettes made from infected leaves (nist.rcsb.org). The viral RNA is infectious by itself, but the addition of a protein coat protects the RNA from enzymes that would destroy it. TMV uses two tricks to release its RNA. As with many viruses, TMV has a chemical switch that causes the proteins to change when the environment changes. The capsid protein has several clusters of acidic amino acids that are stable outside of cells, where calcium levels are high, but repel one another in the low-calcium conditions inside cells. This is enough to loosen the first few capsid proteins, releasing the end of the RNA. TMV then uses ribosomes as the engines to finish the job. As the ribosomes move down the strand, creating the first set of virus proteins, they displace the remaining capsid proteins (nist.rcsb.org). This disease is also transmitted by aphids (Mo et al., 2011).

1.6.2.9 Potato Virus Y (PVY)

PVY mostly infects plants in the family Solanaceae (potatovirus.com). The Solanaceous plants include economically important ones like potato (Solanum tuberosum) and tobacco (potatovirus.com). Aphids can carry viral particles and spread them between plants. PVY can be spread by many species of aphids (potatovirus.com). Within minutes of starting to feed on a PVY-infected plant, the PVY particles get stuck in the aphid's stylet (potatovirus.com). If the aphid then moves to a healthy plant and soon starts to feed, the virus particles are transmitted to the healthy plant. This process is termed "non-persistent" transmission (potatovirus.com). When PVY infects the tobacco plant, it replicates by assuming control of some of the plant's proteins and enzymes to make more PVY viruses. This disrupts the normal functioning of the plant. Tobacco cultivars vary in how susceptible they are to the effects of PVY infection (potatovirus.com). If a plant cannot become infected at all, we call this immunity (potatovirus.com).

1.6.2.10 Migration

Most of the time aphids do not have wings, and in general they move very little, and life consists of feeding and giving birth (www.insect-world.com). The number of winged aphids in a given population is influenced by: death of the host plants which stimulates aphids to migrate in search of the new host, environmental changes and over population also stimulate migration to either primary or secondary host (www.insect-world.com). Aphids are weak fliers, and in still air they move about 1.6 to 3.2 km per hour (www.insect-world.com).

Their migration can be quiet extensive and they often take advantage of favourable wind to enhance their flight efficiency (www.insect-world.com). Aphids land on leaves, because they are attracted to the yellowish light emitted from young actively growing crops or older senescing ones. This is because the flow of nutrients to and from these leaves is greater (www.insectworld.com).

1.6.2.11 Tobacco Aphid Management in Zimbabwe

The control of aphid pests still involves large amounts of pesticides, but other ecologically friendly methods have been in use in other places for some time (www.insect-world.com). Biological and cultural control tools are being extensively researched. Also most aphid populations are controlled or moderated by natural controls that include even stress and natural enemies like lady beetles and other wide variety of Hymenopteran parasitoids (Stary, 1988). When conditions are favourable, aphids' populations grow very rapidly resulting in problems if not controlled. Tobacco varieties resistant to the aphid and which also produce acceptable leaf quality are not yet available in Zimbabwe despite numerous breeding efforts (Cottrell, 1994).

Characteristically, aphid populations normally drop significantly immediately following topping (the removal of the apical region of the plant before flowering structures are developed). Topping increases leaf thickness and the concentration of alkaloids which cause heavy aphid mortality. Consequently, no aphid management is cost-beneficial to a farmer after this operation. Control strategies can be in the form of cultural, biological or chemical techniques (Cottrell, 1994).

1.6.2.12 Cultural control

In Zimbabwe, cultural control is regarded as the most feasible control strategy for the aphid (McDonald et al., 2003). Tobacco aphids primarily come into the crop in late November up till late January which corresponds to the wettest periods of the country (Gates, 1962). Consequently, early-planted tobacco suffers less aphid damage but any tobacco grown after the onset of the rainy season is susceptible to aphid damage. As a result, mandatory cultural stalk destruction dates have been enforced in Zimbabwe to ensure that specific host free periods are created to reduce the populations of the aphid significantly from year to year (Reynolds and Volk, 2007). The effectiveness of this control strategy has been limited by the failure of small-scale farmers to adhere to these stalk destruction dates, thereby creating a continuous host for the parasite and enabling it to establish sufficient population numbers to spread bushy top every season (Tamaki and Weeks, 1968).

Cultural control measures are important in producing strong disease-free seedlings in tobacco arming. Good sanitation such as elimination of weeds which act as reservoir hosts to aphids is also important (Cottrell, 1994). Large aphid populations are most likely to build up on irrigated crops particularly when temperature rises rapidly at the beginning of the rains (September/October). Crops must be scouted once a week and if found infested must be treated immediately (Cottrell, 1994). Legislations are regarded as components of cultural methods and it is important to adhere to legislated plant destruction and sowing dates in order to control transmission of diseases. In Zimbabwe, legislation states that (Tamaki and Weeks, 1968):

- ✓ No sowing of seedbed should be done before June 1 each year,
- ✓ no planting before September 1 and after December 31,
- ✓ destroy seedbeds as soon as possible but before January 1 and
- ✓ destroy stalk/ regrowth in the field by May 15 each year.

1.6.2.13 Chemical control

Aphids may be controlled by a number of insecticides which are classified as contact or systemic aphicides. Stomach poisons which are applied to the surface of the plant are ineffective because the aphids pierce the plant tissue with their specially adapted mouthparts and suck out non-toxic sap (Villacarios, 1987). Among the organophosphates, Parathion and Malathion are contact insecticides which are widely used for aphid control. Demeton-s-methyl, dimethoate and phosdrin are systematic organophosphates which are effective against most aphid species. Work done over the years at Kutsaga to establish a suitable method of applying aphicides for the control of tobacco aphids in the field, shows that aphicides placed in the planting hole, slit-in ridge and slit adjacent ridge, controlled aphids effectively (TRB annual report, 2000). Some of the most widely used aphicides in Zimbabwe are shown in the table below:

Table 1.6.1: Aphicides used in Zimbabwe

Chemical	Active ingredient	Route of entry	Target site of action
Acephate 75 SP	Triazamate	contact	Acetylcholine esterase inhibitors
Dimethoate 40% EC	Triazamate	contact	Acetylcholine esterase inhibitors
Monocrotophos 40% EC	Triazamate	contact	Acetylcholine esterase inhibitors
Methamidophos 60% SL	organophosphate	contact	Acetylcholine esterase inhibitors
Imidacloprid 200 SL	neonicotinoid	systematic	Nicotine acetylcholine antagonists
Thiamethoxam 25% WG	neonicotinoid	systematic	Nicotine acetylcholine antagonists
Aldicarb 15% granules	carbamate	systematic	Acetylcholine esterase inhibitors

Because of the ease with which aphids develop resistance, rotation of aphicides is imperative to prevent resistance by the tobacco aphid. Consequently, growers must take responsibilities to use available products carefully and judiciously (Tamaki and Weeks, 1986).

Chemical control remains the effective means of controlling the aphid pest in Zimbabwe. Several chemicals have been tested and proven efficacious to the control of the pest in both early planted and late planted tobacco. However, continued use of chemicals has led to the resistant strains of the aphid, a case that scholars (McDonald et al., 2003) alike in Zimbabwe agree to be directly linked to improper application rates coupled with failure to adhere to the cultural stalk destruction dates. The issue of chemical residues is also negatively associated with the chemical control strategy. Tobacco companies are increasingly sensitive concerning insecticide residues. Specifically, they do not want Thiodan residues in the tobacco. Some companies spot test for Thiodan residues on the market floors and, if certain markets are exhibiting Thiodan residues, they will shift their buying to other markets. Available alternatives to Thiodan based chemicals include Acephate (Orthene 75S), endosulfan (Phaser 3EC, Thiodan 3EC), Methomyl (Lannate 90 SP, 2.4 LV) and Imidacloprid (Pravado 1.6F).

1.6.2.14 Biological control

Biological control involves the use of predators and parasitoids which feed on aphids. There are many predatory insects which prey or parasitize aphids (www.insect-world.com). These include the bush crickets (Orthoptera), wasps (Vespidae) and spiders (Aranae). Predators of aphids are also sold commercially in the U.S.A and these include ladybird beetles (e.g. Coccinella septempunctata) and laceworms (Chrysoperla cornea) (Stacy, 2003). Aphidus species are important natural enemies of aphids. The adult female lays its eggs in aphids and the larvae develop within the aphid eventually killing it (Miyazaki, 1987).

1.6.2.15 *Beauveria bassiana* for insect management

B. bassiana is a fungus that controls insect pests by causing a disease known as the white muscadine disease in the insects (www.hort.uconn.edu). When the fungal spores of *B. bassiana* come in contact with the cuticle of specific insects that are susceptible to it, they germinate and grow directly through the cuticle to the inner body of the host insect. The fungus then proliferates throughout the insect's body, producing harmful toxins and depriving the insect of nutrients until eventually the insect dies (www.simplynetworking.es).

Unlike bacterial and viral pathogens of insects, *B. bassiana* and other fungal pathogens infect the insect with contact and do not need to be consumed by their host to cause infection (Das, 2013). Once the fungus has killed its host, it grows back out through the softer portions of the cuticle, covering the insect with a layer of white mould (hence the name white muscadine disease). This downy mould produces millions of new infective spores that are released to the environment (www.simplynetworking.es).

1.6.2.16 Efficacy of B. bassiana for insect pest control

B. bassiana can be used as a biological insecticide to control a number of pests such as termites, whiteflies, thrips, aphids and a wide range of beetles (danpritchard.com). *B. bassiana* parasitizes a very wide range of arthropod hosts. However, different strains vary in their host ranges, some having rather narrow ranges (danpritchard.com), it is important to identify the correct strain for a targeted species of pest for efficient control (Xu, Ying and Feng, 2002).

Some strains do have a wide host range and should therefore be considered nonselective biological insecticides (danpritchard.com).

1.6.2.17 Recent work on *B. bassiana* for insect control

B. bassiana was evaluated in combination with insecticides in Maryland USA and a notable reduction in two spotted spider mites, Tetranychus urticae and green peach aphids Myzus persicae by an average of 83% compared with other organic insecticides was observed (www.angr.umd.edu). It was noted in the same trial that certain strains of *B. bassiana* were more efficacious than others and this was more pronounced under rising temperatures. The authors argued that using Azadirachtin, a chemical pesticide, alone or in combination with *B. bassiana* did not increase the control of aphids or mites as compared to using B. *bassiana* alone (www.angr.umd.edu). Furthermore, because of high costs associated with Azadirachtin, it did not appear to be a cost-effective method of controlling aphids and mites at that time (www.angr.umd.edu).

1.6.3. Materials and Methods

1.6.3.1 Study location

The study was conducted at Tobacco Research Board, Kutsaga Research Station in Harare, Zimbabwe. Tobacco Research Board (Kutsaga Research Station) is located 15 km East of Harare at latitude $17^0$55′S and longitude $31^0$08′E with an altitude of 1479 m above sea level. The average temperatures were 32ºC and 18ºC in summer and winter respectively, with annual rainfall between 800-1000mm. The soils at Kutsaga Research Station are on average light textured sandy loam.

1.6.3.2 Beauveria bassiana and aphid cultures used in the experiment

An emulsifiable spore concentration of *Beauveria bassiana* 4 x 109 strains PPRI 5339, a fungal contact insecticide from Broadband® a company in South Africa was used in this study (Fig 1.6.1). It is highly compatible with imidacloprid (Ye et al., 2005; Xu et al., 2002). Serial dilutions were done to reduce the concentrations to low, middle and high (Fig 2). For high, 0.05 ml of 4 x 109 was mixed in 500 ml of water to obtain 4 x 105. For middle, 0.05 ml of 4 x 109 was mixed in 5 L of water to obtain 4 x 104. For low, 50 ml was drawn from 4 x 104 solutions and was added to 450 ml of water obtaining 4 x 103.

1.6.3.3 Beauveria bassiana and aphid cultures used in the experiment

An emulsifiable spore concentration of *Beauveria bassiana* 4 x 109 strains PPRI 5339, a fungal contact insecticide from Broadband® a company in South Africa was used in this study (Fig 1.6.1). It is highly compatible with imidacloprid (Ye et al., 2005; Xu et al., 2002). Serial dilutions were done to reduce the concentrations to low, middle and high (Fig 1.6.2). For high, 0.05 ml of 4 x 109 was mixed in 500 ml of water to obtain 4 x 105. For middle, 0.05 ml of 4 x 109 was mixed in 5 L of water to obtain 4 x 104. For low, 50 ml was drawn from 4 x 104 solutions and was added to 450 ml of water obtaining 4 x 103.

Tobacco aphids used in this study were maintained in screen cages on Chinese cabbage (Brassica pekinensis) (Fig 1.6.1). To prepare age uniformity of aphids, five days before the experiment, six wingless adults were taken from the colony and transferred to six clean potted Chinese cabbage plants. The adults were allowed to freely produce nymphs for 120 hours. Six- hundred, five-day old aphids were taken for the experiment. Ten, five-day old aphids per petri dish (12 cm diameter) were used. New nymphs with age variation of $\leq$ 2 days received the treatments in table 1.6.2.

1.6.3.4 Tobacco leaves used in the experiment

Fifteen upper tender leaves, from healthy tobacco plants (variety KRK 26) were taken from one of Kutsaga fields and were carefully observed to ensure the leaves were clean (have no aphids) (Figure 1.6.3) to produce 60 leaf discs of 5 cm diameter. A pill box of 5 cm diameter was used to cut the leaves into leaf discs, four discs per leaf. The leaf discs were placed in petri dishes, one per petri dish. Twelve treatments were applied to the leaf discs and five blocks were conducted.

Figure 1.6.1: B. bassiana used in the experiment and Chinese cabbage used to rear aphids

1.6.3.5 Imidacloprid used in the experiment

The aphicide used was imidacloprid 200 SL. The insecticide came with the instruction of mixing 220 ml of imidacloprid per 100 litres (L) of water. Imidacloprid had three levels in this experiment low, middle and high. Scaling down was done since the experiment was done at a smaller scale in the laboratory (Fig 1.6.4).

For low, 110 ml imidacloprid was applied to 100 L of water, by scaling down; 0.55 ml imidacloprid per 500 ml of water was used. For middle, instead of 220 ml of imidacloprid per 100 L of water, 1.1 ml of imidacloprid per 500 ml of water was used. For high, instead of 440 ml of imidacloprid per 100 L of water, 2.2 ml of imidacloprid per 500 ml of water was used.

Table 1.6.2: Treatments used in the experiment

Treatment	concentration	treatment	concentration	Coded treatment
1. Control	water	-	-	**Control**
2. Imidacloprid 200 SL	Medium rate (1.1 ml per 500ml water)	-	-	**Imida MR**
3. B. bassiana	Medium rate (4 x 104)			**B. bass MR**
4.B. Bassiana	low rate (4 x 103)	+ **Imidacloprid**	low rate (0.55 ml per 500ml water)	**B. bass LR** + **Imida LR**
5. B. bassiana	low rate (4 x 103)	+ **Imidacloprid**	medium rate (1.1 ml per 500ml water)	**B. bass LR** + **Imida MR**
6. B. bassiana	low rate (4 x 103)	+ **Imidacloprid**	high rate (2.2 ml per 500ml water)	**B. bass LR** + **Imida HR**
7. B. bassiana	medium rate (4 x 104)	+ **Imidacloprid**	low rate (0.55 ml per 500ml water)	**B. bass MR** + **Imida LR**
8. B. bassiana	medium rate (4 x 104)	+ **Imidacloprid**	medium rate (1.1 ml per 500ml water)	**B. Bass MR** + **Imida MR**
9. B. bassiana	medium rate (4 x 104)	+ **Imidacloprid**	high rate (2.2 ml per 500ml water)	**B. Bass MR** + **Imida HR**
10. B. bassiana	high rate (4 x 105)	+ **Imidacloprid**	low rate (0.55 ml per 500ml water)	**B. Bass HR** + **Imida LR**
11. B. bassiana	high rate (4 x 105)	+ **Imidacloprid**	medium rate (1.1 ml per 500ml water)	**B. Bass HR** + **Imida MR**
12. B. bassiana high rate (4 x 105)	+ Imidacloprid	+ **Imidacloprid**	high rate (2.2 ml per 500ml water)	**B. Bass HR** + **Imida HR**

Figure 1.6.2: Serial dilutions for Beauveria bassiana

Figure 1.6.3: Tobacco variety (KRK 26) from land 10 used to make leaf discs

Figure 1.6.4: Dilution of imidacloprid

1.6.3.6 Treatments and inoculation on tobacco disc

For each block, (block 1-5) fungal sprays consisting three concentration treatments, low, medium and high incorporated with one of the three concentration treatments of imidacloprid, low, medium and high were applied to tobacco leaf discs. The three imidacloprid and *B. bassiana* treatment concentrations were, 0.55, 1.1 and 2.2 ml per 500ml of water and 4 x 103, 4 x 104 and 4 x 105, respectively. Water was sprayed in control petri dishes. Treatments with medium rates of imidacloprid only and *B. bassiana* only were also considered (Table 1.6.2). Treatments were collected in 1 L beakers. Leaf disks were dipped in respective beakers with treatments using metal forceps.

Leaf discs were dipped for five seconds to allow the leaf discs to be soaked fully. After five seconds, leaf discs were pulled out of the beaker using metal forceps and allowed to dry on filter papers at room temperature to prevent aphids from drowning. After drying the leaf discs were transferred to the petri dishes with a damp filter paper to keep the leaf disc moist. Five-day old aphids were then infested using a paint brush. Petri dishes were then covered with their lid to prevent aphids from moving out of petri dishes and were maintained in the lab at room temperature at the same regime for seven days. Ten aphids were placed on each leaf disc with a paint brush and mortality recorded daily for seven days.

Figure 1.6.5: Leaf discs used and inoculation of aphids using a paint brush

1.6.3.7 Analysis of aphid mortality

Mortality of aphids was recorded starting from 24 hours after inoculation up to 168 hours after inoculation (7 days). Recordings were done at 8 o'clock in the morning for uniformity. Data sheets indicating the cumulative mortality rates were prepared and recorded. The mortality of the aphids was recorded as a percentage of the initial number infested (10 per petri dish).

ANOVA was used to compare the mortality rates between treatments and blocks and was calculated using Genstat Release 17th edition. For control correction, Abbott's formula was used to calculate corrected efficacy % in pesticide trials.

Abbott's formula

$$\text{Corrected } \% = (1 - x = \frac{\text{n in T afer treatment}}{\text{n in Co after treatment}}) * 100$$

Where: n = Insect population, T = treated, Co = control

1.6.4.0 Results

In the control there were deaths observed (Table 1.6.4), but these did not show any significant difference and they were due to natural variance ($p>0.05$ appendix 2). There were no deaths observed with Imida MR during the first two days (Fig. 1.6.6a). Thereafter deaths started to occur in day 3. B. bass HR + Imida HR showed significantly high rates ($p<0.001$ appendix 3) of mortality in day1 compared to all other treatments except B. bass HR + Imida MR (Table

1.6.3). Deaths in B. bass MR treatment were observed from day one. A white mould was also seen wrapping the aphid body. In treatments B. bass LR+ Imida LR; B. bass LR + Imida MR and B. bass LR + Imida HR mortality was observed from day one.

From day three the mortality rate accelerated in B. bass LR + Imida MR and B. bass LR + Imida HR. A white mould was also seen wrapping the aphid body. The mortality rate was average compared to B. bass LR+ Imida LR; B. bass LR + Imida MR and B. bass LR + Imida HR. Mortality rate also began to accelerate after 72 hours. A white mould was also observed wrapping the aphid body. In B. bass HR + Imida LR; B. bass HR + Imida MR and B. bass HR + Imida HR chemical control, mortality was observed from day one. The mortality rate started to accelerate after twenty- four hours. The mortality rate was high ($p<0.001$ appendix 6) in treatments B. bass HR + Imida MR and B. bass HR + Imida HR. In B. bass HR + Imida HR chemical control (Fig 1.6.6k) by day five almost all aphids were dead.

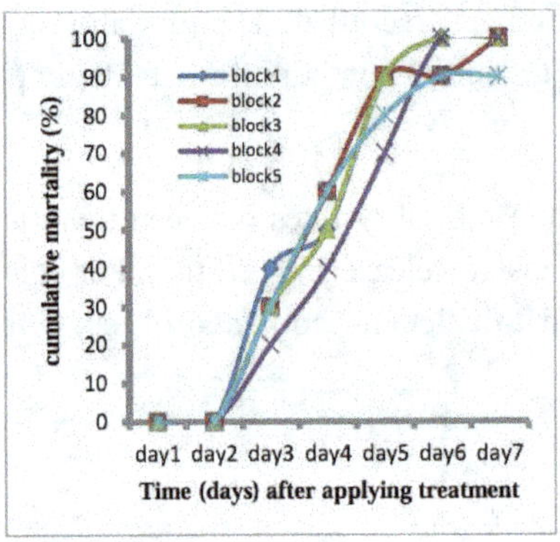

a) Imida MR

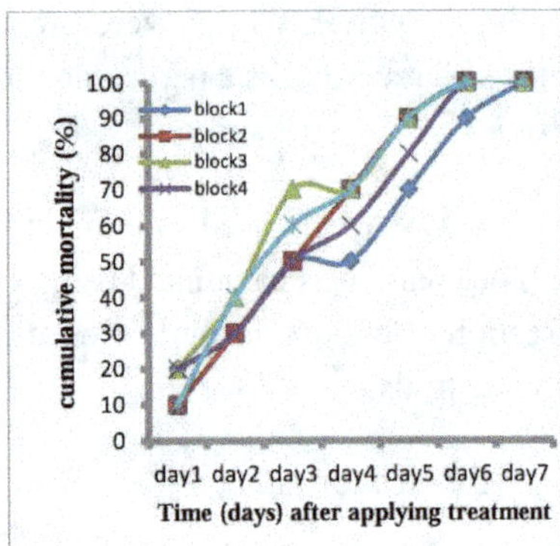

b) B. bass MR

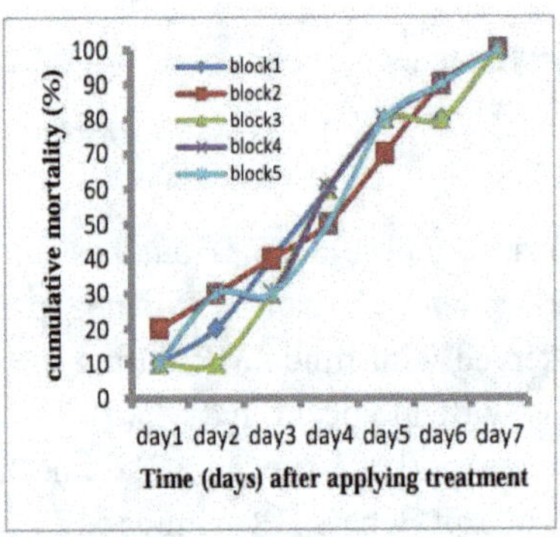

B. bass LR+ Imida LR

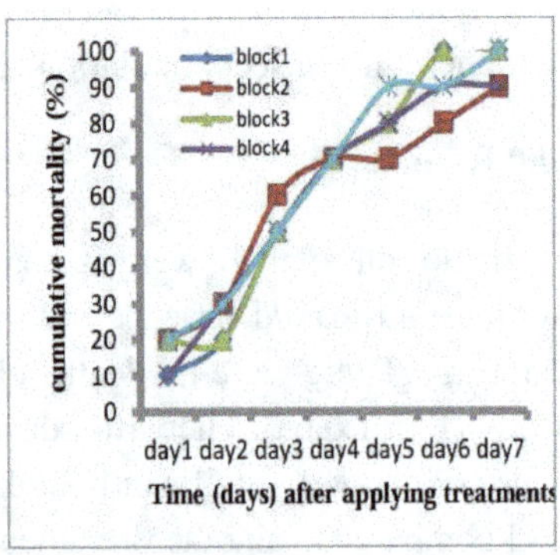

d) B. bass LR + Imida MR

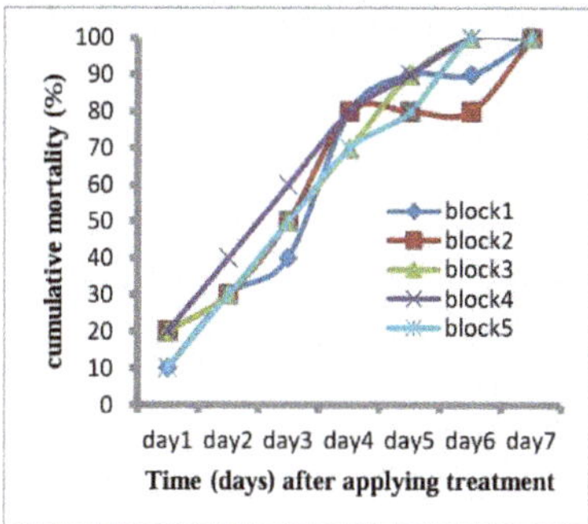

e) B. bass LR + Imida HR

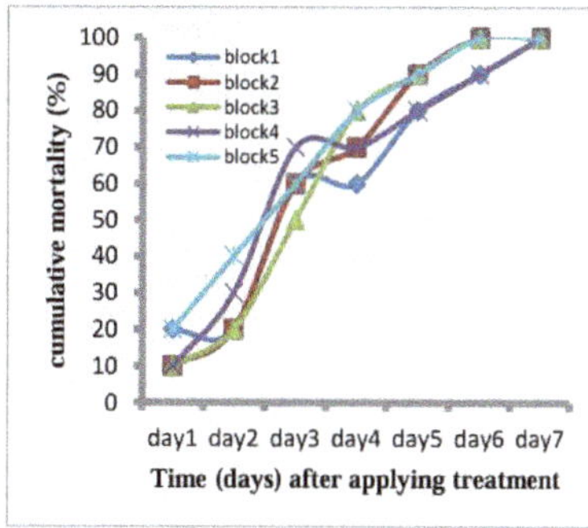

f) B. bass MR + Imida LR

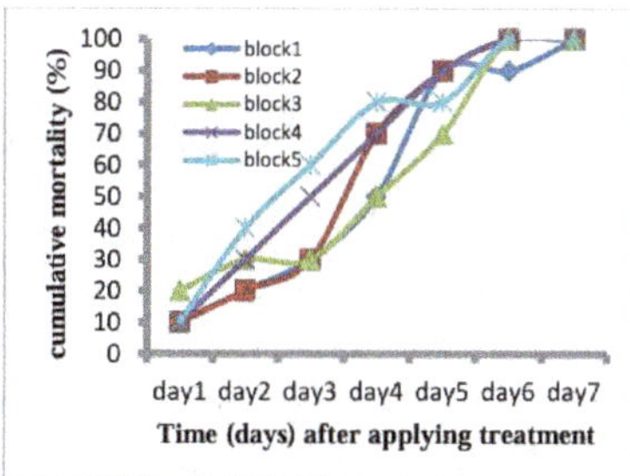

g) B. bass MR + Imida MR

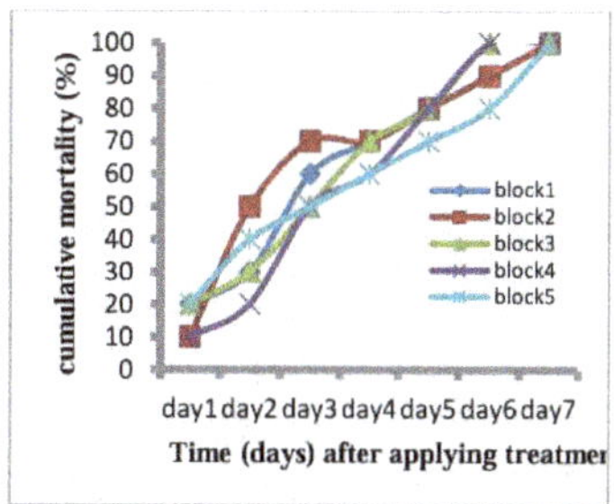

h) B. bass MR + Imida HR

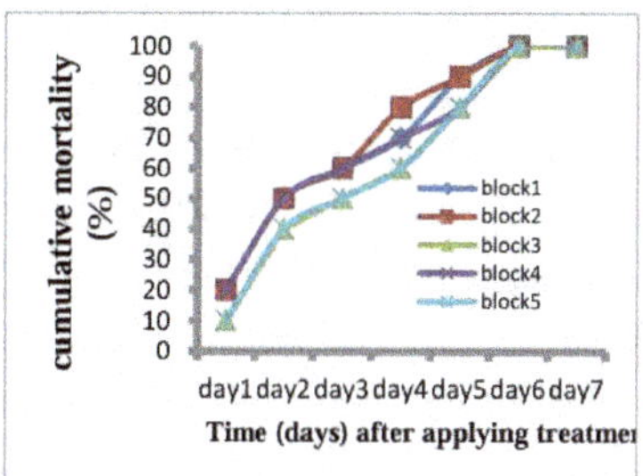

i) B. bass HR + Imida LR

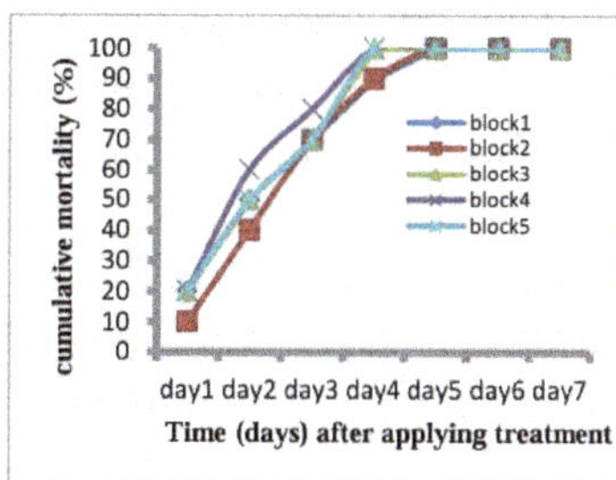

j) B. bass HR

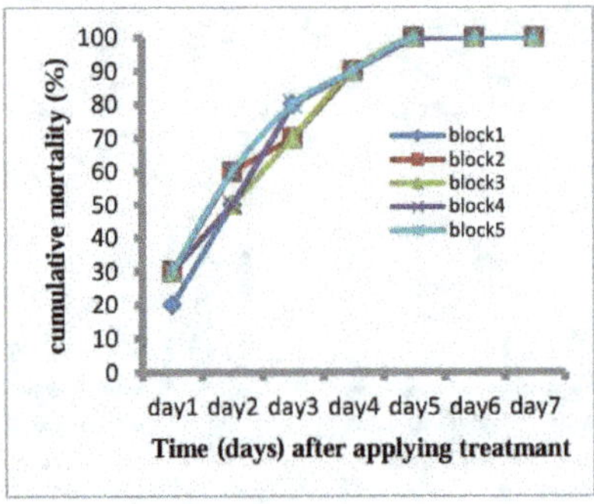

k) B. bass HR + Imida HR

+ Imida MR

Figure 1.6.6: Trends in percent mortalities of M. persicae nicotianae days after exposure to different treatments.

1.6.4.1 Mortality rates

After calculating corrected efficacy percentage in pesticide trials using Abbott's formula, there were significant differences shown among treatments ($p<0.001$ appendix 3-8). On day 1 B. bass HR + Imida HR showed significantly high death rates to all other treatments except B. bass HR + Imida MR (Table 1.6.3). The control and Imida MR showed significantly lower rates compared to the all-other treatments. Day 2, control and Imida MR continued to show significantly lower rates to all other treatments ($p<0.001$ appendix 3). B. bass HR + Imida HR showed high death rates to all other treatments except B. bass HR + Imida LR and B. bass HR+ Imida MR. Day 3, Imida MR started to show death rates but in control plates death rates were still not observed. B. bass HR +Imida HR continued to show high death rates than all other treatments, but death rates were not significantly different with those of B. bass HR + Imida LR and B. bass HR + Imida MR (table 1.6.3). Day 4, B. bass HR + Imida MR showed high death rates though the death rates were not significantly different to B. bass HR + Imida LR and B. bass HR + Imida HR (table 1.6.3). Imida MR still showed lower death rates to all other treatments and in control plates no deaths were recorded.

Day 5, in control plates no death had been recorded, Imida MR showed death rates which had no significant differences to all other treatments except B. bass HR + Imida MR and B. bass HR + Imida HR (table 1.6.3). By day 5, B. bass HR + Imida MR and B. bass HR + Imida HR had reached 100%

mortality rate. Day 6, deaths started to occur in control plates and all other treatments showed high significant death rates with B. bass HR + Imida LR, B. bass HR + Imida MR and B. bass HR + Imida HR showing 100% mortality rates ($p<0.001$ appendix 7). Day 7, an increase in mortality rate in control plates was observed; all other treatments reached 100% mortality rates except in the control, Imida MR and B. bass LR + Imida MR.

Table 1.6.3: Percent mortality after using Abbott's formula to correct for mortality in the control plates

TREATMENT	DAY						
	1	2	3	4	5	6	7
Control	0^{a}	0^{a}	0^{a}	0^{a}	0^{a}	4^{a}	6^{a}
Imida MR	0^{a}	0^{a}	30^{b}	52^{b}	84^{b}	96^{bc}	98^{b}
B. bass MR	16^{b}	34^{bc}	56^{def}	64^{bcd}	84^{b}	97.8^{bc}	100^{b}
B. bass LR+ Imida LR	12^{b}	24^{b}	34^{bc}	56^{bc}	78^{b}	85.6^{b}	100^{b}
B. bass LR + Imida MR	16^{b}	26^{b}	52^{de}	70^{cd}	80^{b}	92^{bc}	96^{b}
B. bass LR + Imida HR	16^{b}	32^{b}	50^{cde}	76^{dc}	86^{b}	93.8^{bc}	100^{b}
B. bass MR + Imida LR	14^{b}	26^{b}	60^{efg}	72^{de}	86^{b}	95.8^{bc}	100^{b}
B. bass MR + Imida MR	12^{b}	28^{b}	40^{bcd}	64^{bcd}	84^{b}	97.8^{bc}	100^{b}
B. bass MR + Imida HR	16^{b}	34^{bc}	56^{def}	66^{bcd}	79^{b}	91.8^{bc}	100^{b}
B. bass HR + Imida LR	16^{b}	46^{cd}	56^{def}	68^{cd}	84^{b}	100^{c}	100^{b}
B. bass HR + Imida MR	18^{bc}	50^{d}	72^{fg}	96^{f}	100^{c}	100^{c}	100^{b}
B. bass HR + Imida HR	28^{c}	54^{d}	76^{g}	90^{ef}	100^{c}	100^{c}	100^{b}
MEAN	13.7	29.5	48.5	64.5	78.7	87.9	91.7
F- PROB	<.001	<.001	<.001	<.001	<.001	<.001	<.001
S.E.D.	3.04	4.05	4.71	4.455	3.88	3.63	1.61
L.S.D.	6.35	8.17	9.49	8.98	7.81	7.32	3.25
CV (%)	35.1	21.7	15.3	10.9	7.8	6.5	2.8

1.6.4.2 Effect of different chemical treatments on cumulative mortality rates of M. persicae nicotianae

Control treatment showed zero cumulative rates up to day 5 (Fig. 1.6.7a). However, the death rates started to accumulate on day 6. In Imida MR death rates started to accumulate at day 3 and reached 98% at day 7 (Fig 1.6.7b). B. bass MR (Fig 1.6.7c) and all other treatments (Fig 1.6.7d- 7i) death rates started to accumulate from day 1 and reached 100% by day 7 except B. bass LR + Imida MR (Fig 1.6.7f) which reached 96% at day 7, B. bass HR + Imida LR (Fig 1.6.7j) which reached 100% mortality rate at day 6 and B. bass HR + Imida MR (Fig 7k) and B. bass HR + Imida HR (Fig 1.6.7l) which reached 100% mortality rate by day 5.

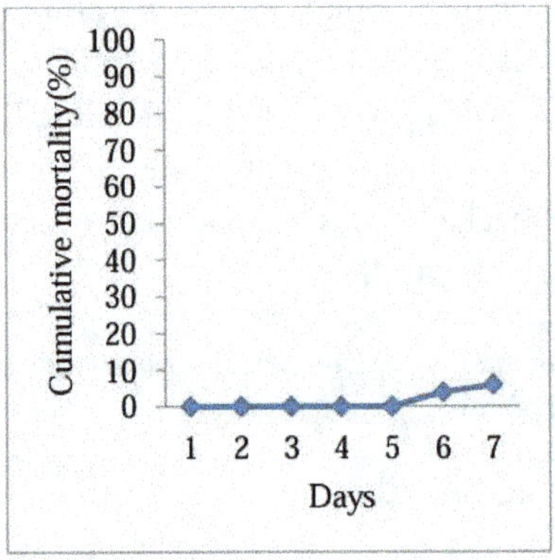

a) Control

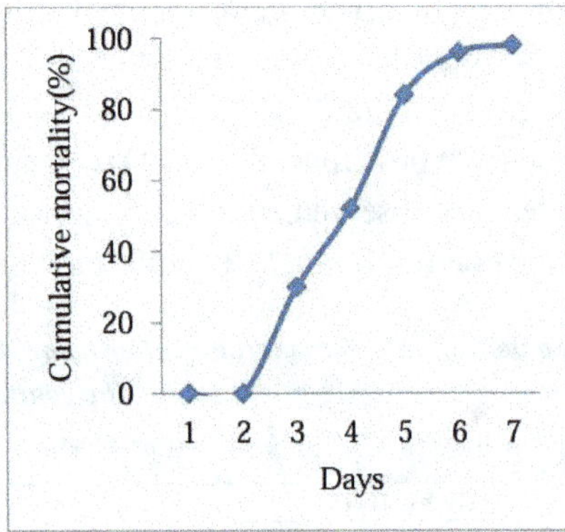

b) Imida MR

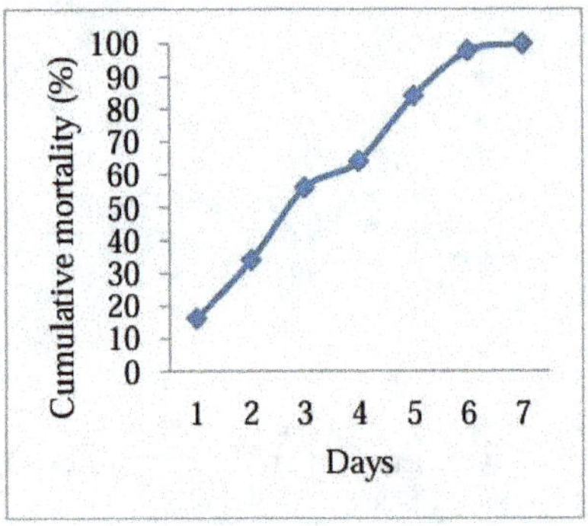

c: B. bass MR

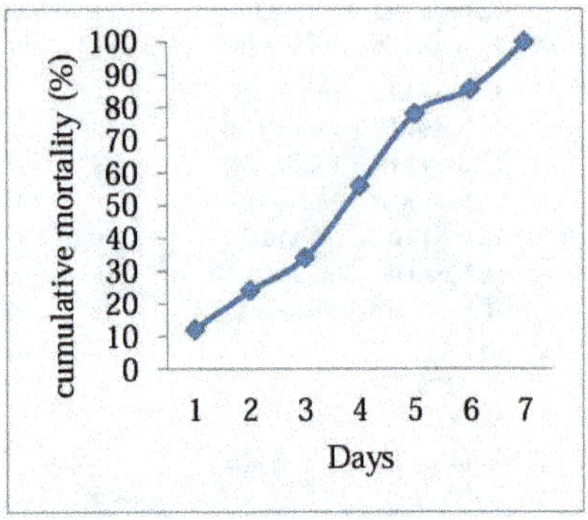

d) B. bass LR+ Imida LR

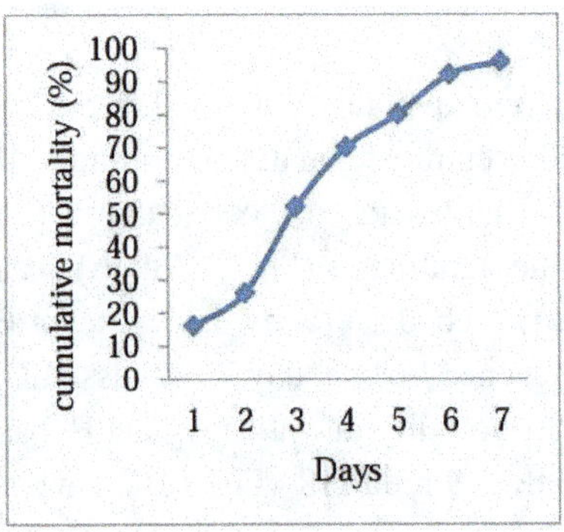

e) B. bass LR + Imida MR

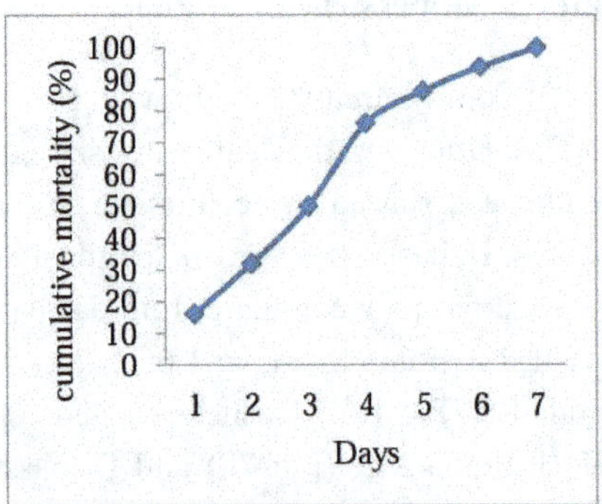

f) B. bass LR + Imida HR

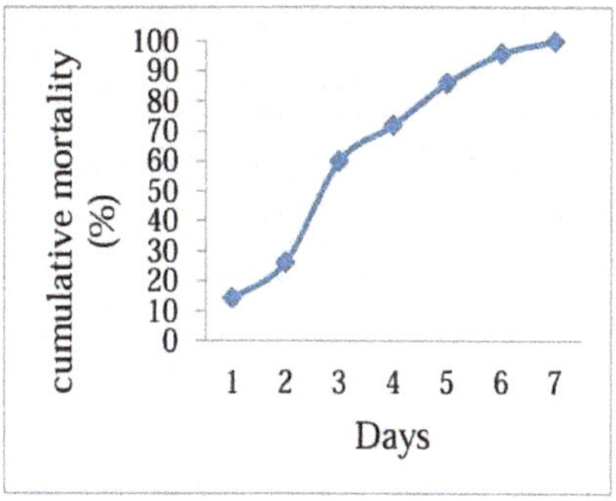

g) B. bass MR + Imida LR

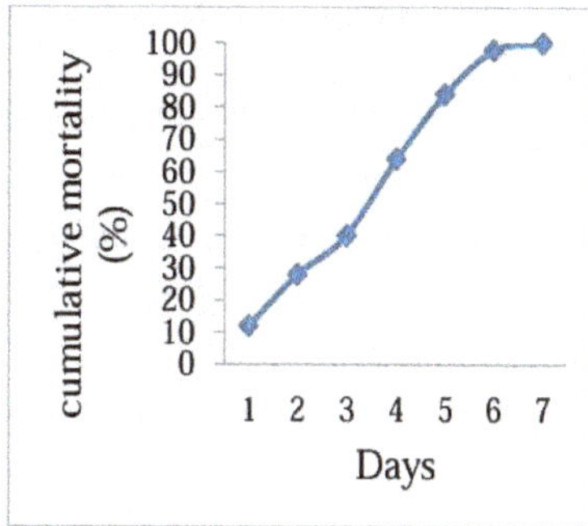

h) B. bass MR + Imida MR

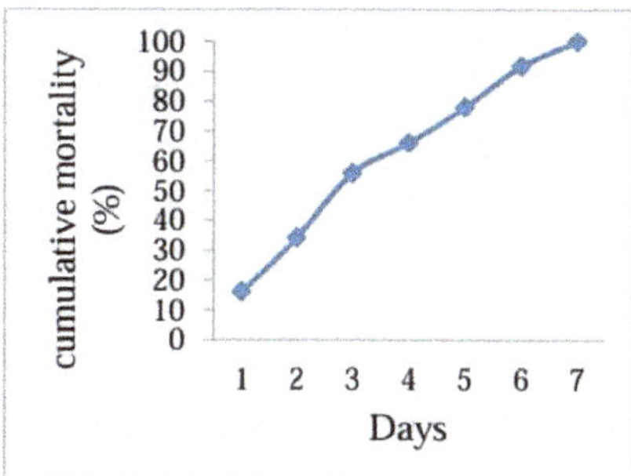

i) B. bass MR + Imida HR

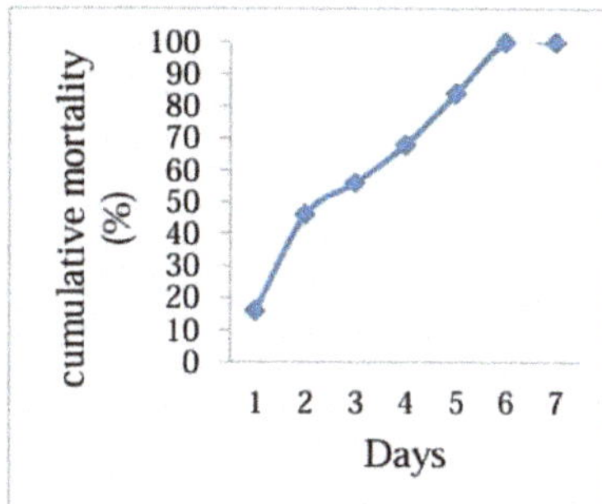

j) B. bass HR + Imida LR

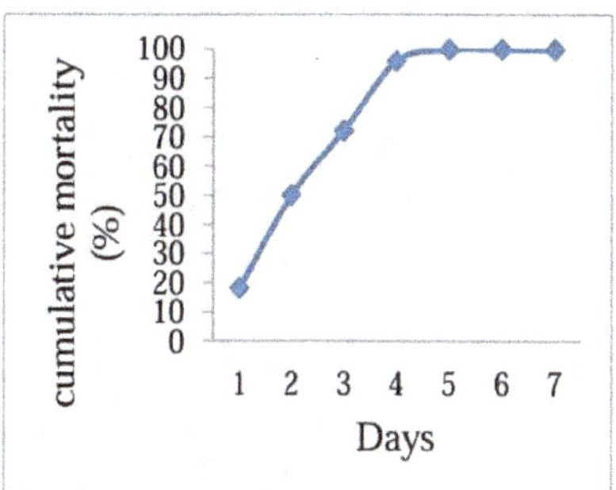

k) B. bass HR + Imida MR

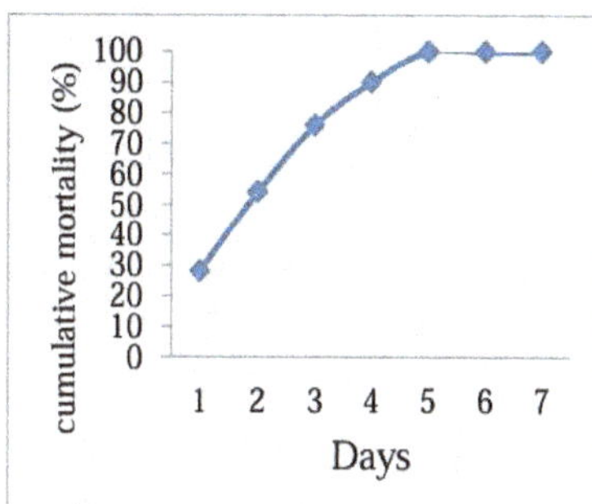

l) B. bass HR + Imida HR

Figure 1.6.7: Effect of different chemical treatments on cumulative mortality rates of M. persicae nicotianae

1.6.5.0 Discussion

The control treatment showed deaths, which were due to natural occurrence because there were no significant differences from day one up to day seven. There were no deaths observed with Imida MR during the first two days. This was because imidacloprid is a systematic insecticide, it should be absorbed first by the plant and this takes about 48 hours. Once the chemical is assimilated, aphids when feeding ingest the chemical and die. This finding clearly suggested that the neonicotinoid insecticides like imidacloprid take slightly longer time for their maximum effect against sap-sucking insect like aphids and could persist up to 10 days (Das, 2013).

Once applied, the insecticide enters into the plant system by translaminar action, after which it enters into the insect body by feeding and finally killing the aphid by binding it with nicotinic acetylcholine receptor (Das, 2013). B. bass HR + Imida HR showed significantly higher rates of mortality on day one compared to all other treatments except B. bass HR + Imida MR. Deaths in B. bass MR treatment were observed from day one, this was because *B. bassiana* is a contact insecticide and it kills aphids on contact. This chemical does not need to be consumed by aphids to cause effect. A white mould was also seen wrapping the aphid body. *B. bassiana* is a fungus that kills aphids by causing a disease known as white muscadine disease. When the spores of *B. bassiana* in contact with the cuticle of specific insects that are susceptible to it, they germinate and grow directly through the cuticle to the inner body of the host (www.hort.uconn.edu

The fungus then proliferates through the insect's body, producing harmful toxins and depriving the insect of nutrients until the insect dies (www.hort.uconn.edu). In treatments B. bass LR+ Imida LR; B. bass LR + Imida MR and B. bass LR + Imida HR mortality was observed from day one because there was a contact insecticide in the formulation, but the mortality rate was slow because the contact insecticide rate used was low. From day three the mortality rate accelerated in B. bass LR + Imida MR and B. bass LR + Imida HR treatments, this was due to low rates of the contact insecticide and medium to high rates of the systematic insecticide which were to be assimilated by the plant first then fed by the aphid eventually killing the aphid by binding it with nicotinic acetylcholine receptor.

This number of deaths on day one indicated clearly that the level of mortality was dose dependent and the presence of the contact insecticide (Ye et al., 2005). The higher rates of B. bass and Imida MR and HR indicated that by four to five days cumulative mortality had reached 100% unlike higher rate of B. bass with lower rate of imidacloprid which took 6 days to reach 100%. A white mould was also seen wrapping the aphid body in this treatment. The mortality rate was moderate compared to B. bass LR + Imida LR; B. bass LR + Imida MR and B. bass LR + Imida HR. Mortality rate also began to accelerate after 72 hours because the systematic insecticide had been absorbed into the plant system and also absorbed by the system of the pest after feeding.

In B. bass HR + Imida LR; B. bass HR + Imida MR and B. bass HR + Imida HR chemical control, mortality was observed from day one because the contact insecticide concentration was high. The mortality rate was high in treatments B. bass HR + Imida MR and B. bass HR + Imida HR this was because the systematic insecticide rate used was medium to high. In B. bass HR + Imida HR chemical control (Fig 1.6.71) by day five cumulative mortality had reached 100%.

1.6.6 Conclusion and Recommendation

1.6.6.1 Conclusion

Formulations of fungal and chemical agents are sustainable options for *Myzus persicae nicotianae* control. Adopting combined formulations of fungal and chemical agents can effectively control *Myzus persicae nicotianae* and sustain pesticide usage at lower levels. B. bass HR + Imida MR had significantly high rates of mortality in day one and by day five, mortality had reached 100%. However, there would be no need to use the higher rates if the same results could be attained without using higher rates of Imidacloprid. Uses of this formulation will benefit agrochemical users economically, while maintaining a friendly environment and reducing exposure when handling the pesticides.

1.6.6.2 Recommendation

I strongly recommend the use of B. bass HR + Imida MR because this treatment had significantly high rates of mortality in day one and by day five, mortality had reached 100% which was the same as B. bass HR + Imida HR. Such a combination undoubtedly increases practicability of fungal formulation and may help to manage aphid resistance to chemical insecticides. It is also economical viable to use this formulation compared to B. bass HR + Imida HR because efficacy was the same. Experiments should be conducted further with the aim of developing our own strain of B. bassiana suitable for tobacco aphid control. *B. bassiana* is environmentally friendly and it has worked well on controlling aphids, experiments on the control of other pests using *B. bassiana* should be conducted since its wide application as a biological pesticide could be taken up after exploring its toxicity and field trials.

1.6.7 Reference

1. Banga, N., (2012). Zimbabwe tobacco output seen up, but shy of target. The Zimbabwean Mail, 16 April 2012.

2. Blackman, R. L., (1987). Morphological discrimination of a tobacco-feeding from Myzus persicae (Sulzer) (Hemiptera: Aphididae), and a key to New World Myzus (Nectarosiphon) species. Bulletin of Entomological Research, 77: 713-730.

3. Blair, B. W., (1990). Insect and Mite Pests of Tobacco in Zimbabwe: Description, Biology and Damage. TRB Technical Bulletin.

4. Breen, T. H. (1985). Tobacco Culture. Princeton University Press.

5. Cabrera-Brand, M. A., Fuentes-Contreras, E. and Figueroa, C. C. (2010). Detoxification metabolism between two clonal linages of the aphid Myzus persicae (Sulzer) (Hemiptera: Aphididae) reared on tobacco (Nicotiana tubacum L). Chilean journal of Agricultural Research, 70 (4): 567- 575.

6. Capinera, L. J., (2005). Green Peach Aphid, Myzus persicae (Sulzer) (Insecta: Hemiptera: Aphididae). University of Florida, IFAS Extension.

7. Chamberlin, F. S., (1958). History and status of the green peach aphid as a pest of tobacco in the United States. USDA Technical Bulletin, 1175: 12.

8. Chandler, D. (1997). Selection of an isolate of the insect pathogenic fungus Metarhizium anisopliae virulent to the lettuce root aphid, Pemphigus bursarius. Biocontrol Science and Technology, 7: 95–104.

9. Clements, K. M., Sorenson, C. E., Wiegmama, B. M., Neese, P. A. and Roe, R. M. (2000). Genetic, Biochemical, & Behavioural uniformity among populations of Myzus persicae. Entomologia Experiments et Applicata, 95: 269-281.

10. Cottrell, C. B. (1994). Aphids and virus diseases in tobacco. Tobacco Research Board Journal, 15: 12-13. danpritchard.com

11. Das, G. (2013). Efficacy of imidacloprid, a nicotinoid group of insecticide against the infestation of chilli aphid, Myzus persicae (Hemiptera: Aphididae). International Journal of Biology and Biological Sciences 2 (11): 154-159, Available online at ISSN 2327-3062 ©2013 Academe Research Journals.

12. DiFonzo, C. D., Ragsdale, D. W., Radcliffe, E. B., Gudmestad, N. C and Secor, G. A. (1997). Seasonal abundance of aphid vectors of potato virus Y in the Red River Valley of Minnesota and North Dakota. Journal of Economic Entomology, 90: 824-831.

13. Dixon, A. F. G. (1977). Aphid ecology: life cycle, polymorphism and population regulation. Annual Review of Ecology and Systematics, 8: 329-353.

14. Dixon, A. F. G. (1998). Aphid ecology, 2nd edn. London: Chapman and Hall.

15. Dorschner, K. W., Feng, M. G. and Baird, C. R. (1991). Virulence of an aphid-derived isolate of Beauveria bassiana (Fungi: Hyphomycetes) to the hop aphid, Phorodon humuli (Homoptera: Aphididae). Environmental Entomology, 20: 690–603.

16. England, A and Hawkins, T. (2011). Zimbabwe tobacco farmers yield success available at www.cfuzim.org (accessed on Friday 19 December 19, 2014)

17. FAO, (2010): Issues in global tobacco economy. Tobacco in Zimbabwe.

18. Feng, M. G. and Johnson, J. B. (1990). Relative virulence of six isolates of Beauveria bassiana (Fungi: Hyphomycetes) on the Russian wheat aphid, Diuraphis noxia (Homoptera: Aphididae). Environmental Entomology, 19: 785–790.

19. Feng, M. G., Johnson, J. B. and Kish, L. P. (1990a). Virulence of Verticillium lecanii and an aphid-derived isolate of Beauveria bassiana (Fungi: Hyphomycetes) for six species of cereal aphids (Homoptera: Aphididae). Environmental Entomology, 19: 815–820.

20. Feng, M. G., Poprawski, T. J. and Khachatourians, G. G. (1994). Production, formulation and application of the entomopathogenic fungus Beauveria bassiana for insect control: current status. Biocontrol Science and Technology, 4: 3–34.

21.Francis, F., Gerkens, P., Harmel, N., Mazzucchelli, G., DePauw, P. and Haubruge, E. (2006). Proteomics in Myzus persicae : effect of aphid host plant switch. Insect Biochemistry and Molecular Biology 36: 219-227.

22. Gates, L. F., (1962). A virus causing axillary bud sprouting of tobacco in Rhodisia and Nyasaland. Annals of Applied Biology, 50 (1): 169-174.

23. Hatting, J. L., Wraight, S. P. and Miller, R. M., (2004). Efficacy of Beauveria bassiana (Hyphomycetes) for control of Russian wheat aphid (Homoptera: Aphididae) on resistant wheat under field conditions. Biocontrol Science and Technology, 14: 459–473.

24. Masuka, A. G., Cole, D. L. and Mguni, C., (1998). List of plant diseases in Zimbabwe. Plant protection research institute. Zimbabwe.

25. McDonald, S. A., Halbert, S. E., Tolin, S. A and Nault, B., (2003). Seasonal Abundance and Diversity of Aphids in a pepper Production Region in Jamaica. Environmental Entomology, 32 (3): 499-509.

26. Milner, R. J. (1997). Prospects for biopesticides for aphid control. Entomophaga, 42: 227–239.

27. Miyazaki, M. (1987). Forms and morphs of aphids. In: Minks, A. K., and P. Harrewijn (eds), Aphids: Their Biology, Natural Enemies and Control, 2A, World Crop Pests. Elsevier, Amsterdam. pp. 27-50.

28. Mo, X. H., Qin, X. Y., Tan, Z. X., Li, T. K., Wu, J. Y and Chen H. R. (2002). First report of Tobacco Bushy-top disease in China. Plant disease, 8b (1): 74.

29. Mo, X. H., Qin, X. Y., Wu. J., Yang, C., Wu, J. Y., Duan, Y. Q., Li, T. F and Chen H. R. (2011). Complete nucleotide sequence and genome organisation of a Chinese isolate of Tobacco Bushy top virus. Archives of virology, 148: 389-397.

30. Moran, N. A. (1990). Aphids life cycle: two evolutionary steps. Am Nat 136: 135-138. nist.rcsb.org

1.7 Evaluation of planting time, pest incidence and yield for 5 (Five) Industrial Hemp varieties (Cannabis Sativa.L) in Zimbabwe.

Beloved Mtetwa

A thesis submitted in partial fulfilment of the requirements of the degree of Master of Science in Crop Science

Department of Agronomy and Horticulture

Faculty of Agriculture, Environment and Natural Resources Management

Midlands State University

DECEMBER 2021

Acknowledgements

- Ms Katy D. Percival for working with me through-out the project.

- Zimbabwe Prisons & Correctional Services: the Commission General and the Officers in Charge for giving access to the secure plot of land for the trial, supplying staff and overseeing all daily work required for this trial to run efficiently.

- Department of Research & Specialist Services (Agronomy): Dr M.Chiduwa, Mr B. Neurashe & Ms M. Mahovo for their tireless assistance with data collection, analytical testing of soils and plant disease sampling.

- DripTechfortheirfinancialsupportatsupplyingthein-lineirrigationdrip system used

- Vollig Private Limited management for their financial support through tuition fees payments, travel costs and other financial obligations.

- My family members Godwin Mtetwa, Mirriam Mtetwa, Answer Mtetwa, Desire Mtetwa and Evidence Mtetwa who gave me support in motivating me to complete this research project.

- My wife Kudakwashe Pswatai Mtetwa who stood by me all the time when I was facing hardships during my examination and research project period.

1.7.1.1 Introduction

Industrial hemp (Cannabis sativa L., Cannabaceae) is a versatile herbaceous crop that has been used for fiber, food, and medicinal purposes (Andre, 2016, Vonapartis, 2015). Industrial hemp is a variety of *Cannabis sativa L.* and is one of the marijuanas with lower levels of the psychoactive compound THC (tetrahydracannabinol); < 0.3%, while marijuana has approximately 17% (Johnson, 2013). (Johnson, 2013). The cultivation of hemp dates back to China around 2700 BC and is believed to have then expanded across Asia, making its way to Europe 2000–2200 years ago (Callaway, 2004, Struik, 2000).

There are global estimates that the hemp market consists of more than 25,000 products (Johnson, 2013). Hemp can be grown as a fiber crop for many products like fabrics and textiles, yarns and raw or processed spun fibers, paper, carpeting, home furnishings, construction and insulation materials, auto parts, and composites. The stalk has two different fibers; hurd, the interior of the stalk and the bast, the outer fibers. The hurd can be used for animal bedding, raw material inputs, low-quality papers, and composites while the bast is among the strongest and most durable natural fibers (Brook, 2008). Oil and "milk" from crushed hemp seed are ingredients in a number of body care products, nutritional supplements, cosmetics, pharmaceuticals, and food products (Johnson, 2013).

Precise data are not available on the size of the market for hemp-based products; however, current industry estimates report that United States retail sales of all hemp-based products may be nearly $500 million per year (Johnson, 2013). By comparison, wheat in Oregon had a farm gate value of $503 million in 2011 and all of Oregon agricultural and fisheries commodities had an aggregate value of $5.4 million in 2011 (Oregon Department of Agriculture; 2012).

Industrial hemp (Cannabis sativa L.) has recently been the subject of increasing interest in Zimbabwe, following the gazzeted statutory Instrument 218/20 of the Agricultural Marketing Authority (Industrial Hemp) Regulations, of September 2020 (SI 2020-218 Agriculture Marketing Authority). In terms of these 2018 Regulations, persons interested in the production of the cannabis plant must apply for the producer's license which will only be valid for 5 years and may be renewed after that before its expiry. It is noteworthy that the 2018 Regulations allow for the production of cannabis, which is defined to include fresh or dried cannabis, cannabis oil, cannabis plants or cannabis seeds; the cannabis plant also includes the whole or any portion, whether green or dry, of any plant of the genus cannabis (Nyamidzi, 2020).Further, in September 2020 a new law being Statutory Instrument, S.I 218 of 2020 Agricultural Marketing Authority (Industrial Hemp) Regulations 2020, 'the Regulations' was gazetted. These regulations were put in place to regulate and to provide a conducive environment in the production, procurement, distribution possession, sale, provision, transportation of industrial hemp.

Industrial hemp (Cannabis sativa L.), like all new crops, needs proper agronomic understanding so as to optimize production. There is need to identify which pests are of economic importance to Industrial hemp. There is also need to investigate content that can be attained by the crop in Zimbabwe. In the current project, we sought to understand the agronomy, and pests affecting industrial hemp. This is the first industrial hemp study which will help farmers with information on how to grow Industrial hemp in Zimbabwe.

1.7.1.2 Overall Aim/Objective

The objective of the project is to evaluate the effects of planting date, pest incidence fibre and seed yield on five industrial hemp varieties.

1.7.1.2.1 Specific objectives

i. To investigate the planting time of Cannabis sativa L

ii. To identify the pests of concern on Cannabis sativa L

iii. To investigate the effect of Cannabis sativa L on yield.

1.7.2. Literature Review

Hemp (Cannabis sativa L.) is an annual herbaceous plant with a slender stem, ranging in height from 1.2 to 4.5 m and a diameter from 0.6 to 1.9 cm. The innermost layer is the pith, surrounded by woody material known as hurds. Outside of this layer is the growing tissue which develops into hurds on the inside and into the bast fibers on the outside. The stem is more or less branched, depending on the crop density. When sown thickly the stems do not branch (Ariskan, 2010). The leaves are of a palmate type and each leaf has 7 to 11 leaflets, with serrated edges. The strong tap-root penetrates deep into the soil.

However, if the soil conditions are unfavorable, the main root remains short, while lateral roots become more developed (Ariskan, 2010). Hemp (Cannabis sativa L.) is also known as industrial hemp, plant of the family Cannabaceae and is cultivated for its bast fibre its seeds. Hemp is usually confused with the cannabis plants that serve as source of drug (Bocsa, 1998).

1.7.2.1 Plant densities

Economic efficiency of planting a crop such as industrial hemp (Cannabis sativa L) from seed requires the establishment of ideal sowing rates (kg seed ha-1) which determine the final plant population (plants m-2). Under-sowing may result in undesirable product qualities and poor yield (Ranalli.1999). Excessive planting density for hemp leads to increased self-thinning and slower plant growth in later growing stages (Van der Werf 1996). Appropriate planting density will vary depending on the variety, season, soil type and the extent of other agronomic practices applied to the crop. Therefore, seeding densities for any variety should be optimized for the specific sites of interest. Industrial hemp is also sown at different densities depending on the purpose of the crop: fiber or seed. The recommended seeding rate for the following crop (one seed crop) is usually around 30 kg ha1 or between plants 100-150 m2 (Bocsa 1998). Low seeding densities allow for more tillering, shorter plant heights and higher individual plant weights than fiber crops sown at higher densities. The latter prevents branching, producing individual trees that are taller and lighter. Harvesting hemp for mature fibers is much easier when the plant is upright and has few branches, the fiber quality is better than in unbranched plants. The planting density of hemp is approximately twice that of hemp seed. Early research (Bocsa 1998) suggested that no more than 80 kg of ha1 seed should be sown for hemp, as there is little difference in final yield between 60–100 kg ha1 (300–500 m2). However, these ratios are considered excessive for nonwoven hemp yarn produced because of yarn volume rather than quality. The seeding rate for nonwovens may be appropriate at 30 kg ha1 (Burczyk 2009). A high percentage of seeds often results in shorter, thinner stems (Amaducci 2008) with a higher fiber-to-stem ratio, which is desirable for hemp fibers (Van der Werf 1996). The seeding rate depending on the intended use may not vary much between varieties. In a study done in Wales by (Bennet,2006), five varieties were sown in trees covering 150 and 300 m2. The results indicated that although the total final fiber yield increased at the highest densities, no interaction between cultivars with seeding rate was observed. The percentage of fibers in harvested straw increased slightly at the highest density although the percentage of long fibers remained the same. Hemp fiber production around

1.7.2.2 Environmental benefits

Hemp has environmental benefits and it is recognized as one of the crops that can be used for soil improvement due to its strong growth ability especially for high yielding fiber varieties (Piotrowski 2011). It has been shown that hemp has the ability to treat plants, restore soil contaminated by heavy metals (Askew, 2002). The hemp plant grows for tall fibers; the roots go deep into the soil about 45 to 90 cm. This allows the roots of the plant to deepen and increases the efficiency of removing contamination over a large area compared to other plants with shallow roots & # 40; Citterio, 2003 & # 41; Hemp has a high potential for absorbing and accumulating heavy metals like lead (Pb), nickel (Ni), cadmium (Cd), etc., through its roots and then storing them, thus allows plants to be harvested from hemp in addition to dangerous compounds. A variety of hemp cultivars are good candidates for plant treatment and they are highly resistant to Cd stress (Linger, 2002). The same study also showed that the highest concentrations of heavy metal accumulation were found in the leaves, even though they were distributed throughout all parts of the plant. Hemp leaves collected from a heavy metal contaminated site in Pakistan showed heavy metal accumulation; Cu (1530 mg kg-1), Cd (151 mg kg-1) and Ni (123 mg kg-1), respectively (Ahmad et al., 2016). In 1986, hemp was planted to help decontaminate the soil around the site of the Chernobyl nuclear disaster (Citterio, 2003).

The rapid growth and development of hemp make it one of the fastest sources of CO_2 biomass conversion. Hemp has proven to be an ideal carbon sink as it can capture more CO_2 per hectare than other cash crops or even forests. For example, one hectare of hemp can absorb 22 tons of CO2 per hectare. High biomass crops such as hemp, which are grown for fiber, can sequester larger amounts of carbon through photosynthesis and then store it in the plant's body and roots through photosynthesis. biological response. Most of the carbon is stored in the harvested hemp stalk and less in the roots and leaves. Hemp can produce at least 13 tons of biochar per hectare per year (Günther, 2019). One of the other potential uses of hemp biomass is the production of biochar for soil applications that has the potential to improve soil carbon sequestration and reduce greenhouse gas emissions (Lehmann, 2006 and Andreae 2001).

Hemp has been recognized as one of the energy plants due to its high biomass and energy concentration per hectare (Kraszkiewicz, 2019). Hemp's ability to suppress weeds, low pesticide requirements and soil health-promoting properties make it even more energy efficient (Kraszkiewicz, 2019 and Kolodziej, 2012). The energy properties of hemp are similar or superior to those of other solid biofuels such as grain straw, wood, etc. (Prade, 2012). For example, the heat of combustion of hemp (18.4-19.1 MJ/kg) is comparable to that of corn Zea mays (18 MJ/kg), which is slightly higher than that of Jerusalem artichoke Helianthus tuberosus (16.5 MJ/kg) and slightly lower than Miscanthus sp. (19.8 MJ/kg) (Kolodziej; et al 2012 and Prade et al 2012). In addition, wet and silage hemp biomass can be converted to biogas and ethanol (Kraszkiewicz; et al, 2019 and Prade; et al, 2011). Hemp also emits compounds that are low in sulfur (Kolodziej, 2012) and have low ash content (Kraszkiewicz, 2019). Therefore, hemp has the potential to contribute to renewable energy.

1.7.2.3 Soil condition

Hemp can be grown on a variety of soils, but hemp grows best in loose loamy soils that are well drained and rich in organic matter (Van der Werf, 2019). The most suitable soil for growing hemp should have a pH between 6.0-7.5 (Amaducci; et al, 2015) and according to (Bócsa, 1998) the optimal soil pH for hemp production is between 5 and 5.8 to 6.0 because it does not grow well in acidic soil. Soil should be deep, open, rich in nutrients and have good water holding capacity. The texture of sandy clay is ideal for hemp growth followed by clayey loam, but heavy clayey and sandy soils are less suitable as they are too moist or too low (Li)., 1982). Soil preparation is an essential process in growing cannabis, and flooding kills plants, especially young plants, so hard soil needs to be crushed for free drainage. The presence of compacted layers can impede root development, especially if compacted bread is the result of inadequate soil preparation. Hemp is a taproot plant, and in fine soils the taproot is usually L-shaped, which adversely affects the plant's uptake of nutrients and water (Desanlis2013). One study concluded that neutral alkaline fertile loam or siltrome soils are optimal for cannabis cultivation (Dewey, 1913). Hemp does not germinate well on acidic sandy, heavy clay, or gravel soils. Hemp plants can grow in Pete's wetlands, but with lower fiber quantity and quality (Dewey, 1913). Hemp is very sensitive to soil moisture and should not be exposed to drought. Good soil drainage is important for maximizing its production, as it grows well in highly water-retaining soils, and most cannabis plants do not grow on poorly drained soils (Vessel, 2019).

Since hemp is a short-day plant, it is very sensitive to day length. It has been reported that the length of the day affects the amount of light that hemp plants receive and has a significant impact on the productivity of different types of hemp. The transition from the vegetative stage to the flowering stage of hemp depends on the length of the day and the variety. However, the Canadian Hemp Trade Alliance (2019) found that some varieties started flowering regardless of the length of the day, while others took a short time to transition to the flowering period. Did. Data on the specific number of hours of light required per day is limited and needs further investigation.

1.7.2.4 Plant

The distance between cannabis plants depends on the type of cannabis cultivated. H. Fiber, seeds, or CBD. In general, densely planted hemp promotes higher altitudes and limits flowering. Hemp grown primarily for textiles is grown close to each other, reducing branching and promoting stem elongation. This results in longer and stronger fiber yields (Amaducci; et al., 2002). High-density cannabis plants are also produced because the dense stands control weeds and eliminate the need for herbicides. On the contrary, hemp planted for seeds and CBD is well spaced to maximize desirable flowering and branching.

Hemp is usually grown on seed drills with 3–7-inch row spacing, especially when grown for textile and seed products. However, the recommended sowing amount will vary greatly depending on the optimum sowing depth of 1.9 to 3.2 cm. These recommendations ensure that the competition for Amaducci space required for vegetative growth and root development is minimized. et al., 2002). Another study reported that the distance between hemp cultivated for textiles was 20-40 cm (Bócsa, 1998).

High plant populations of fiber hemp varieties close the canopy faster and minimize weed growth (Lotz, 1991). For cannabis grown for CBD, the highest yield of flowers or buds was achieved with a plant density of 15 plants m-2. It has been suggested that similar planting densities are required for the highest cannabis oil production from seed-containing varieties (Meije; et al., 1995). One study reported that the row spacing of CBD hemp was similar to the row spacing of hemp strains cultivated for seeds (Williams, 2019). In another study, the optimal density of seed production varied significantly from

30 to 75 plants m-2 (Hennink; et al., 1994). The most important variation is found in hemp cultivated for fiber production, and one study reports that it is a plant with a seeding density of 50-750 m-2 (Dempsey, 1975). Appropriate plant spacing to achieve high overall yields of stems, seeds and inflorescences was found in 120 plant m-2 with 0.5 m row spacing (Campiglia, 2017). Another study reported that the optimal density for growing CBD hemp was 10 plants m-2 (Ivonyi; et al, 1997). The results of these studies show that optimal cannabis spacing is a function of the type of cannabis cultivated and the soil environment. However, information on optimal CBD and recommended intervals for cannabis seed production in different soils and climates is very limited and clearly needs to be addressed.

1.7.2.5 Temperature

Temperature plays a clear and important role in the production of cannabis at various stages of growth. Hemp grows in a variety of environmental conditions, and many studies have shown that it is better suited for temperate climatic zones. Hemp grows best when the average daily temperature is 16-27 ° C, but it can withstand colder and warmer conditions. For example, at low temperatures of 8-10 ° C, it takes 8-10 days for seeds to germinate. Young saplings with 8-10 leaves can withstand exposure to frost, usually up to -5 ° C. Plant height reached in 90 days in the field can be reached in 40 days by plants grown at 19 ° C under controlled conditions by adjusting the number of growing days (GDD). Cannabis grown in the Mediterranean region requires approximately 1900-2000°C GDD for fiber production and approximately 2700-3000°C GDD for seed production (Merfield, 2019). However, the ideal GDD requirements for high-yielding CBD hemp need further investigation.

1.7.2.6 Water requirement

Hemp requires high levels of water throughout its growing season, especially during the first 6 weeks of growth when young plants are beginning to settle (Dewey, 1913). When roots are firmly rooted, plants can tolerate drier conditions, but severe droughts can negatively accelerate maturation and cause stunted crop production. Ideally, hemp requires 63 to 75 cm of rainfall per year. The need and consumption of plant water depend on the type of hemp, soil, climatic conditions and management methods.

Cannabis roots can penetrate to a depth of 2-3 m and remove water, as long as the soil is not soaked and the pot does not block the roots of the plant. According to a study conducted in Europe, cannabis yields are highly dependent on precipitation from June to July and can reach up to 700 mm (Amaducci; et al., 2002). According to other studies, cannabis requires 25-35 cm of water and 50-70 cm of available water at the nutritional stage for optimal yields.

1.7.2.7 Fertilizer requirement

Nitrogen (N) is the most important nutrient in cannabis production as it plays an important role in plant nutrition and yield. This fact is well illustrated by studies reporting the response of cannabis to N fertilization. The use of N fertilizer has a positive effect on the plant height of cannabis, the biomass of fiber cannabis varieties, and the grain and protein content of varieties including seeds. Due to the high N requirement during the first month of cannabis growth, most field trials usually apply N at the time of sowing. One study reported that 79% of total N intake occurred within the first month and daily N intake was 3-4 kg ha-1 (Ivonyi; et al., 1997). Another study reported that post-sowing N or split application did not increase stem yield compared to the N distribution at sowing (Finnan and Burke, 2013). We found that N fertilization increased from 0 to 120 kg Nha-1. However, no further increase in yield above the N fertilization rate of 150 kg ha-1 has been reported (Finnan and Burke, 2013). Excessive application of N stimulates rapid stem growth and facilitates the sedimentation of hemp plants (Desanlis; et al., 2013). On the other hand, if N is too low, the yield will be low, and if it is too high, the quality of the fiber will be poor. Various experimental results confirm that N fertilization should be determined based on the initial soil fertility. Struik et al. A study conducted by 2000 also concluded that the reaction of cannabis growth to fertilizer-N was negligible in N-rich soils, so further investigation is needed. Recent studies have reported that N-deficiency symptoms were observed in CBD greenhouse hemp when the N content in leaf tissue analysis was 1.62% (Cockson; et al., 2019). However, more research is needed to assess the effect of N application on the yield of seeds and CBD cannabis varieties under field conditions.

Phosphorus (P) is an essential major nutrient required during the growth stage of hemp. However, as the plant matures, the need gradually increases. P is also important in the early stages of cannabis growth, as it plays an important role in imparting strength and resistance to pests. The effect of P on hemp production has been studied only to a limited extent. It was concluded that the application of P increases plant height, but the effect of cannabis on biomass and seed yield is inconsistent and therefore minimal and therefore highly negligible (Vera; et al. 2004). Previous studies have also shown that the response of hemp to P may vary depending on growth conditions (Vera; et al, 2010), and its availability also plays an important role in the elasticity and tensile strength of hemp fibers. I did. One study reported that P uptake from cannabis ranged from 52 to 67 kg ha-1 and that fertilization of P did not affect stem yield (Finnan and Burke, 2013). Treatment with P-enrichment resulted in a 16% reduction in THC concentration, but did not affect flower CBD concentration (Bernstein; et al., 2019). P deficiency can also affect the ability of hemp to absorb other essential nutrients, which in turn can reduce plant health, disease immunity, crop quality, and yield. Cannabis plants suffering from P deficiency exhibit visual symptoms such as stunting and reddish-purple development of leaves due to the formation of anthocyanin pigments (Abbott; et al. 2019). A recent study reported a P leaf tissue analysis of P-deficient CBD hemp plants grown in greenhouse conditions at 0.09% (Cockson; et al., 2019). Hemp is less responsive to potassium (K) than N and P fertilization, and few experiments have been conducted to investigate the effect of K on hemp. One study reported that K did not significantly affect cannabis biomass and seed yields (Cockson; et al., 2019). K is a major nutrient and is needed in large quantities by hemp. K uptake increases as the plant grows. This is similar to the trend observed in P (Aubin; et al., 2015). However, for fiber hemp varieties, peak uptake occurs during fiber development. K has been reported to have a greater impact on fiber quality than P (Merfield, 2019). The K fertilization test suggested that cannabis should be ingested 175 kg K ha-1 (Neenan, 1969). A recent study of CBD hemp under greenhouse conditions found that the leaf tissue of K-deficient hemp plants had a K content of 0.41% (Cockson; et al., 2019).

However, the effect of K fertilization on the seeds and CBD yields of cannabis varieties in the field is unknown to our knowledge. Crop rotation is essential to disrupt the pest cycle, maintain and improve soil health, and promote crop diversification. Crop rotation also increases the levels of organic carbon in soil and organic matter in soil (Land; et al., 2017). Fiber hemp can produce 25 tha^{-1} of terrestrial biomass and large amounts of root biomass and can be distributed deeper in the soil compared to corn and wheat (Struik; et al., 2000). Roasting fibrous hemp returns more than two-thirds of the organic matter to the soil, increasing the porosity and fragility of the soil (Kraenzel; et al., 2019). Fiber hemp plants are also deeply rooted and have been shown to affect soil structure. Its taproot penetrates deeply into the soil, aerating it and at the same time forming soil aggregates and preventing erosion (Amaducci; et al, 2008).

A group of researchers evaluated cannabis grown in single crops and in alternation with wheat (Gorchs; et al., 2017). They concluded that cannabis can be cultivated for several years in traditional single cultivation without sacrificing yield. Cannabis increases wheat yields in two of the three growing seasons, and researchers conclude that cannabis has great potential as a precursor to wheat in rainwater Mediterranean conditions. Another study reported a 10-20% increase in wheat yield, followed by cannabis growth (Bosca and Karus, 1998). Recent studies of growing soybeans as a single crop have shown that previously grown cannabis had a positive effect on soybean growth (Liu; et al., 2012).

Hemp plant has been reported to decrease the population of nematodes and pathogenic fungi in the soil, and it has been stated that hemp has the potential to be grown without the use of pesticides, herbicides, or fungicides (Van Der Werf; et al, 1995). Three soil pathogens (the fungus Verticillium dahlia, rootknot nematodes Meloidogyne chitwoodi, and Meloidogyne hapla) were suppressed by hemp, and it was concluded that introducing hemp in a crop rotation might enhance soil health (Kok; et al, 1994) Hemp`s high planting density, rapid soil surface coverage ability and fast growth (especially the fiber yielding) varieties after emergence render it very competitive against weeds (Lotz; et al, 1991). This makes it potentially one of the greatest agronomic and environmental benefits of cultivating hemp in rotation with other crops. An experimental study also demonstrated hemp was able to suppress weed effectively without the use of herbicide, except for places where plant densities were relatively low, between 10 and 30 plants per m2 (Van Der Werf; et al, 1995)In order to introduce hemp successfully into a crop rotation, information on the optimal sequencing of hemp with other crops, the effect of integrating hemp into rotations on pests, and potential effects of hemp on soil health is necessary and needs to be further investigated across diverse growing conditions in the US.

For generations, animal manures and composts have been identified as soil builders because of their enormous impact on improving soil health (Bhowmik; et al, 2016 and Bhowmik; et al, 2017). The nutritional content of animal excrement depends on various factors such as the source and type of feed, the age of the animal, handling and storage techniques, temperature and water content. Organic additives are rich in nutrients essential to the nutritional needs of plants, such as N, one of the essential nutrients for growing cannabis. They also add carbon to the soil and improve soil biodiversity. Organic additives also help reduce nutrient spills and nitrate leaching into the soil. There are many environmental benefits to be gained from fertilizer application, but these benefits are optimized when applied at the right time and in the right amount and technology. Improper fertilizer management can contribute to increased greenhouse gas emissions (Bhowmik; 2018). There are several types of organic additives that can be used based on location and availability. The most common types of fertilizers used in hemp production are horse, cattle and chicken fertilizers. Hemp has been reported to naturally adapt to mammalian manure use. B. Horse manure is indicated as a fertilizer and can also be used efficiently for livestock manure (Kraenzel et al. It has been reported to improve manure quality and soil health characteristics. Manure composting is recommended (Bhowmik; et al., 2017). To reduce the potential for pathogens, it is advisable to compost chicken, horse and cow manure before use. It is reported to contain about 1-2% N and 1-3% K, but the amount may vary depending on the type of animal excrement used. It is very likely that animal excrement will be used for hemp production, but it is rarely reported.

Another cause of ecological changes is the use of mulching in cannabis cultivation. Organic mulches, namely chopped leaves, straw, cutouts, compost, wood chips, shredded bark, sawdust, pine needles, etc., have several advantages. Mulching should improve soil structure by mitigating fluctuations in soil temperature and promoting the ingress of water and oxygen, making it a beneficial habitat for useful soil organisms. When cannabis is cultivated for textiles, most of the leaves are returned to the field and act as a mulch. This maintains soil moisture and increases the biodiversity of microorganisms in the soil (Kraenzel; et al., 2019). Also, roasting cannabis in the field leaves stalks that can be used as soil roots. After that, much of the residual nutrients from the stems are returned to the soil for use in the next year's harvest. Most studies on CBD hemp use black or white plastic mulches, and few reports have been made on the use of organic or biodegradable mulches. Catch crops are crops that are cultivated primarily for the ecological benefits of the soil, not for the crop yield (Daryanto; et al, 2018).

Catch crops are usually grass or legumes and are cultivated primarily in the winter before summer crops. Inclusion of legume capture crops with biological nitrogen fixation capacity may also meet additional nitrogen requirements for key crops in the cultivation system. Catch crops also bring environmental benefits to the soil by controlling weed and pest pathogens, controlling soil erosion, increasing and improving soil fertility and increasing biodiversity (Bhowmik et al 2017. and Liu et al 2012). Cannabis is an annual crop that blends well with crop rotation, but little research has been done on the effects of cannabis on crop rotation. Also, some strains are very sensitive to light, so using hemp as a cover crop presents challenges. As a catch crop, fiber hemp in crop rotation has been reported to be able to effectively control weeds in the following crops (Struik; et al. 2000)]. It improves soil health by reducing the need. A report from Canada's Hemp Trade Alliance suggested the use of fixed-crop crops each year. H. Sweet clover, broad beans, peas, etc. Some studies have shown that the inclusion of hemp in crop rotation has an allelopathic effect and reduces the pest nematode. Therefore, hemp can act as a nematode-killing agent for certain crops. B. Potatoes, corn, peas, etc. that are susceptible to nematode invasion (Rothenburg, 2018). Certain hemp residues are also suitable for use as botanical pesticides (Beneli; et al, 2018).

1.8.1 A Survey of Pests of Cannabis Sativa in Zimbabwe

1.8.1.1 Abstract

Industrial hemp (Cannabis sativa L.) was first introduced in Zimbabwe in September 2020. As a new crop, studies to understand its agronomy and biotic factors affecting its cultivation were carried out. A study was carried to identify pests attacking Industrial Hemp at Harare Central Prison. Sampling was done when 50% of the crop had reached key physiological stages i.e germination, vegetative stage, flowering, milk dough stage, dough stage, physiological maturity and at harvest. Field scouting of ten (10) tagged plants per plot was done using canopy observation. The most abundant pests observed were aphids, whitefly, cutworm, leafhopper, looper, hairy caterpillar. Natural enemies observed were black ants, Honeybee ladybird beetles. Insect damage observed include weakening of head, chlorosis of leaves, and sooty mould development on the panicles. The study showed that Hemp attracts a diversity of leaf eating and sap sucking insects. This is the first report of insect pests of quinoa in Zimbabwe, which indicate the need to develop management strategies for pests of Hemp in the country.

1.8.2 Introduction

Industrial hemp (Cannabis sativa L.), like all new crops, hemp needs proper agronomic understanding so as to optimize production. There is need to identify which pests are of economic importance to Industrial hemp. There is also need to investigate yield content that can be attained by the crop in Zimbabwe. In the current project, we sought to understand the pests affecting industrial hemp. This is the first industrial hemp study which will help farmers with information on which pest is of economic importance to Industrial hemp.

1.8.3 Study Site and Materials

The experiment was conducted at Harare Central Prison protected site (17.8°89° S, 31.07°4° E) Harare, Zimbabwe. The trial was carried out in a piece of land comprising brown medium grained sand clay loam soils, at an elevation of 1494 meters above sea level. The location of the trial site falls under Zimbabwe's agro-ecological region II, which receives annual rainfall ranging between 831-950mm and a mean annual temperature of 12° to 25°C.

Industrial Hemp varieties were planted in plots measuring 1.5m x 5m. The seeds were planted by way of drilling in open furrows. There was no pest control measure which was used on the trial. Insect sampling was done when 50% of the crop had reached physiological stages i.e germination and vegetative stage. From each plot, ten (10) plants were randomly selected, tagged and used as observation plants. Cut worms were identified and manually counted from the plots.

1.8.4 Field Scouting and Insect identification

Scouting was done from 14 days after planting. Scouting involved carefully examining the presence of pests and damages caused on panicles, leaves and stems of the selected plants. Scouting was done on 10 consecutive rows. Observed insect pests were counted and data recorded.

1.8.5 Data Analysis

The insect data was collected. Data was transformed using square root transformation. Transformed data was analyzed using Genstat 14th Edition. Means were separated using Dancan multiple range test.

1.8.6 Results

1.8.6.1 Effect of cutworm infestation on time of sowing

Statistical analysis showed that transformed data did not have any significant differences p=0.23 at 14 days after sowing and p=0.094 at 21 days after sowing table 1.8.1. This means that cutworm did not have any preferences in respect of sowing time. The number of cutworms did not increase significantly with different planting time.

Table 1.8.1: Effect of cutworm infestation on time of sowing

	Transformed 14 days	Transformed 21 days
January	2.07	3.66
October	1.50	2.96
November	1.78	3.73
Mean	1.78	3.45
CV%	49.1	29.7
LSD	0.655	0.766
SE	0.876	1.024
P-value	0.23	0.094

**Figures separated by the same letter do not have significant differences*

Plants were scouted daily and observations noted for their impact on plant growth and yield. Photographs of some of the pest and beneficial insects observed are shown in Figure 1. Descriptions of pests observed are listed below in order of impact to the Hemp plants (most to least).

1.8.6.2 Birds

Pigeons were one of the more prominent pests. They ate the seed sown and young sprouts at planting, as well as the seed harvest maturing on the plants. This problem was more prevalent by the time of the harvest of the second planting and sowing of the third planting, which was over the same timeframe. There was significant seed loss of yield of planting 3 from birds. In part this was from the harvesting date of P3 being delayed due to a countrywide COVID lockdown in April 2020, thus allowing the birds to feast on the crop.

1.8.6.3 Cutworm

Cutworms were present at every planting. The greasy looking, grey caterpillars either ate through the hemp stems at ground level or just below.

1.8.6.4 Aphids

Heavy infestations were present from full flowering to seed formation, and were particularly prevalent on Santhica 27. Some plants were black with infestation, with nearly every plant of Santhica having aphids to a greater or lesser degree. These small green or grey-brown delicate insects occurred in colonies on the flowering heads and leaves. Futura 75 and Fedora 17 had the least infestation of aphids.

1.8.6.5 Caterpillars

Several different species of caterpillar and loopers were observed eating hemp leaves and / or young shoot stalks.

1.8.6.6 Leaf Feeders

These include transitory pests such as rose beetles, chafer beetles and grasshoppers. They caused some leaf damage, and ate into stems.

1.8.6.7 Black Ants

These were present around the site and climbing up and down plant stems, but no actual plant damage was observed. Ants appeared to be feeding on plant secretions and honeydew excreted by aphids. There were some ants infesting the soil around the roots of young seedlings in P3.

1.8.6.8 Beneficial Insects

Bees were prolific. They were attracted to the nectar produced at flowering and would have assisted with pollination of flowers. There were many ladybirds seen when the aphids were present.

Figure 1.8.1: Insects observed in the hemp fields

1. Aphids and black ants.	2. Aphids and whiteflies.	3. Cutworm.
4. Leafhopper.	5. Looper.	6. Hairy caterpillar.
7. Leaf miner damage.	8. Ladybird.	9. Honeybee carrying pollen.

1.8.7 Conclusion and Recommendation

Our results are the first report of insect pests of *Cannabis sativa L* in Zimbabwe. This study is important in giving clues and guidelines to the management of pests of *Cannabis Sativa L.* It shows the onset of pest infestation and therefore guides as to management of the pests. Therefore, below are the general recommendations for *Cannabis sativa L* cultivation.

1.8.7.1 General Growing Recommendations for Cultivation

- Check for cutworm and white grub (can be localized problem). Treat land if found.
- Apply a broad-spectrum fungicide to seed. This is important when the rains have started and soil remains moist for long periods to prevent rot.
- Scout regularly for potential problems

- Over or under irrigation
- Pests
- Diseases
- Weed pressure
- Crop appearance for nutritional problems

Pigeons were one of the more prominent pests. They ate the seed sown and young sprouts at planting, as well as the seed harvest maturing on the plants. This problem was more prevalent by the time of the harvest of the second planting and sowing of the third planting, which was over the same timeframe.

1.9. Effect Of Hemp Variety (*Cannabis sativa L*) On planting time and Yield

1.9.1 Abstract

A trial to investigate the effect of Hemp varieties on planting time and yield was carried out at Harare Central Prison protected site (17.8089°S, 31.0704°E) Harare, Zimbabwe. The experimental design was a 3x5 Factorial in a Randomized Complete Block Design Structure, with factor 1 being the variety and factor 2 being the planting date. The experiment consisted of 5 varieties (Fedora 17, Felina 32, Futura 75, Santhica 27, and USO31) and 3 planting dates (September, November and January). Cultural and chemical pest control measures were used on the trial. There were significant plant height differences among sowing dates p<0.001 at 14 days after sowing (table 4.1). January plant heights (16.55 cm) showed significantly taller plant compared to October (11.31 cm) and November (3.22 cm) sowing dates. October sowing dates showed significant plant height differences to November sowing date. Significant plant height differences p<0.001 were shown between sowing dates at 21 days after sowing. November sowing date showed significant taller plants p<0.001 compared to October and January at 21 days after sowing. Results from this study showed that Futura 75 had significantly the highest seed yield per hectare compared to all other varieties. Felina 32, Santhica 27 and Fidora 17 did not have significant yield seed differences. Felina 32 and Santhica 27 showed significant seed yield differences to Uso 31. However, Fidora 17 and Uso 31 did not show any significant seed yield differences to each other.

1.9.2 Introduction

Industrial hemp (*Cannabis sativa L.*, Cannabaceae) is a versatile herbaceous crop that has been used for fiber, food, and medicinal purposes (Andre, 2016, Vonapartis, 2015). Industrial hemp is a variety of *Cannabis sativa* L. and is one of the marijuanas with lower levels of the psychoactive compound THC (tetrahydracannabinol); < 0.3%, while marijuana has approximately 17% (Johnson, 2013). (Johnson, 2013). The cultivation of hemp dates back to China around 2700 BC and is believed to have then expanded across Asia, making its way to Europe 2000–2200 years ago (Callaway, 2004, Struik, 2000).

There are global estimates that the hemp market consists of more than 25,000 products (Johnson, 2013). Hemp can be grown as a fiber crop for many products like fabrics and textiles, yarns and raw or processed spun fibers, paper, carpeting, home furnishings, construction and insulation materials, auto parts, and composites. The stalk has two different fibers; hurd, the interior of the stalk and the bast, the outer fibers. The hurd can be used for animal bedding, raw material inputs, low-quality papers, and composites while the bast is among the strongest and most durable natural fibers (Brook, 2008). Oil and "milk" from crushed hemp seed are ingredients in a number of body care products, nutritional supplements, cosmetics, pharmaceuticals, and food products (Johnson, 2013).

Precise data are not available on the size of the market for hemp-based products; however, current industry estimates report that United States retail sales of all hemp-based products may be nearly \$500 million per year (Johnson, 2013). By comparison, wheat in Oregon had a farm gate value of \$503 million in 2011 and all of Oregon agricultural and fisheries commodities had an aggregate value of \$5.4 million in 2011 (*Oregon Department of Agriculture; 2012*).

Industrial hemp (*Cannabis sativa* L.) has recently been the subject of increasing interest in Zimbabwe, following the gazzeted statutory Instrument 218/20 of the Agricultural Marketing Authority (Industrial Hemp) Regulations, of September 2020 (SI 2020-218 Agriculture Marketing Authority).In terms of these 2018 Regulations, persons interested in the production of the cannabis plant must apply for the producer's license which will only be valid for 5 years and may be renewed after that before its expiry. It is noteworthy that the 2018 Regulations allow for the production of cannabis, which is defined to include fresh or dried cannabis, cannabis oil, cannabis plants or cannabis seeds; the cannabis plant also includes the whole or any portion, whether green or dry, of any plant of the genus cannabis (Nyamidzi, 2020).Further, in September 2020 a new law being Statutory Instrument, S.I 218 of 2020 Agricultural Marketing Authority (Industrial Hemp) Regulations 2020, 'the Regulations' was gazetted. These regulations were put in place to regulate and to provide a conducive environment in the production, procurement, distribution possession, sale, provision, transportation of industrial hemp.

Specific Objective

To investigate the planting time of *Cannabis sativa L* on yield

1.9.3 Materials and Methods

1.9.3.1 Trial Site and Experimental Treatments

The experiment was conducted at Harare Central Prison protected site (17.8°89°S, 31.07°4°E) Harare, Zimbabwe. The trial was carried out in a piece of land comprising brown medium grained sand clay loam soils, at an elevation of 1494 meters above sea level. The location of the trial site falls under Zimbabwe's agro ecological region II, which receives annual rainfall ranging between 831-950mm and a mean annual temperature of 12° to 25°C.

Hemp seed of five (5) cultivars was procured from the French breeders into Zimbabwe in July 2019 (Table 1.9.2). The first planting was sown mid-October 2019 (planting 1 = P1). Additional plantings occurred end of November 2019 (planting 2 = P2) and end of January 2020 (planting 3 =P3).

Standard agricultural practices were used for soil preparation. The beds were pegged out (1.5m width x 5m length) and completed by hand using hoes. Beds were irrigated to field capacity the day before sowing. A 100cm plank with 25 nails spaced at 4cm was used to make holes for seed placement. Seeds were firmed into soil, covered lightly with sand and irrigated. Care was taken to ensure seeds planted were placed at 10-25mm in the soil. Wheat-straw mulch was applied over the plants after germination counts were completed, on day 14 after sowing. Beds were kept moist for germination and were irrigated daily with 7.3mm water.

1.9.3.2 Data Collection

Seed stalks, by trial plot, were harvested by hand and deposited onto large sheets for weighing (the wet weight reading). Afterwards the sheets were placed on drying tables in full sunlight. Seed and stalks were dried for 3 to 7 days, depending on the amount of plant material and humidity levels at the time, to get a seed moisture content of +/- 9%. Care was taken to ensure no moisture came onto the hemp whilst drying. Seed was threshed, winnowed, weighed (seed yield) and stored in cloth bags in a cool room. Each plot's stalks were bundled together and a dry weight measurement taken.

1.9.3.3 Statistical Analysis

The experimental design was a 3x5 Factorial in a Randomized Complete Block Design Structure, with factor 1 being the variety and factor 2 being the planting date. The experiment consisted of 5 varieties (Fedora 17, Felina 32, Futura 75, Santhica 27, and USO31) and 3 planting dates (September, November and January). Cultural and chemical pest control measures were used on the trial. Analysis of variance (ANOVA) was done using GenStat 18th edition to analyze data from the treatments. Treatment means were separated using LSD at 5 % level of significance.

1.9.4 Results and Discussion

1.9.4.1 Effect of time on plant heights

There were significant plant height differences among sowing dates $p<0.001$ at 14 days after sowing (table 1.9.1). January plant heights (16.55 cm) showed significantly taller plant compared to October (11.31 cm) and November (3.22 cm) sowing dates. October sowing dates showed significant plant height differences to November sowing date. Significant plant height differences $p<0.001$ were shown between sowing dates at 21 days after sowing. November sowing date showed significant taller plants $p<0.001$ compared to October and January at 21 days after sowing. January sowing dates showed significant taller plants compared to October sowing date. Significant plant height differences were shown between sowing dates $p<0.001$ 9 days after sowing. November plant heights were significantly taller than October and January sowing dates at 30 days after planting. January sowing dates had significant taller plants compared to October at 30 days after sowing. There were significant plant height differences between sowing dates $p<0.001$ at 60 days after sowing. November sowing dates showed significant taller plants compared to October and January at 60 days after sowing. January sowing showed significantly taller plants compared to October sowing dates. Plants sown in November showed significantly taller plants $p<0.001$ compared to October and January sowing at 90 days after sowing. November sowing dates showed significantly taller plants compared to October and January and January sown plants showed significantly taller plants than October. At 120 days after sowing January sown plants showed significantly taller plants compared to October and November sowing dates. There were no significant plant height differences between October and November at 120 days after sowing. At Harvest, Day after sowing November sown crop was significantly taller than January and October sown. January sown crop also showed significantly taller plants compared to October crop.

Table 1.9.1: Effect of time on plant heights

	14 days	**21 days**	**30 days**	**60 days**	**90 days**	**120 days**	**Harvest day**
January	16.55^{a}	57.5^{b}	96.9^{b}	121.8^{b}	132.7^{b}	128.71^{a}	131.9^{b}
October	11.31^{b}	43.4^{c}	82.2^{c}	95.5^{c}	91.5^{c}	103.81^{b}	99.4^{c}
November	3.22^{c}	83.8^{a}	117.9^{a}	141.2^{a}	148.8^{a}	100.85^{b}	150.2^{a}
Mean	10.36	61.6	99.0	119.5	124.4	111.12	127.1
CV%	30.2	13.9	9.7	8.0	6.8	6.8	7.3
LSD	1.998	5.45	6.10	6.1	5.38	4.796	5.9
SE	3.131	8.54	9.56	9.56	6.94	7.515	9.24
P-value	<0.001	<0.001	<0.001	<0.001	<0.001	<0.001	<0.001

Figures separated by the same letter do not have significant differences

1.9.4.2 Effect of variety on plant heights

Analysis showed that there were significant plant height differences $p<0.001$ among varieties at 21 days after sowing (table 1.9.2). Fidora 17 (13.83 cm) had significantly the tallest plants compared to all other varieties. There were no significant plant height differences between varieties Felina 32, Futura 75 and Santhica 27. There were no significant plant height differences between Futura 75, Santhica 27 and Uso 31. However, there were significant plant height differences between Felina 32 (11.27 cm) and Uso 31 (8.25 cm). There were no significant plant height differences $p=0.134$ on varieties at 30 days after sowing. Varieties showed significant plant height differences $p<0.001$ at 60 days after sowing. Future 75 showed significant taller plants at 60 days after sowing compared to all other varieties.

There were no significant plant height differences at 21 days after sowing between Uso 31, Santhica 27 and Selina 32. However, Fidora was significantly shorter than Futura 75 and USo 31 at 60 days after sowing. At 90 days after sowing Futura 75 showed significantly taller plants $p<0.001$ compared to other varieties. There were no significant plant height differences between Felina 32, Santhica 27 and Uso 31. Fedora 17 showed significantly the shortest plants compared to the other varieties at the same sowing date. At

120 days after sowing Futura 75 showed significantly taller plants compared to all other varieties. Santhica and Felina had no significant differences on plant height, but significantly taller than Fedora 17 and Uso 31. However, Fedora 17 had significantly taller plants compared to Uso 31 at the same sowing date. At 120 days after sowing there were interactions on variety and sowing dates $p<0.001$. Futura75 sown in January had significantly the tallest plants compared all varieties planted in January, October and November. Future 75 sown in January showed significantly taller plants compared to the same variety sown in October and November. There were significant plant height differences $p<0.001$ at harvest day after sowing. Futura 75 showed significantly taller plants compared to all other varieties. Felina 32, Santhica 27 and Uso 31 did not show significant differences among each other but had significant differences with Fedora 17 which was significantly shorter.

Table 1.9.2: Effect of variety on plant heights

	21 days	30 days	60 days	90 days	120 days	Harvest
Fedora 17	13.83^{a}	66.2	112.4^{d}	114.0^{c}	92.96^{d}	115.4^{c}
Felina 32	11.27^{b}	63.8	116.6^{cd}	122.3^{b}	114.99^{b}	123.6^{b}
Futura 75	8.89^{bc}	60.2	130.0^{a}	137.6^{a}	125.02^{a}	144.1a
Santhica 27	9.56^{bc}	58.1	117.3^{cd}	124.0^{b}	118.36^{b}	126.4^{b}
Uso 31	8.25^{c}	59.4	121.2^{bc}	123.9^{b}	104.28^{c}	126.3^{b}
Mean	10.36	61.6	119.5	124.4	111.12	127.1
CV%	30.2	13.9	8.0	6.8	6.8	7.3
LSD	2.579	7.03	7.88	6.94	6.192	7.61
SE	3.131	8.54	9.56	8.43	7.515	9.24
P-value	<0.001	<0.134	<0.001	<0.001	<0.001	<0.001

Figures separated by the same letter do not have significant differences

1.9.4.3. Interaction of sowing time and variety on plant heights

There significant interaction $p=0.027$ on sowing date and variety on plant height. Fedora 17 (22.82 cm) sown in January had significantly the tallest plants compared to all other varieties and sowing dates (table 1.9.3). Feline 32 (18.27 cm) sown in January did not show any significant plant height differences $p=0.027$ with varieties Futura 75 (14.82 cm, Santhica 27 (15.98 cm) sown in the same months and Fedora (14.11 cm) sown in October. November sowing date showed significantly shorter plants $p=0.027$ ranging from 1.14 cm – 4.57 cm compared to January sown ranged from 10.87 cm-22.82 cm. However, in October there were no significant plant height differences among varieties Felina 32 (11.85 cm), Futura 75 (7.86 cm),

Santhica 27 (9.97 cm) and Uso 31 (12.74 cm). There were significant sowing date and varieties interactions p=0.009 on plant heights. Varieties Fedora 17, Felina 32, Futura 75 and Uso 31 sown in November had significant taller plants compared January and October sowing dates. Santhica 27 sown in Novembers had significant shorter plant height differences compared to all other varieties sown the same month. Uso 31 sown in January and October showed no significant plant height differences p=0.009 compared to other varieties sown in October. At y days after sowing there were interactions on variety and sowing dates p<0.001. Futura75 sown in January had significantly the tallest plants compared all varieties planted in January, October and November. Future 75 sown in January showed significantly taller plants compared to the same variety sown in October and November.

Table 1.9.3: Interaction of sowing time and variety on plant heights

	30 days after planting			90 days after sowing			Day of harvest		
Variety	January	October	November	January	October	November	January	October	November
Fidora 17	22.82^{a}	14.11^{bcd}	4.57^{fg}	65.1^{bc}	46.4^{ef}	87.2^{a}	119.72^{bcd}	95.07^{f}	64.1^{h}
Felina 32	18.27^{b}	11.85^{cde}	3.70^{fg}	62.8^{bcd}	45.1^{f}	83.6^{a}	121.74^{bcd}	104.41^{ef}	118.82^{d}
Futura 75	14.82^{bcd}	7.86^{ef}	4.00^{fg}	59.5^{cde}	36.6^{f}	84.5^{a}	142.58^{a}	115.13^{de}	117.33^{cd}
Santhica 27	15.98^{bc}	9.97^{de}	2.72^{fg}	59.4^{cde}	39.5^{f}	75.5^{b}	130.26^{b}	102.63^{f}	122.18^{bcd}
Uso 31	10.87^{cde}	12.74^{cde}	1.14^{g}	41.0^{f}	49.2^{def}	88.1^{a}	129.23^{bc}	101.82^{f}	81.80^{g}
Mean	10.36			61.6			111.12		
CV%	30.2			13.9			6.8		
LSD	4.47			12.18			10.72		
SE	3.131			8.54			7.515		
P-value	0.027			0.009			<0.001		

Figures separated by the same letter do not have significant differences

1.9.4.4 Effect of Sowing Date on Seedling Survival, Seed Yield g/Plant and seed yield kg/ha

1.9.4.4.1 Seedling survival

There was significant differences p<0.001 on seedling survival on sowing dates (table 1.9.4). January sowing showed significantly higher seedling survival compared to October and November sowing dates. November sowing dates showed significantly higher survival rates compared to October sowing date.

1.9.4.5 Effect of sowing date on seed yield g/plant

There were significant seed yield differences per plant p<0.001 (table 1.9.5). November sowing date had significantly the highest seed yield per plant. October sowing date exhibited significant yield differences compared to January.

1.9.4.6 Effect of sowing date on seed yield kg/ha

There was significant yield differences p<0.001 between sowing dates (table 1.9.5). November sowing date showed significantly higher seed yield p<0.001 compared to January and October sowing. October sowing had significant higher seed yield differences compared to January sowing.

Table 1.9.4: Effect of sowing date on seedling survival, seed yield g/plant and seed yield kg/ha

	Seedling survival	Yield g/plant	Yield kg/ha
January	98.37^{a}	1.16^{c}	931^{c}
October	81.41^{c}	1.79^{b}	1432^{b}
November	91.24^{b}	4.89^{a}	3893^{a}
Mean	90.34	2.61	2085
CV%	6.9	35.7	35.7
LSD	3.98	0.594	475.4
SE	6.23	0.931	745.0
P-value	<0.001	<0.001	<0.001

Figures separated by the same letter do not have significant differences

1.9.5 Effect of Variety on Seedling Survival, Number of Seeds Per Plant Yield g/plant and Seed Yield Kg/Ha

1.9.5.1 Seedling survival

There were no significant differences among varieties on seedling survival p=0.451 (table 1.9.5).

1.9.5.2 Number of seeds per plant

There was significant differences p<0.001 on number of seeds per plant on varieties (table 1.9.6). Futura 75 showed significantly the highest number of seeds per plant compared to all other varieties. There were no significant differences between Fidora 17, Felina 32 and Uso 31. However, Fidora 17, Felina 32 and Uso 31 showed significantly higher seed per plant compared to Santhica 27.

1.9.5.3 Seed yield g/plant

There significant yield differences per plant p<0.001 among varieties (table 1.9.5). Futura 75 produced significantly higher seed yield per plant compared to all other varieties. Felina 32, Santhica 27 and Fidora 17 had significantly higher yield per plant compared to Uso 31.

1.9.5.4 Seed yield kg/ha

There were significant yield differences p<0.001 shown on seed yield per hectare (Table 1.9.5). Futura 75 had significantly the highest seed yield per hectare compared to all other varieties. Felina 32, Santhica 27 and Fidora 17 did not have significant yield seed differences. Felina 32 and Santhica 27 showed significant seed yield differences to Uso 31. However, Fidora 17 and Uso 31 did not show any significant seed yield differences to each other.

Table 1.9.5: Effect of variety on seedling survival, number of seeds per plant yield g/plant and seed yield kg/ha

	seedling survival	Number of seeds per plant	Yield g/plant	Seed yield kg/ha
Fidora 17	90.0	3.95^{b}	1.88^{bc}	1504^{bc}
Felina 32	91.1	3.91^{b}	2.61^{b}	2086^{b}
Futura 75	92.7	6.71^{a}	4.59^{a}	3671^{a}
Santhica 27	89.87	2.37^{c}	2.43^{b}	1944^{b}
Uso 31	87.99	3.42^{b}	1.53^{c}	1222^{c}
Mean	90.34	4.07	2.61	2085
CV%	6.9	15.9	35.7	35.7
LSD	5.13	0.999	0.767	613.8
SE	6.23	0.648	0.931	745.0
P-value	0.451	<0.001	<0.001	<0.001

1.9.6 Interaction of Sowing Time and Variety on Seed Survival, Seed Yield g/Plant and Seed Yield Kg/Ha.

1.9.6.1 Seed survival

There were significant interaction differences on seedling survival on variety by sowing date (table 1.9.6). January sowing date showed significant high survival rates compared to October.

1.9.6.2 Seed yield g/plant

There was significant interaction yield differences $p<0.001$ (table 1.9.6). Futura 75 planted in November produced significantly higher seed yield per plant compared to all the other varieties and sowing dates. Fedora 17, Felina 32, Santhica 27 planted in November and Futura 75 sown in October had the second significant seed yield per plant.

1.9.6.3 Seed yield kg/ha

There was significant interaction $p<0.001$ between variety and sowing date (table 1.9.6). Futura 75 planted in November had significantly the highest seed yield compared to the all-other varieties and sowing date. Fedora 17, Felina 32 and Santhica 27 planted in November and Futura 75 planted in October had significantly the second highest seed yield.

Table 1.9.7: Interaction of sowing time and variety on seed survival, seed yield g/plant and seed yield kg/ha

	Seed survival			Seed yield g/plant			Seed yield kg/ha		
Variety	January	October	November	January	October	November	January	October	November
Fidora 17	100.0^{a}	78.03^{f}	92.0^{abcd}	1.13^{de}	0.59^{e}	3.92^{b}	902^{c}	471^{e}	3138^{b}
Felina 32	100.0^{a}	82.84^{de}	90.46^{bcd}	1.12^{de}	1.88^{de}	4.83^{b}	895^{c}	1503^{cde}	3862^{b}
Futura 75	100.0^{a}	84.77^{de}	93.41^{abc}	1.92^{de}	3.90^{bc}	7.94^{a}	1536^{cde}	3124^{b}	6353^{a}
Santhica 27	93.67^{abc}	88.90^{cd}	87.06^{cde}	0.68^{e}	1.41^{d}	5.21^{b}	541de	1126^{cde}	4167^{b}
Uso 31	98.19^{ab}	72.51^{f}	93.26^{abc}	0.98^{e}	1.17^{d}	2.44^{cd}	783^{de}	935^{cde}	1948^{c}
Mean	90.34			2.61			2085		
CV%	6.9			35.7			35.7		
LSD	8.887			1.329			1063.1		

SE	6.228	0.931	745
P-value	<0.001	<0.001	<0.001

Figures separated by the same letter do not have significant differences

1.9.7 Conclusion and Recommendation

1.9.7.1 Conclusion

None of the cultivars tested can be grown as a fibre crop, with the exception of Santhica 27, without the introduction of supplementary lighting. Crops grown for their fibre need to be 3-4m tall in order to be economically viable. The mean maximum plant height obtained was 1.7m for Futura 75 (P2), with all other cultivars being between 1.38-1.49m (P2). Longer length of stem will only be achievable of temperate bred varieties with the use of additional lighting in field to increase the number of hours of day-length, and therefore the period of growth before flowering is initiated. More varieties need to be trialed, in particular from sub-tropical latitudes such as Malawi, to see if they will grow better under our latitude for fibre production without use or with minimal use of supplementary lighting.

Regardless of Santhica 27's relatively short height of 1.47m (P2), its extrapolated dry weight yield per Hectare (9.8 tons/ Ha) was within the expected yields for a fibre crop. Futura 75's yield (7.2 tons/Ha) was just below the lower limit of the expected dry mass yield of 8 to 12 tons per Hectare. Santhica 27 had longer internode lengths, with not as many leaves present as other cultivars trialed; all being desirable characteristics for a fibre crop. If it is grown solely for fibre, there won't be an issue with aphid infestation as the crop would be harvested before flowering when the first aphids were observed. The potential of Santhica 27 for fibre should be looked into further, particularly using fresh seed to see if germination issues can be overcome and that its THC content doesn't exceed 0.3%.

Futura 75 had the greatest potential to be grown as a dual-purpose crop (for seed and fibre), or for its seed alone. The seed yields obtained from Futura 75 at all planting dates was greater than 1,500Kg per Hectare. The second planting date yield was exceptional though, being the equivalent of 6.3 tons/Ha. With the high seed yield and good dry fibre yield, it makes the growing the crop for dual purpose appealing. The key to success will be getting the seed crop off the plants before birds consume it. Other cultivars trialed, and did not perform to the same extent as Santhica 27 or Futura 75, and as such they are not recommended to take to commercial production.

Without use of lighting, the naturally short growing season of hemp means it could potentially be grown in more crop cycles within the year; i.e. 3-4 crops rather than 2-3. This requires further investigation though to see if yields are higher planting more crops in a year, or just by using supplementary lighting and growing less crop cycles.

1.9.7.2 Fibre Crops

In order to grow hemp commercially for its fibre it will be important to either provide additional lighting to keep the crop actively growing for the 3–4-month period, or to obtain cultivars that are genetically less sensitive to day-length. Without the plants reaching their optimum height, it will be difficult to get the yields required to be commercially viable.

Plant densities are determined by the end product required and the growth characteristics of the cultivar being grown. Fibre production requires high plant densities (up to 150 plants per m^2), which encourages growth of single straight stems with minimal branching. Planting densities necessary are between 40-60Kg/Ha, with a narrow row spacing of 150mm.

1.9.7.3 Seed Crops

Grain or seed production is planted at 70-80 plants per m^2. This gives more space for plants to branch, and maximize seed yields. Seed rates of 25-35Kg/Ha are used, with 150-800mm row spacing (depending on cultivar and seed size). Branching of stems can be induced by nipping main stem between the 4th and 7th nodes. This also reduces the height of the plants, which makes seed collection easier.

1.9.7.4 General Growing Recommendations for Cultivation

- Obtain the necessary licenses from Ministry of Agriculture.
- Check the cultivar being grown is certified as low THC. The THC level must be <0.3% to be grown as industrial hemp in Zimbabwe.
- Hemp is tolerant to a wide variety of soil types. Select land with good moisture and nutrient holding capacity, such as in deep, well-drained loam soils. Hemp is not suited to waterlogged, badly

structured or compacted soils.

- Clear land of weeds. If weeds are a known issue, irrigate lands before sowing hemp to induce weed germination and treat/remove. Chemical treatment of weeds with herbicides will be specific to the prevalent weeds, and dependent on what the crop is being grown for (i.e. human consumption and medicinal use *etc*., as opposed fibre for clothing, construction *etc*.). This will significantly reduce weed pressure at germination. Once hemp is established, it easily outgrows and out competes weeds. Hemp growth can be up to 10cm per day.
- Take full soil analysis. Hemp prefers a soil pH of 6.2A6.8 ($CaCl_2$ scale), but will grow in slightly acidic soil. Cations should be present in healthy proportion to each other. Correct any soil imbalances through addition of fertilizer during land preparation.
- Depending on the time of year of planting and the variety being grown, it may be advisable to apply crop nutrition simultaneously. The fertilizer can therefore be applied at land preparation and be thoroughly incorporated into the soil. A slow-release fertilizer is preferable. Hemp is nitrophilic, and removes from the soil the following nutrients during a growing season:
    - Nitrogen (N) 100Kg/Ha – required between weeks1A8
    - Phosphate (P_2O_5) 70Kg / Ha – required at flowering and seed production
    - Potash (K_2O) 110Kg / Ha– required at flowering and seed production

  Broadcasting fertilizer after planting can cause tissue burn, which does impact the growth of the plants.

- Deep rip where soil compaction and soil pans occur. Addition of organic matter (compost, green manure, animal manures) will improve and create a healthy soil structure.
- Plough and furrow the land to give a seedbed of loose, even and level soil. Remove all clods. A well-prepared, fine and firm seedbed is critical to achieving a uniform germination of seed. Planting on beds / ridges will assist water drainage in areas that drain poorly.
- Mark out beds/lands for sowing. The planting rows width is dependent on the cultivars growth characteristics as well as what product the crop is being grown for (flower / seed / fibre etc).

- Take a full agricultural water analysis. Identify any potential issues, and attend to them.
- Determine irrigation application and plan layout. Overhead irrigation is the easiest and cheapest application method. Over watering can cause root rot.
- Do a seed germination test and if necessary, alter the planting density accordingly to achieve the density required. It is important to sow fresh seed as viability of seed rapidly decreases the older it is.
- Irrigate to field-capacity the day before planting.
- Sowing from September through to November gives maximum dry matter fibre production potential. Flowering is initiated by the start of shorter days (after 21st December in Zimbabwe), so the earlier the planting, the more vegetative growth, and increase in plant height, will occur.
- Apply a broad-spectrum fungicide to seed. This is important when the rains have started and soil remains moist for long periods to prevent rot.
- Soil temperatures of 10°C plus (ideally 15°C) is necessary for germination.
- In commercial plantings the seed is sown using precision planters or air seeders. Seed planting rates depend on the variety (northern or southern originating cultivar) and the type of yield the crop is being grown (seed, fibre or dual crop). Planting depth is 10-25mm. Good seed to soil contact is necessary for germination, using a roller press.
- Germination of seed occurs 4-7 days after sowing. Soils should be moist during germination and not dry out completely.
- Be prepared to deal with flocks of birds eating the seed. If a problem, provide an easier source of food for birds away from the crop for the period of germination. Once the crop has fully germinated, the birds are no longer a considered a pest as they do not consume the seedlings.
- Hemp requires 300-400ml irrigation during the season, with 75% in the first 6 weeks. This is important if the sub-soil has no moisture reserve. Overhead irrigation is best. Do not overwater as this facilitates root rot.
- Monitor flower development, which will be dependent on where the cultivars grown have been sourced, and if in-field lighting has been used.

1.9.7.5 Fibre Crop Harvest

Harvest stems when 50% of male plants are flowering, or when female plants have just started flowering. After this stage, plants become lignified and are more difficult to harvest.

A. Use of mechanized harvesters cut stalks to 300-700mm lengths, which are left in field to dry. Drying can take 2-3 days, depending also on the present of heavy morning dews. Hempisraked2-3timestodryevenlyand remove leaves. Moisture content should be 12% before baling and transferring to mill. OR

B. At onset of female flowering, spray crop with a defoliant. Leave crop to stand in field for up to 3 weeks. When stem moisture content is correct, direct harvest and transfer to mill.

Stalks used for construction industry are not retted. Retting is done for textiles, fiberglass replacement, plastics, papers, and a multitude of other uses. To ret hemp, the crop is left in the field with dew on it or irrigated to rot. This causes a separation of the outside fibre to the inside fibre. This process can be sped up using steam explosion methods, which are very expensive, and also using new enzyme methods.

Seed Crop Harvest

- Seeds mature 4-8 weeks after the peak of flowering. Cut down on nitrogen and water at this time. The seed heads mature from the bottom of the plant upwards. The seed coat will harden up, and the bracts dry out.
- At 60-70% seed head maturation, a desiccant can be applied to the crop to dry out the leaves and stalk, which will assist in ease of handling at harvest.
- Implement bird control measures. Birds can consume up to 50% of the seed yield if not prevented from doing so.
- Check weather reports to determine the best time to harvest. The weather should be sunny and dry, with low humidity. If you have no alternative but to harvest in wet weather, driers will need to be utilized.
- Harvesting is done using rotary headers or conventional headers. Small areas can be cut with brush-cutters. Seed moisture should be

around9%.

- Drying can take 2-3 days. Separate seed from hurd. Hemp seeds thresh easily.
- Collected seed needs to be stored away from moisture as it can absorb it. With storing seed, rats can be a problem.

1.10 References

1. Abbott, G. Making Cannabis Plants Thrive with Phosphorus. Available online: https://www.ganjapreneur.com/making-cannabis-plants-thrive-with-phosphorus/

Accessed October 2019

2. Ahmad, R.; Tehsin, Z.; Malik, S.T.; Asad, S.A.; Shahzad, M.; Bilal, M.; Shah, M.M.; Khan, S.A. 2016. Phytoremediation potential of hemp (*Cannabis sativa* L.): Identification and characterization of heavy metals responsive genes. *Clean Soil Air Water*, *44*, 195–201.

3. Aladić, K.; Jarni, K.; Barbir, T.; Vidovic, S.; Vladic, J.; Bilic, M.; Jokic, S. 2015Supercritical CO_2 extraction of hemp (*Cannabis sativa* L.) seed oil. *Ind. Crop Prod.*, *76*, 472–478.

4. Amaducci S, Colauzzi M, Bellocchi G and Ventu G.2008; Modeling post emergent hemp phenology (Cannabis sativa L.): Theory and evaluation. European Journal of Agronomy. 28:90-102.

5. Amaducci S, Zata A, Pelai F and Venturi G. 2008; Influence of agronomic factors on yield and quality of hemp (Cannabis sativa L.) fibre and implication for an innovative production system. Field Crops Research. 107:161-169.

6. Amaducci, S.; Errani, M.; Venturi, G. 2002Plant population effects on fibre hemp morphology and production. *J. Ind. Hemp*, *7*, 33–60.

7. Amaducci, S.; Scordia, D.; Liu, F.H.; Zhang, Q.; Guo, H.; Testa, G.; Cosentino, S.L. 2015Key cultivation techniques for hemp in Europe and China. *Ind. Crop Prod.*, *68*, 2–16.

8. Amaducci, S.; Zatta, A.; Raffanini, M.; Venturi, G. 2008. Characterization of hemp (*Cannabis sativa* L.) roots under different growing conditions. *Plant Soil*, *313*, 227–235.

9. Andre, C.M.; Hausman, J.F.; Guerriero, G. Cannabis sativa, 2016: The plant of the thousand and one molecules. Front. Plant Sci., 7, 19.

10. Andreae, M.O.; Merlet, P. 2001Emission of trace gases and aerosols from biomass burning. *Glob. Biogeochem. Cycles*, *15*, 955–966.

11. Anwar, F.; Latif, S.; Ashraf, M. 2006Analytical characterization of hemp (*Cannabis sativa*) seed oil from different agro-ecological zones of Pakistan. *JAOCS*, *83*, 323–329.

12. Ariskan, S., 2010. The Benefits of Hemp.

13. Askew, M.F. 2002 Interactive European Network for Industrial Crops and their Applications. In *Trends in New Crops and New Uses*; Janick, J., Whipkey, A., Eds.; ASHS Press: Alexandria, VA, USA, pp. 55–61.

14. Aubin, M.P.; Seguin, P.; Vanasse, A.; Tremblay, G.F.; Mustafa, A.F.; Charron, J.B. 2015. Industrial hemp response to nitrogen, phosphorus, and potassium fertilization. *Crop Forage Turfgrass Manag.*, *1*, 1–10.

15. Bassei P, koneman M, Mediavilla V and Scheid-slembrouck L.1998; Optimization of nitrogen fertilization and seed density in hemp crop. Agraforschung. 5:241-244.

16. Bennet S. J, Snell R and Wright D. 2006. Effect of variety, seed rate and time of curing on fibre yield of dew-retted hemp. Industrial Crops and Products. 24:79-86.

17. Bernstein, N.; Gorelick, J.; Zerahia, R.; Koch, S. 2019Impact of N, P, K, and humic acid supplementation on the chemical profile of medical cannabis (*Cannabis sativa* L.). *Front. Plant Sci.*, *10*, 736.

18. Bhowmik, A.; Fortuna, A.-M.; Cihacek, L.; Bary, A.I.; Carr, P.; Cogger, C.G. 2017 Potential Carbon Sequestration and Nitrogen Cycling in Long-Term Organic Management Systems. *Renew. Agric. Food Syst.*, *32*, 498–510.

19. Bhowmik, A.; Fortuna, A.M.; Cihacek, L.; Rahman, S.; Borhan, M.S.; Carr, P. 2017 Use of Laboratory Incubation Techniques to Estimate Green House Gas footprints from Conventional and No-Tillage Organic Agroecosystems. *Soil Biol. Biochem.*, *112*, 204–215.

20. Bhowmik, A.; Fortuna, A.-M.; Cihacek, L.J.; Bary, A.I 2016.; Cogger, C.G. Use of biological indicators of soil health to estimate reactive nitrogen dynamics in long-term organic vegetable and pasture systems. *Soil Biol. Biochem.* , *103*, 308–319.

21. Bocsa I and Karus M. 2012 The cultivation of hemp: botany, varieties, cultivation and harvesting. Sebastopol, Hemptech. California. 186.

22. Bócsa, I.; Karus, M. 1998*The Cultivation of Hemp: Botany, Varieties, Cultivation and Harvesting*; Hemptech: Sabastopol, CA, USA.

23. Bouloc, P. 2013The uses of hemp for domestic animals. In *Hemp: Industrial Production and Uses*; Bouloc, P., Allegret, S., Arnaud, L., Eds.; CAB International: Wallingford, UK; Boston, MA, USA,; pp. 260–262.

24. Brook, G. 2008. "National Industry Hemp Strategy." Manitoba Agriculture, Food and Rural.

25. Bullard M. J, Musill S. J, Carver, P and Nixon P. M. I. 2002; Yield improvements through modification of planting density and harvest frequency in short rotation coppice Salix spp.-2. Resource capture and use in two morphologically diverse varieties. Biomass and Bioenergy. 22:27- 39.

26. Burczyk H., Grabowska L., Strybe M and Konczewicz W.2009; Effect of sowing density and date of harvest on yields of industrial hemp. Journal of Natural Fibers. 6:204-218.

27. Callaway, J.C.2004, Hempseed as a nutritional resource: An overview. Euphytica, 140, 65–72.

28. Campiglia, E.; Radicetti, E.; Mancinelli, R. 2017Plant density and nitrogen fertilization affect agronomic performance of industrial hemp (*Cannabis sativa* L.) in Mediterranean environment. *Ind. Crops Prod.*, *100*, 246–254.

29. Canadian Hemp Trade Alliance (CHTA). Available online: http://www.hemptrade.ca/eguide/production/fertility-in-organic-production (accessed on 23 December 2019).

30. Cerdan PD and Chory J. 2003; Regulation of lowering time by light quality. Nature. 423:881-5.

31. Chimwamurombe F 2021. Legal framework on Industrial Hemp farming.

32. Citterio, S.; Santagostino, A.; Fumagalli, P.; Prato, N.; Ranalli, P.; Sgorbati, S. 2003 Heavy metal tolerance and accumulation of Cd, Cr and Ni by *Cannabis sativa* L. *Plant Soil*, *256*, 243–252.

33. Cockson, P.; Landis, H.; Smith, T.; Hicks, K. 2019; Whipker, B.E. Characterization of nutrient disorders of *Cannabis sativa*. *Appl. Sci.*, *9*, 4432.

34. Crescente, G.; Piccolella, S.; Esposito, A.; Scognamiglio, M.; Fiorentino, A.; Pacifico, S. 2018 Chemical composition and nutraceutical properties of hempseed: An ancient food with actual functional value. *Phytochem. Rev.*, *17*, 733–749.

35. Daryanto, S.; Fu, B.; Wang, L.; Jacinthe, P.A.; Zhao, W. 2018 Quantitative synthesis on the ecosystem services of cover crops. *Earth Sci. Rev.*, *185*, 357–373.

36. Dempsey, J.M. 1975 *Fiber Crops*; University of Florida Press: Gainesville, FL, USA. pp. 46–89.

37. Desanlis, F.; Cerruti, N.; Warner, P. 2013Hemp agronomics and cultivation. In *Hemp: Industrial Production and Uses*; Bouloc, P., Allegret, S., Arnaud, L., Eds.; CAB International: Wallingford, UK; Boston, MA, USA. pp. 98–124.

38. Dewey, L. 1913A purple-leaved mutation in hemp. USDA. *Plant Ind. Circ.*, *113*, 23–24.

39. Dingha, B.; Sander, L.; Bhowmik, A.; Akotsen-Mensah, C.; Jackai, L.; Gibson, K.; Turco, R. 2019. Industrial hemp knowledge and interest among North Carolina organic farmers in the United States. *Sustainability*, *11*, 2691.

40. Erasmus, U., 1993. *Fats that heal, fats that kill.* Alive Books, 7436 Fraser Drive,Burnaby, B.C., Canada V5J 5B9.

41. Getmanov, P.I. 1967 Effect of trace fertilizers during hemp cultivation of peat-humus soils. *Khim. Sel. Khoz. 5*, 412–413.

42. Gorchs, G.; Lloveras, J.; Serrano, L.; Cela, S. 2017 Hemp yields and its rotation effects on wheat under rainfed mediterranean conditions. *Agron. J.*, *109*, 1551–1560.

43. Gryndler, M.; Sudova, R.; Puschel, D.; Rydlova, J.; Janouskova, M.; Vostka, M. 2008 Cultivation of high-biomass crops on coal mine spoil banks: Can microbial inoculation compensate for high doses of organic matter? *Bioresour. Technol.*

44. Günther, F. Carbon Sequestration for Everybody: Decrease Atmospheric Carbon Dioxide, Earn Money and Improve the Soil. Available online: https://terrapreta.bioenergylists.org/files/Terra%20pretav1_0.pdf (accessed on 26 December 2019).

45. Hall J, Bhatarai S. P and Midmore D.J. 2013. The effects of photoperiod on phenological development and yields of industrial hemp. Journal of Natural Fibers.

46. Hall J, Bhatarai S. P and Midmore D.J.2013; The effects of different sowing times on maturity rates, biomass, and plant growth of industrial fiber hemp. Journal of Natural Fibers. 10:40-50.

47. Hall J., Bhatarai S.P and Midmore D. J. 2012; Review of lowering control in industrial hemp. Journal of Natural Fibers. 9:23-36.

48. Initiative Agriculture and Agri-Food Canada. Accessed October 2013.

49. Ivonyi, I.; Izsoki, Z.; van der Werf, H.M. 1997 Influence of nitrogen supply and P and K levels of the soil on dry matter and nutrient accumulation of fiber hemp. *J. Int. Hemp Assoc.*, *4*, 84–90.

50. Johnson, R. 2013. "Hemp as an Agricultural Commodity." Congressional Research Service.

51. Jones, K., 1995. *Nutritional and medicinal guide to hemp seed.* Rainforest BotanicalLaboratory, P.O. Box 1793, Gibsons, B.C., Canada VON1VO.

52. Kaskey, J. Pending Legislation Makes Insurance Available to Hemp Growers. Available online: https://www.propertycasualty360.com/2018/11/29/pending-legislation-makes-insurance-available-to-hemp-growers/?slreturn=20191128144209 (accessed on 23 December 2019).

53. Kluwer Academic/Plenum Publishers: New York, NY, USA, 2004; pp. 32–83. Schultes, R.E. *Random Thoughts and Queries on the Botany of Cannabis*; J &A Churchill: London, UK, 1970; pp. 11–33.

54. Kok, C.J.; Coenen, G.C.M.; de Heij, A. 1994 The effect of fibre hemp (*Cannabis sativa* L.) on selected soilborne pathogens. *J. Int. Hemp Assoc.*, *1*, 6–9.

55. Kolodziej, J.; Wladyk-Przybylak, M.; Mankowski, J.; Grabowska, L. 2012 *Heat of Combustion of Hemp and Briquettes Made of Hemp Shives*; Renewable Energy and Energy Efficiency: Cairo, Egypt, pp. 163–166.

56. Kraenzel, D.G.; Petry, T.; Nelson, B.; Anderson, M.J.; Mathern, D.; Todd, R. Industrial Hemp as an Alternative Crop in North Dakota. Available online: http://www.industrialhemp.net/pdf/aer402.pdf (accessed on 6 December 2019).

57. Kraszkiewicz, A.; Kachel, M.; Parafiniuk, S.; Zapac, G.; Niedziolka, I.; Sprawka, M. 2019 Assessment of the possibility of using hemp biomass (*Canabis Sativa* L.) for energy purposes: A case study. *Appl. Sci.*, *9*, 4437.

58. Land, M.; Haddaway, N.R.; Hedlund, K.; Jorgensen, H.B.; Katterer, T.; Isberg, P.E. 2017 How do selected crop rotations affect soil organic carbon in boreo-temperate systems? A systematic review protocol. *Environ. Evid.* , *6*, 9.

59. Lehmann, J.; Gaunt, J.; Rondon, M. 2006 Bio-char sequestration in terrestrial ecosystems—A review. *Mitig. Adapt. Strateg. Glob. Chang. 11*, 403–427.

60. Li, Z. 1982 *The Theory and Techniques for Bast Fibre Crops*; Shanghai Science and Technology Press: Shanghai, China, pp. 332–387.

61. Linger, P.; Mussig, J.; Fischer, H.; Kobert, J. 2002 Industrial hemp (*Cannabis sativa* L.) growing on heavy metal contaminated soil: Fibre quality and phytoremediation potential. *Ind. Crop Prod.*, *16*, 33–42.

62. Lisson S. N and Mendham N. J.2000; Cultivar, sowing date and plant density studies of fibre hemp (Cannabis sativa L.) in Tasmania. Australian Journal of Experimental Agriculture. 40:975-986.

63. Liu, A.; Ma, B.; Bomke, A. 2005 Effects of cover crops on soil aggregate stability, totalorganic carbon, and polysaccharides. *Soil Sci. Soc. Am. J.*, *69*, 2041–2048.

64. Liu, X.; Li, Y.; Han, B.; Zhang, Q.; Zhou, K.; Zhang, X.; Hashemi, M. 2012 Yield response of continuous soybean to one-season crop disturbance in a previous continuous soybean field in Northeast China. *Field Crop Res.*, *138*, 52–56.

65. Lotz, L.A.P.; Groeneveld, R.M.W.; Habekotte, B.; Van Oene, H. 1991 Reduction of growth and reproduction of *Cyperus esculentus* by specific crops. *Weed Res.*, *31*, 153–160.

66. Mathieu, J.-P., 1980. Chanvre (hemp). *Techn. Agric.* 5: 1-10

67. Matthäus, B.; Brühl, L.2008. Virgin hemp seed oil: An interesting niche product. Eur. J. Lip. Sci. Technol. 110, 655–661.

68. Meijer W. J. M, Van der Werf H. M. G, Mathijssen E. W. J. M and Van den Brink P. W. M.1995; Constraints to dry mater production in fibre hemp (Cannabis sativa L.). European Journal of Agronomy. 4:109-117.

69. Merfield, C.N. Industrial Hemp and Its Potential for New Zealand. Available online: https://researcharchive.lincoln.ac.nz/bitstream/handle/10182/4801/Industrial_hemp.pdf?sequence=1 (accessed on 23 December 2019).

70. Miller, R.L. Hemp as a Crop for Missouri Farmers: Markets, Economics, Cultivation, Law. Available online: http://www.druglibrary.org/olsen/hemp/crop/hemp-01.html (accessed on 23 December 2019).

71. Mitch McConnell Pushed for Hemp Legalization. Now Kentucky Farmers Are Tripling Down on the Crop. Available online: https://www.cnbc

.com/2019/03/28/kentucky-farmers-bet-on-hemp-as-new-cash-crop.html (accessed on 23 December 2019).

72. Mosjidis J. A and Wehtje G.2011; Weed control in sunn hemp and its ability to suppress weed growth. Crop Protection. 30:70-73.

73. Neenan, M. 1969The cultivation of hemp in Ireland. *FIBRA*, *14*, 23.

Nyamidzi T., 2020, Business and Investment opportunities for cannabis production in Zimbabwe.

74. Olschewski, M. 1995. Umweltvertragliche Tenside fur Wasch- und Reingungsmittel auf Naturstoffbasis. Pages 562-567 in *Bioresource Hemp*, 2nd edition. nova-Institute, Rosenstr. 53, 50678 Cologne, Germany.

75. Pari, L.; Baraniecki, P.; Kaniewski, R.; Scarfone, A. 2015 Harvesting strategies of bast fiber crops in Europe and in China. *Ind. Crop Prod.*, *68*, 90–96.

Pertwee, R.G. *The Pharmacology and Therapeutic Potential of Cannabidiol, in Cannabinoids*.

76. Piomelli, D., Russo, E.B. 2016 The *Cannabis sativa* versus *Cannabis indica* debate: An interview with Ethan Russo, MD. *Cannabis Cannabinoid Res.*, *1*, 44–46.

77. Piotrowski, S.; Carus, M. *Ecological Benefits of Hemp and flax Cultivation and Products*; Nova Institute: Hurth, Germany, 2011; Available online: http://eiha.org/attach/643/110513_Ecological_benefits_of_hemp_and_flax.pdf/ (accessed on 5 December 2019).

78. Prade, T.; Finell, M.; Svensson, S.E.; Mattsson, J.E. 2012 Effect of harvest date on combustion rleated fuel properties of industrial hemp (*Cannabis sativa* L.). *Fuel* , *102*, 592–604.

79. Prade, T.; Svensson, S.E.; Andersson, A.; Mattsson, J.E. 2011 Biomass and energy yield of industrial hemp for biogas and solid fuel. *Biomass Bioenergy*. *35*, 3040–3049.

80. Ranalli, P.; Venturi, G. 2004, Hemp as a raw material for industrial applications. Euphytica 140, 1–6.

81. Rausch, P. 1995. Verwendung von Hanfsamen–l in der Kosmetik. Pages 556-561 in*Bioresource Hemp*, 2nd edition. nova-Institute, Rosenstr. 53, 50678 Cologne, Germany.

82. Small, E., 1979. Practical and natural taxonomy for*Cannabis.*Volume I, Pages 171-211 in Small E, ed., *The species problem in* Cannabis. Corpus, Toronto, Canada.

83. Small, E.; Naraine, S.G. 2016. Size matters: Evolution of large drug-secreting resin glands in elite pharmaceutical strains of *Cannabis sativa* (marijuana). *Genet. Resour. Crop Eviron*, *63*, 349–359.

84. Stevulova, N.; Cigasova, J.; Estokova, A.; Terpakova, E.; Geffert, A.; Kacik, F.; Singovszka, E.; Holub, M. 2014Properties characterization of chemically modified hemp hurds. *Materials 7*, 8131–8150.

85. Struik, P.C.; Amaducci, S.; Bullard, M.J.; Stutterheim, N.C.; Venturi, G.; Cromack, H.T.H 2000. Agronomy of fibre hemp (*Cannabis sativa* L.) in Europe. *Ind. Crop Prod.*, *11*, 107–118.

86. Struik, P.C.; Amaducci, S.; Bullard, M.J.; Stutterheim, N.C.; Venturi, G.; Cromack, H.T.H.2000Agronomy of fibre hemp (Cannabis sativa L.) in Europe. Ind. Crops Prod., 11, 107–118.

87. United States Department of Agriculture. Organic Regulation. Available online: https://www.ams.usda.gov/rules-regulations/organic (accessed on 23 December 2019).

USDA, 2000. Industrial Hemp (Cannabis sativa)-Economic Viability and Political Concerns.

88. Van der Werf H. M. G, Mathijssen E. W. J. M and Haverkort A. J. 1996; Crop physiology of Cannabis sativa L.: A simulation study of potential yield of hemp in Northwest Europe. In: RANALLI, P. (ed.) Advances in Hemp Research. New York: Food Products Press.

89. Van der Werf, H.M.G. Agronomy and Crop Physiology of Fibre Hemp: A Literature Review. Available online: https://edepot.wur.nl/346939 (accessed on 23 December 2019).

Van der Werf, H.M.G.; 2008 Turunen, L. The environmental impacts of the production of hemp and flax textile yarn. *Ind. Crop Prod.*, *27*, 1–10.

90. Van der Werf, H.M.G.; Mathussen, E.W.J.M.; Haverkort, A.J. 1996 The potential of hemp (*Cannabis sativa* L.) for sustainable fibre production: A crop physiological appraisal. *Ann. Appl. Biol.*, *129*, 109–123.

91. Vera, C.L.; Malhi, S.S.; Phelps, S.M.; May, W.E.; Johnson, E.A. 2010N, P, and S fertilization effects on industrial hemp is Saskatchewan. *Can. J. Plant Sci.*, *90*, 179–184.

92. Vera, C.L.; Malhi, S.S.; Raney, J.P.; Wang, Z.H. 2004. The effect of N and P fertilization on growth, seed yield and quality of industrial hemp in the Parkland region of Saskatchewan. *Can. J. Plant Sci.*, *84*, 939–947.

93. Vessel, A.J.; Black, C.A. Soil Type and Soil Management Factors in Hemp Production. Available online: https://lib.dr.iastate.edu/researc hbulletin /vol28/iss352/1/ (accessed on 23 December 2019).

94. Vonapartis, E.; Aubin, M.P.; Seguin, P.; Mustafa, A.F.; Charron, J.B.2015. Seed composition of ten industrial hemp cultivars approved for production in Canada. J. Food Composite. Anal., 39, 8–12.

95. Zhukov, M.S.; Bedak, G.R., 1963. Boron fertilizer and its application on hemp. *Len Konop.* 1963, *8*, 24–25

PART 2: Water Harvesting for Sustainable Agriculture Development

2.0 Ground Water Resource in Semi-Arid Region for Climate Resilience

2.1. Small scale irrigation using collector wells in Zimbabwe

Published: Environmental Science, Hydrology, Agriculture and Soil Science Water for Agriculture in Zimbabwe

C.J. Lovell, C.H Batchelor, D.K Waughray, A.J. Semple[1]

E. Mazhangara, G. Mtetwa, M. Murata, M.W. Brown, T. Dube[2]

D. M. Thompson, P.J. Chilton, D.M.J. Macdonald[3]

D. Conyers[4] *and O. Mugweni*[5]

1. Institute of Hydrology, Wallingford, Oxon, OX10 8BB, UK

2. Lowveld Research Stations, P. O. Box 97, Chiredzi, Zimbabwe

3. British Geological Survey, Wallingford, Oxon, OX10 8BB, UK.

4. P.O. Box 83, Kariba, Zimbabwe

5. Agritex, P.O Box 354, Masvingo, Zimbabwe

ACRONYMS AND ABBREVIATIONS

Agritex	Agricultural Technical and Extension Services
R&SS	Research and Specialist Services
DWD	Department of Water Development
IH	Institute of Hydrology
BGS	British Geological Survey
NR	Natural Region
BGS	British Geological Survey, Wallingford, Oxon, OX10 8BB, UK.
LVRS	Lowveld Research Stations
NRI	Natural Resources Institute
DWD	Department of Water Development

NGO	Non-Governmental Organisation
NGADI	Nutrition Garden development Initiative
CSO	Central Statistical Office
ODA	Overseers Development Administration
VIDCO	Village development committee
WARDCO	Ward development committee
GoZ	Government of Zimbabwe
DDF	District Development Fund
TWD	Traditional dug well
LDW	Large diameter well
CDB	Conventional deep borehole
SRB	Screened Regolith Borehole
DEB	Deep exploratory borehole
Kg	Kilograms
N	Nitrogen
R&D	Research and Development

2.1.1 Collector well summary

Throughout Africa, many communities rely on crystalline basement rocks as the primary source of water supply. Well yields are often low and shortage of water remains the principal constraint on development in many areas. Recent studies in southern Zimbabwe have focused on the water resource potential of the crystalline basement rocks and in particular, on the potential for using collector wells to abstract sufficient water from the regolith aquifer for both domestic use and small-scale irrigation by rural communities.

This was the final report in the Government of Zimbabwe/British Overseas Development Administration- funded pilot project. The project was implemented by the Zimbabwe Department of Agricultural Technical and Extension Services (Agritex), Research and Specialist Services (R&SS) and Department of Water Development (DWD), in collaboration with the Institute of Hydrology (IH) and British Geological Survey (BGS). It was completed in Masvingo Province, one of the driest parts of Zimbabwe, and has allowed implementation of one of the most comprehensive comparisons of alternative well designs and detailed assessments of social, economic, institutional and environmental viability of community-based irrigation ever undertaken in Africa.

Currently, groundwater in southern Zimbabwe is abstracted by dug wells or by deep boreholes sited using geophysics. Low yields are typified by a recent Drought Relief Programme-only 21 per cent of boreholes gave yields of 0.6 l/s or higher and 18 per cent were dry. A collector well is designed for use in crystalline basement rock. It is a large diameter well sited by exploratory drilling whose yield is enhanced by lateral drilling at the base to a distance of 30 metres in several directions (harvesting water). Results are presented of collector well performance at nine locations, and include the effect of lateral drilling on well yields, measured aquifer properties, and comparison of collector well performance with that of traditional dug wells, large diameter wells, and boreholes screened in the regolith, and conventional deep boreholes cased in the regolith.

Exploratory drilling has been found to be vital to locate optimum well sites in this region as present geological methods do not have sufficient resolution in terrain of such high spatial variability. Rainfall and geology influence choice of well design and in southern Zimbabwe (average annual rainfall 400-800 mm) collector wells have been found to be well suited in areas of younger undifferentiated gneiss and intrusive granite, less suited in areas of Karoo basalt, and unsuitable in areas of older gneiss because either depth of saturated weathering was insufficient or depth of water table exceeded 15 metres. At all project sites, however, a consistently adequate supply of water for domestic use and small-scale irrigation has been obtained from the shallow regolith aquifer, both by large diameter wells (17.6m^3/day, on average) and collector wells (25.9 m^3/day on average). The increase in suitability yield obtained by lateral drilling is viable but on average is significant at 38 per cent improvement.

Economic viability of siting deep boreholes to support community gardens depends entirely on success rate of drilling. Some screened regolith boreholes drilled by the project gave higher yields than existing deep boreholes. This is important, as it suggests that: a) siting deep boreholes by shallow exploratory drilling in the regolith to identify the greatest depth of saturated weathering may be a reasonable approach for helping to overcome the spatial variability found in the deep aquifer and b) boreholes in this region would generally be better if screened in the regolith rather than cased. The challenge for future groundwater development in the communal lands of southern Zimbabwe is suggested to lie in quick and cost-effective selection of the appropriate well design and location in a given area. Practical examples illustrate that this can be achieved by rapid exploratory drilling, simple pump tests of exploratory holes and existing water points, and limited geophysical surveys, used in conjunction with a hydrogeological decision tree developed during the pilot project. The results of comprehensive baseline surveys, routine monitoring and return-to-household surveys highlight the positive impact on rural economies and quality of life that small-scale irrigation using groundwater can have. They also highlight the importance of a range of human and institutional factors that can influence this type of development. The schemes are highly valued by the communities both for their provision of reliable domestic water and sufficient irrigation water. An estimated 3882 people obtain their domestic water from the six project wells, consuming approximately 13 litres per person per day. The estimated population of households directly involved in the six community gardens is 4461 people with a mean income generated per member per year of Z$255.20 The schemes are economically viable and water use efficient, with an average IRR in 1994 of 19 per cent, average total gross margin per hectare per year of Z$24,000, and average gross margin per unit of water (Z$/ha/m) of $43,800 per scheme. They provide extensive marketing opportunities for the sale of dry season vegetables to surrounding townships and play an important role in improving the welfare of women and children by making more efficient use of labour time, improving nutrition, and providing an important source of disposable income for household items and school fees. In essence, the collector well gardens have reduced the period of scarcity of fresh vegetables that communities in the area face by four to five months, lowered the number of people who feel there is a period of fresh vegetable scarcity by about 25 per cent, and decreased the time during scarce periods that people miss out on ting fresh vegetables by about four days in every week. Price increases for plots in the gardens indicate the degree of welfare improvement the schemes are now bringing to members. Many people, originally disinterested are now wishing to join, to the extent that there is a willingness to pay up to 16 per cent of a year's income from rain-fed crops to join the schemes and benefit from the steady revenue they can provide. It should be noted that the implementation of community-based development initiatives is not easy, and many factors can influence scheme performance. Local ownership of the resource is one vital ingredient.

Local communities are much more likely to look after and pay for the upkeep of their water points and gardens if they know that they belong to them and not to another agency. An important corollary of this is that the community is involved at all stages of the resource development, from its inception, through planning and construction, to subsequent maintenance and management. An inter-disciplinary project team is vital to success, recognising that decision making at all stages is an inter-disciplinary process. Valuable lessons learnt at first schemes allowed key steps to be identified that help to promote full community involvement and ensure schemes that are more likely to be sustainable from a social perspective. Among these, an informal community contract considered prior to construction is a potent means of outlining obligations, clarifying misconceptions and creating sense of ownership. Leadership remains the main social factor influencing scheme performance, principally because decision making in a community-based project is slow compared to a private or family-based project and more easily disrupted by leadership disputes. Whether these disputes can be overcome depends on whether the potential benefits of the development are sufficient to ensure that the community maintain interest in the development during a dispute and have sufficient incentive to overcome a dispute. Fortunately, group ownership of a resource is not new in Zimbabwe, potential benefits in this case are significant, and in most communities, there appears to exist the necessary qualities of leadership, experience of gardening, and local institutional structures to make this type of community project successful. Where such qualities or structures are lacking or temporarily fail, trained facilitators can help enormously in resolving difficulties. Specific environmental benefits of community gardens using groundwater include the reduction in pressure to cultivate marginal land, particularly stream banks, and the promotion of longer-term management strategies due to decreased risk and increased security of tenure that the schemes bring. There remains, however, a critical need to ensure that any new investment in water and garden facilities does not fail through failure to protect the ground water resource. Community participation in development of small-scale irrigation using groundwater can have the major benefit of providing the ideal springboard to improved resource management at the village or catchment scale. In many cases, it gives rural communities their first experience of the institutions needed to make community-based activities successful. Future development of wells and gardens should proceed hand-in-hand with programmes of community-based natural resource management, building on the opportunity and incentive that this initial type of water development project provides. An increasing number of communities and agencies are requesting to replicate the pilot schemes. A steering committee has been found to develop a framework for the next phase of work. Details of the proposed NGADI project: "Nutrition Gardens and Groundwater Development in Zimbabwe" are included here. Initially, the project will focus on the critical need to enhance institutional capacity in Zimbabwe to implement sustainable water points and community gardens. It will include programmes of village level training and permit extended monitoring required to confirm scheme viability. Later, as this initial form of project support is withdrawn, it will provide a sound basis for replication of schemes on a wider scale on behalf of local authorities and in response to community and NGO requests. In summary, the pilot project: "Small-scale irrigation using collector wells" has been highly successful. The quantified benefits to rural communities recorded at pilot schemes and expected at future schemes, -the anticipated benefits of enhanced institutional capacity to implement sustainable water points and community gardens in the future, and the far-reaching benefits of using this type of first water development project as a springboard to other community-focused projects in health, agriculture and the environment, may be considered sufficient to justify project replication on a wider scale. The development of community gardens using groundwater has little risk of negative consequences, while offering numerous additional positive externalities, such as improved productivity due to better nutrition and health (both to scheme members and surrounding populations), reduced demands on women's time and energy spent fetching water and in providing relish, the provision of employment and skills development at the local level, the development of community spirit and collective action, and the contribution to development of local institutional structures. For much of Zimbabwe and many parts of Africa underlain by basement rock, this offers a possible way in which communities threatened by aridity and land degradation can stabilise or improve living standards and quality of life.

2.1.2 Background

2.1.2.1 Semi-Arid Areas of South-East Zimbabwe

Communal areas of southern Zimbabwe typify the problems now facing people and the environment in many semi-arid parts of Africa. Population densities are high - up to 61 persons per square kilometre, and increasing at an average rate of 2.2 per cent per annum (CSO, 1993). The majority of people live in rural areas and depend on subsistence farming. They are struggling to survive on a poor and degrading resource base. Drought is recurrent and agricultural production is low and highly variable. Land degradation and desertification are taking place at an alarming rate. The extent of land degradation in south-east Zimbabwe has been described and discussed in a number of reports (e.g. Whitlow 1983; Whitlow and Campbell 1989; Abel and Blaikie, 1989; Campbell et al, 1989; Tagwira, 1992; Scoones, 1992). The main processes of land degradation in the region are deforestation, overgrazing and soil erosion. These result from poor management of natural resources in a region characterised by high population growth, normally high livestock numbers, a lack of financial and manpower resources for sustainable land management and a land tenure system which promotes overgrazing (lied, 1992). Since 1980, the Government of Zimbabwe has adopted policies aimed at improving the management of land resources in rural areas. However, the successful implementation of these policies remains elusive, as does the goal of sustainable agricultural development. Prime constraints on sustainable development in south-east Zimbabwe are the low and erratic rainfall and the availability of ground and surface water resources. Rain-fed crop production provides the main source of staple foodstuffs. However, increasing population and declining productivity of existing croplands have led to cultivation of more marginal terrain better suited to other less intensive forms of land use. Extension of cropland is often at the expense of grazing lands, thus placing even greater pressures on the remaining grazing land (Whitlow, 1989). It should be noted that there is no obvious evidence in south-east Zimbabwe of environmental recovery being promoted by increase in population as has been reported for some semi-arid areas of Kenya (Tiffen et al (1994).

In areas where sufficient water resources are available, large-scale irrigation projects have been and continue to be constructed. However, such schemes have often proven to be expensive, relatively inefficient and difficult to reconcile with traditional farming systems. Some large irrigation schemes have also caused a range of health-related and environmental problems and necessitated considerable social adjustments. In contrast, experience in the region has shown that informal or garden irrigation can be economically viable and appropriate to farmers, especially women fanners, for whom it is already a traditional component of the farming system. The unavailability of reliable safe water is one of the main problems confronting rural populations in south-east Zimbabwe. In Masvingo Province for example, 32% of people still rely on relatively unsafe water from unprotected wells, rivers, streams and dams (CSO, 1993). There is a recorded density of only one protected water point (well or borehole) per 17 km^2. Of these, 40% are known to go dry during drought (Anon, 1992). For some communities the priority is for cleaner and more reliable sources of water. For many it is for sufficient water to allow vegetables to be grown.

In conclusion, south-east Zimbabwe is typical of much of semi-arid Africa. The many attempts that have been made to alleviate the serious social and environmental problems in the area have met with only limited success. The reasons for this lack of success are varied and complex. Some interventions have been inappropriate; some have been introduced without. the participation of stakeholders, and others have not been economically viable. Some interventions, although relatively successful, have been introduced at such an isolated and local scale that they have gone only a short way to alleviating the multiple problems of the local communities let alone the region as a whole. There is therefore, a fundamental need for the implementation of projects that are appropriate, economic, involve the participation of stakeholders, and address the problems of the people and the environment of the whole region. The pilot project described here has these wider objectives.

2.1.2.2 Collector Well Research

Basement aquifers are of particular importance in tropical and sub-tropical regions because of their widespread extent and accessibility and because there is often no readily available alternative source of water supply, particularly for rural communities (Wright et at, 1989; Wright, 1992; Howard and Karundu, 1992; Howard et at, 1994). Figure 2.1.1 shows the extent of crystalline basement aquifers worldwide. In general, water in basement aquifers occurs within the weathered residual overburden (or regolith) and the fractured bedrock. In Zimbabwe, current development is generally by digging wells within the regolith or drilling boreholes to intercept fractures in the bedrock. However, work during the last twelve years has shown that collector wells can be used to increase groundwater abstraction from regolith aquifers (Wright, 1992; Anon, 1994). A collector well is a shallow, hand-dug well of large diameter with horizontal boreholes drilled radially from the bottom of the well to a distance of approximately 30 m (see Figure 2.1.1).

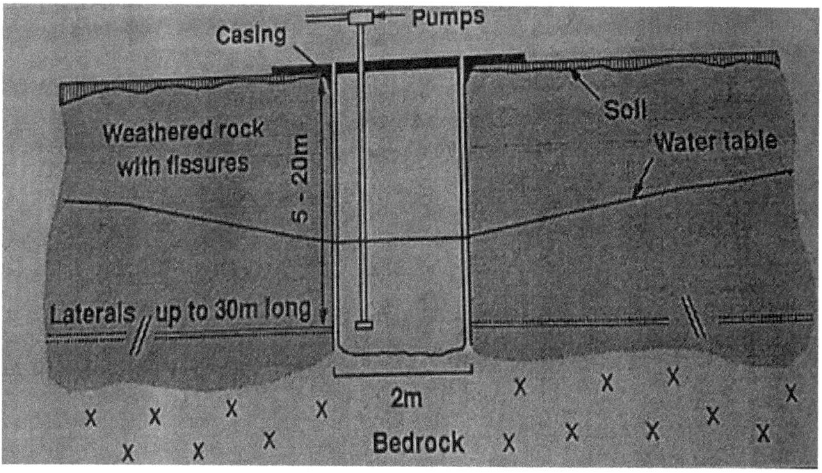

Figure 2.1.1: Schematic diagram of a collector well

It has been demonstrated that in some areas higher average yields can be obtained from collector wells than from slim boreholes with the added advantage of small drawdowns. Safe yields from eight wells in Zimbabwe and Malawi and 20 in Sri Lanka were calculated at 2.7 I/s-1 (ranges of and I/s-1 respectively) with drawdowns of 2-3 m, and were compared with typical yields in the range I/s-1 for slim boreholes at pumping drawdowns in excess of 30 m (Wright, 1992).

2.1.2.3 First Collector Well Gardens

In 1988, the Institute of Hydrology began a collaborative research project with the British Geological Survey and the Lowveld Research Stations. The main objectives of this project were to study the feasibility of using regolith aquifers as a source of water for small-scale irrigation and to compare and develop methods of low-cost, high efficiency irrigation which would be suitable for use on small irrigated gardens. In 1989, a collector well was constructed at the Chiredzi Research Station. Construction of laterals in this well led to a 150% improvement in yield (Chilton and Talbot, 1990). Water from this well has been used to irrigate a 1 ha demonstration garden that has been the location for a series of trials that have evaluated different approaches to improving irrigation efficiency for a range of crops. The results of these trials can be found in a series of interim reports (Batchelor et al, 1990; Lovell et al, 1990; Lovell et al, 1992; Murata et al, 1995).

In 1991, the collaborative research project completed the construction of a first off-station collector well garden at Romwe Catchment in Chivi District. Although the yield improvement due to laterals was relatively low at 10% (Chilton and Talbot, 1992), this well with its two hand pumps was able to provide domestic water for over 1200 people and water for a 0.5 ha community garden. At the outset, 46 families had plots or allotments on this garden. The results of the first cropping seasons are presented in the interim reports listed above. After initial social and institutional "teething" problems this garden has flourished most notably during the 1991/92 drought when many boreholes and traditional dug wells dried up and when all irrigation schemes using surface water failed. The success of this collector well in providing domestic water and a valuable and continuous supply of vegetables during the 1991/92 drought went a long way to overcoming residual scepticism and was also instrumental in securing Government of Zimbabwe and British Overseas Development Administration support for the pilot project reported here.

2.1.3. The pilot project

2.1.3.1 Wider Objectives

The pilot project "Small scale irrigation using collector wells -Zimbabwe" had the following wider objective: i) To improve the long-term well-being of rural communities and the rural environment by effective use of groundwater.

2.1.3.2 Specific Objectives

The project contained the following main elements:

i) Selection of suitable sites and the installation of six small irrigation schemes using water from collector well in south-east Zimbabwe;

ii) The assembly and collection of adequate baseline data and design of a monitoring system to facilitate the assessment required at iv below;

iii) The regular collection of data through the monitoring system;

iv) production of a final integrated report on the scheme's technical, economic, financial, institutional, social and environmental Viability, with recommendations for future development.

Specific objectives of the project were:

i) To field test the validity of small-scale irrigation and collector well research results obtained at the Lowveld Research Station (LVRS);
ii) To identify ways of improving the operation of the schemes, for example by identifying and overcoming constraints;
iii) To identify a basis for replicating the schemes on a wider scale.

(Ref. Project Data Sheet, BDDSA March 1992) The project started on the 1 October 1992 and finished on 16 January

2.1.4. Project implementation

2.1.4.1 Activities and Staffing

In order to fully evaluate collector well garden viability, monitoring of the initial scheme at Romwe continued during the pilot project, and the six new pilot project schemes were planned and sited to broaden the range of physical, social, economic and institutional settings considered. Practical constraints of distance and equipment available to construct and monitor all schemes in a limited period of time meant that, logistically, this objective was best achieved by siting the new schemes in Zaka District, on a line running north of Chiredzi Research Station and reasonably parallel to the main Buffalo Range - Zaka road., In this way, four principal geologies (younger mobile belt gneiss, older gneiss complex, paragneiss and younger intrusive granite) and corresponding terrain types likely to be important in future development could be studied. With success of the initial scheme at Romwe and increasing interest as first pilot project schemes were completed in Zaka District, two additional schemes were also requested and funded by Plan International in Chiredzi District. These schemes, at Machoka and Masekesa, enabled study of a further geology (basalt) and social setting (Shangaan, as opposed to Ndebele and Karanga at previous sites). Results from these two schemes are included here.

Figure 2.1.3 shows staff involved and key activities during the pilot project. Figure 2.1.4 shows regional geology and eventual location of the nine schemes completed during this pilot phase.

2.1.4.2 Baseline Data Collection

Socio-economic aspects of the schemes funded by GoZ/ODA have received considerable attention. Comprehensive baseline data at each scheme has been collected in four parts:

i) Rapid rural appraisal to identify main characteristics and structure of each community, existing water and garden resources, community problems and aspirations;
ii) Formal surveys by questionnaire of 30 households randomly selected from the whole community to assess local institutions important at district and village level, key people and leadership patterns in the community, both modern (Vidco/Wadco) and traditional (Chief/Kraalhead), the range and pattern of income and expenditure, class structure and wealth present in the community, family circumstances including decision making, division of labour and problems faced, water sources past and present, current patterns of vegetable consumption, health, farming assets and practices; gardening activities past and present, contact with extension staff, experiences of development;
iii) Case studies of 4 households selected to represent different income groups and provide more detailed information on particular aspects such as income and expenditure, household budgets, savings and credit, patterns of vegetable consumption, decision making within the household and within the community, gender divisions and attitudes, and specific effects of the collector wells and gardens;
iv) Informal meetings with groups of women only and groups of men only to discuss issues of gender, equity, health, leadership and rural development.

2.1.4.3 Monthly Monitoring

Following baseline data collection, regular data collection throughout the project has been achieved by monthly monitoring visits to all schemes. In this way, information has been obtained not only on well and garden performance but also on the dynamics of change that have occurred at each scheme. In particular, routine monitoring has been achieved in four main ways:

i) collection by garden committee secretaries at each scheme of routine data on inputs to and outputs from the gardens, including cropping patterns, irrigation and water use, crop protection, fertilisation, crop yields and market prices;
ii) collection by an appointed member of each garden committee of weekly data on well water levels and abstraction measured by a

Munro Chart Recorder and two water meters fitted to each collector well;

iii) monthly visits by project staff to collect data (i) and (ii), observe progress and to discuss with members and extension staff issues that have included the role of local institutions and leadership patterns within the community and their impact on community efforts, and the degree and form of community efforts, problems, successes and failures and the differences and similarities between schemes;

iv) Regular analysis of all data to provide detailed accounts of both financial performance and social dynamics at each scheme on a season-by-season basis (see project progress reports Lovell et al, 1995 and Mazhangara et ai, 1995).

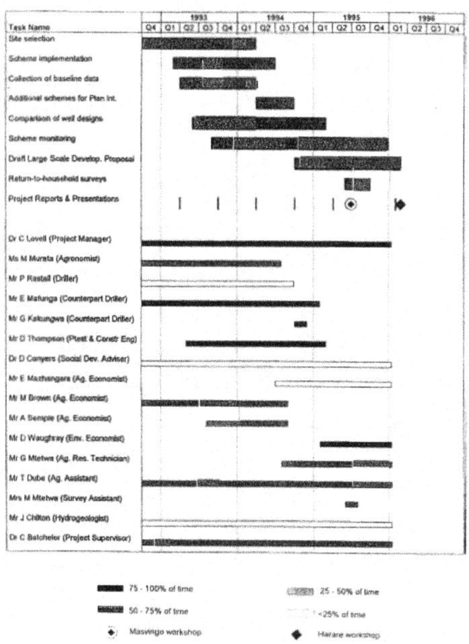

Figure 2.1.3 Key activities and staff involved in the pilot project

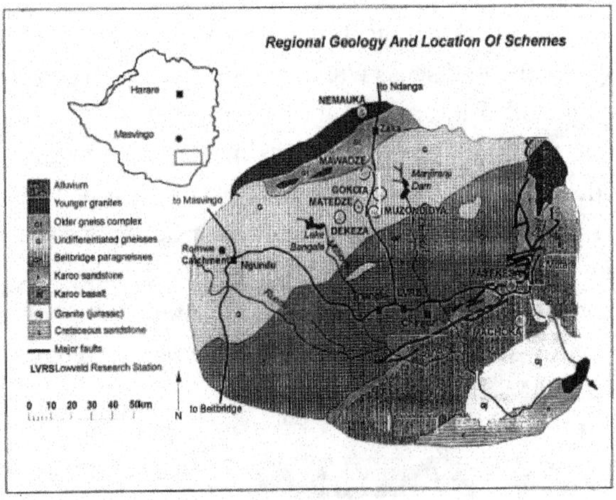

Figure 2.1.4: Regional Geology and Location of Schemes

2.1.4.4 Return-to-Household Survey

To help in overall evaluation of scheme impact on both members and non-members of each community, a socio-economic return-to-household survey was completed towards the end of the project in August using a similar questionnaire to that used during baseline data collection, the 30 households at each scheme were revisited. As well as information listed in Section 3.2 (ii) above, emphasis was placed on:

i) issues of equity, particularly looking at who are and who are not members of the schemes and reasons why;
ii) willingness to join, using proxies for money in order to better assess the welfare improvement that the schemes bring;
iii) differences between vegetable availability and consumption and money spent on vegetables before and after the schemes; wealth breakdown of members and non-members, including the potential created by the schemes for employment and income generation and paying special attention to the money that women are now saving or using in revolving funds or generating from other new small scale income generation activities.

2.1.5 Scheme implementation

2.1.5.1 Methods of Site Selection

During the first two years of the project, various methods of site selection were tried in order to represent ways in which site selection might be undertaken in a larger development programme

2.1.5.1.1 Formal nomination by Agritex staff.

Requests were made to Agritex staff at ward level to suggest potential areas for collector well gardens based upon geographical, hydrogeological, socio-economic and agricultural selection criteria shown in Annex 1.

A total of thirty-eight prospective sites were identified in this way and all were visited. Twelve were shortlisted but following a team visit only four were considered to satisfy most of the selection criteria. Table 1 provides an example of rapid rural appraisal data collected at this stage. Three of the sites were nominated for exploratory drilling. At the first (Chindondodza) saline water was found, at the second insufficient water, and at the third a successful site (Muzondidya) was located.

It should be noted that the beginning of the project coincided with the worst drought recorded in Zimbabwe. In the project area, rainfed crops had failed, surface water supplies had all but failed, and relief programmes of well drilling and well deepening were on going. The situation encountered at all sites visited was one of critical need first and foremost for adequate supplies of domestic water. Only if this could be supplied could people begin to consider the possibilities of garden irrigation. By formal Agritex nomination, numerous communities with dire need for water and with great enthusiasm for gardening could be identified. The value of local knowledge provided by the Agritex field staff could also not be bettered. But the problem remained at all sites to know whether groundwater of suitable quality and quantity existed at a depth and in weathering appropriate for collector wells or other well designs. Unfortunately, in many instances the hardships faced by the communities nominated were found to be directly related to unfavourable hydrogeological conditions in these areas

2.1.5.1.2 Formal nomination by DDF staff

Hydrogeological records of completed wells and boreholes are kept by district and provincial offices of the District Development Fund (DDF) and Department of Water (DWD). Making use of these records and, in effect, considering only the geographical and hydrogeological criteria initially, eight potential areas were quickly identified. Following subsequent liaison with staff of Agritex and the local communities, this approach improved success rate. Of six areas with vehicular access, five were found hydrogeologically worthy of exploratory drilling and at the first selected, a successful site (Gokota) was located. It should be noted, however, that the more favourable hydrogeological conditions identified in this way meant that a(several locations the communities were better off than communities visited by the previous method of site selection, at several locations the hydrogeological conditions found were different to those on record, and even at the site where hydrogeological evaluation was successful (Gokota) the location of the best exploratory hole was not in the immediate area of the existing DDF well but in a neighbouring valley, a finding that highlighted the extreme spatial variability of aquifer properties in this region that was to become a key factor throughout all subsequent exploratory drilling.

2.1.5.1.3 Site selection by geology

The principal physical factors influencing choice of well design are geology and rainfall. These determine the relative positions of the base of the regolith (weathered zone) and the water table (Chilton and Foster, 1995). Depth of weathering is a complex function of geology and rainfall; in general, the former determines the degree and type of weathering and the subsequent properties of the aquifer, and the latter may have the strongest influence on the depth of weathering. Geology and rainfall may also influence groundwater quality. Thus, only where geology and rainfall produce a saturated regolith that is relatively thick (greater than 5m) and a water table that is relatively shallow (less than 15 m below ground level) is a collector well likely to be viable.

Five principal geologies and corresponding terrain types are of importance in the communal lands of southern Zimbabwe, namely younger undifferentiated gneisses, younger granites, Beitbridge paragneiss, older gneiss complex and Karoo basalt (Figure 2.1.4). The first pilot project schemes at Muzondidya and Gokota and the initial scheme at Romwe were all successfully sited on the first of these geologies. Thereafter, attention was placed on siting schemes on the remaining geologies.

On younger granites, exploratory drilling was successful and one pilot project collector well was sited at Nemauka. On Beitbridge paragneiss, depth of weathering and rest water levels were generally found to be suitable for collector wells but reconnaissance of existing water points in the southern parts of Zaka District and Nyajena Communal Land revealed groundwater salinity to be a serious problem on this geology, consistent with the findings of first exploratory drilling at Chindondodza. Consequently, a scheme was not sited on this geology. On the older gneiss complex around Zaka, exploratory drilling revealed either small amounts of water in shallow depths of weathering on top of fresh rock, or no water present in depths of weathering often greater than 20 mbgl. Despite concerted effort, a suitable location for a collector well could not be found on older gneiss complex, although a number of locations suitable for deep conventional boreholes were identified and an ideal site for a collector well was subsequently identified at Mawadze on the transition between older and younger gneiss complexes

2.1.5.1.4 Site selection by request

As first pilot project schemes came into operation, interest from neighbouring communities and from aid agencies working in the region increased. Members of the community at Dekeza, for example, made a request for a collector well garden through their Agritex Extension Worker when hearing of exploratory drilling at nearby Makwevere, and members of the community at Matedze made a request for a collector well garden through the Headman of one of their kraals after observing success of the scheme at Dekeza. Similarly, two additional schemes were requested and successfully sited (on basalt) for Plan International at Machoka and Masekesa in Sangwe Communal Land. Potential site identification in this way, based on initial interest expressed either by a community or by a donor on behalf of a community, is likely to occur again in a larger development programme particularly if work is concentrated on a ward-by-ward or district-by-district basis. Where well siting is found possible, this approach represents the best way to identify prospective locations because it creates a community sense of ownership of the project from the outset.

2.1.5.1.5 Site selection using geophysics

Geophysical methods have been extensively used in a wide range of hydrogeological environments to assist in the selection of optimum sites for borehole drilling. Indeed, in areas underlain by basement rocks, considerable reliance has been placed on such methods. Varying degrees of improvement in "success rate", have been claimed, but the published cases of comprehensive and thorough. Post-drilling comparison of geophysics and drilling success are few, and comparisons of different methods under the same conditions even rarer. It is, for example, not often easy to evaluate the additional benefits of geophysics, as it is rarely possible to compare on a site-by-site basis the "with" and "without" geophysics drilling result. In most rural water supply projects either one or other method is employed, and the best that can be expected is overall statistical comparisons of similar projects using different approaches in the same region.

Geophysical methods have been deployed in basement areas with three main objectives:

- To eliminate negative sites with hard bedrock at very shallow depth rather than waste time on unproductive drilling. Experience shows this can be achieved by a variety of geophysical methods.
- To locate simply, quickly and cheaply the precise spot where the weathering is of the optimum depth (usually maximum) for the type of abstraction point proposed, relatively permeable and where the water table is relatively shallow. This is much more difficult to achieve.
- To locate productive fractures in the unweathered bedrock, where the weathered zone is either very thin and/or is above the water table, and the bedrock is therefore the target for groundwater exploration.

It should be remembered that, to be effective, geophysical methods need strong contrasts in the physical properties of rocks, and need these to be closely related to hydrogeology and groundwater occurrence. Basement rocks, are highly variable in both weathered and unweathered states, and many of these variations are gradational and small in scale. A great deal of effort over many years has been expended by the hydrogeological community in the development of instrumentation and field methods to address the three objectives outlined above. Although promising results have been reported, methods which are scientifically effective but easy and cheap to operate are not yet widely available. Extensive work has been carried out by BGS in

Zimbabwe, including some field testing of the shallow seismic method in the present project. While it was able to delineate areas of shallow bedrock to be avoided, it was not able to pinpoint consistently preferential locations for shallow groundwater supplies (Davies, 1994). While further work on the development of geophysical methods is underway, its routine use in projects such as this one cannot be recommended as cost-effective at present.

2.1.5.2 Methods of Scheme Construction

Originally it was envisaged that the project would identify and construct six collector well gardens during the first year (PDS, BDDSA 1992). In the event, this proved not to be possible if the schemes were to represent a range of physical, social, economic and Institutional settings. Instead, schemes were implemented two at a time. This had the added advantage that first schemes completed provided examples of problems that could occur and an adaptive approach to scheme implementation became possible.

2.1.5.2.1 Key steps identified

Valuable lessons were learnt at first schemes and these led to significant improvements in implementation of later schemes. The key steps now known to be important in development of community gardens using groundwater are shown in Figure 2.1.5. These are not intended as a prescription but rather as a checklist. The exact order and importance of the steps will vary to some extent between different areas and between different communities, but in southern Zimbabwe the sequence shown has been found to be helpful in promoting successful collaboration with communities and in implementing schemes more likely to be sustainable from a social perspective.

The importance of community meetings in this sequence cannot be overstated. They are the key to success. No work should be done in an area until local leaders have been consulted. Sufficient time must be allocated to allow these meetings to be well organised, attended by all leaders and interested members of the community. Sufficient time must also be allowed for reflection and consideration by both community and project staff.

In southern Zimbabwe it will be rare to meet a community that does not have need for or express interest in a community garden particularly as this will also offer a source of clean domestic water. Thus, at the first community meeting it is important not to raise hopes too high. It should be explained from the outset that the scheme will be possible only if hydrogeological evaluation is positive, a step described in detail later in Section If hydrogeological evaluation is successful, the second community meeting provides the venue for more detailed discussion of the prospective project.

2.1.5.2.2 Informal contract with community

Community problems which occurred at the first schemes could be linked primarily to a poorly developed sense of ownership of the project, lack of self-confidence, leadership problems, and misconceptions about the project regarding ownership, responsibility for management and maintenance, and distribution of benefits.

In order to help develop a sense of community ownership at the earliest opportunity and to avoid misconceptions, project staff introduced at the final two schemes (Mawadze and Matedze) an informal contract between the community and project staff. This was prepared in Shona and signed at the time of the second community meeting. A copy in English is shown at Annex 2. The contract outlines the potential project and stages of scheme construction, lists contributions expected from both parties, and recognises explicitly community ownership of the project and community responsibility for scheme management and maintenance.

Use of the contract was well received and proved to be successful at both locations. The people appreciated the clarification which the document provided. Well construction proceeded at record pace, with all work performed on a voluntary basis and in high spirits by teams of five men working for five days each under supervision of a project foreman. Similarly, completion of the gardens was undertaken voluntarily by both male and female members as a group activity. Most importantly, there has been increased reference to project ownership noted during interviews with case study families and the incidence to date of social teething problems has been less at these two sites.

2.1.5.2.3 Garden membership

The pilot investigated a number of different social settings. This has involved garden membership being decided by communities in a number of different ways, with resulting differences both in the performance of the schemes and in non-equitable distribution of benefits between members and non-members. The consensus made by project staff at this stage is that institutional structures are still evolving and it is too early to draw conclusions on the 'best' or preferred mechanism by which scheme membership can and should be decided.

2.1.5.2.4 Pump maintenance

In a project design to implement relatively high yielding water points and provide rural communities with water sufficient for both domestic use and irrigation, it is vital that appropriate, reliable and affordable components and sources of inputs are used in scheme design. Choice of pump type and pump maintenance thus become two key issues. In Zimbabwe, a locally manufactured type 'B' bush pump has proved to be robust and can be repaired at relative low cost. During the project two were fitted per collector well. In addition to meeting the total pumping capacity of 15-20 m^3/day, fitting two pumps has provided a safety net in the event of one pump breaking down. A steel gantry was fitted above each well to facilitate lifting of the pump, and training days held for 8-10 members of each community at the time of pump installation when the necessary tools and repair manual were also donated. All schemes were registered with DDF in order to benefit from the third tier of their maintenance system that provides assistance in cases of emergency.

This approach has worked well, and its success and popularity are illustrated by the continuing ability and willingness of the project communities to repair not only their own pumps but also those of neighbouring communities. At one site (Mawadze), members have gone to an extent of purchasing spare parts in anticipation of future breakdowns. Use of a motorised pump is not advisable. It has higher costs and difficulties of repair that can be beyond the means of poor rural villages. It also lacks the control against ground water depletion inherent in the use of hand pumps, an important consideration in these dry areas.

2.1.5.2.5 Community ownership

It was found helpful upon completion of each scheme to hold an official opening ceremony and handover. This event can be attended by local and foreign dignitaries as appropriate. It helps to seal the sense of ownership felt by the communities, it allows the community to show with pride what they have already achieved in their garden, and it is an appropriate for a celebration. Community participation in development of small-scale irrigation using groundwater also provides the ideal springboard to other community projects, especially in health, agriculture and the environment. Community workers in the area are now able to advice on nutrition at the new garden. Agricultural extension staff and NGO's are providing advice on agronomy and garden management and are promoting low water use methods of irrigation. Perhaps most exiting is the opportunity this type of first water development project creates to initiate community-based management of soil and water resources in each small catchment considered. In this way, the people themselves become responsible for recharge to their well and the life if their scheme, and in so doing they also begin to address the problems of environmental degradation often too daunting when considered on a larger scale.

2.1.6 Technical and economic viability of collector wells

2.1.6.1 Background

Appraisal of the first collector well garden completed in Romwe catchment in 1991 gave an indication of the economic viability of small-scale vegetable production when water for irrigation could be made available to communities in this dry region (Brown and Dube 1992).

2.1.6.2 Collector Well Performance

The monitoring confirms that the peak water required to irrigate a 0.5 ha garden and satisfy local domestic need is presently 14-15 m^3/day. However, the initial scheme at Romwe also demonstrates that a viable (if somewhat smaller) community garden is possible using 10-11 m^3/day. Over the three-year period 1993-95, a slight net fall in the well levels at sites 1-6 can be detected. However, the process of ground water recharge in the garden depends heavily on the pattern of rainfall distribution during any year, and results of the complementary study of groundwater recharge on going in Romwe Catchment (Butterworth et al, 1995) indicate that 1994/95 was particularly a poor year. Recharge in this catchment during 1994/95 (total rainfall 735 mm) was very low due and even distribution of rainfall and significantly lower than during 1993/94 (total rainfall 661 mm) when consecutive heavy rains fell early in the season. It is encouraging therefore that all collector wells have provided a continuous supply of water even through years of low recharge, but it is also clearly important to monitor recharge and well performance over the longer-term.

2.1.6.3 Effect of Lateral Drilling

At the ninth site (Machoka), pumping tests of the large diameter well prior to laterals showed that it could satisfy the required target of 15 m^3 of water per day without drilling. Only small improvement in well yield by lateral drilling was recorded at site 2 (Gokota) and at Romwe. Despite this, the large diameter wells at these two sites have since proved able to support viable community gardens. At the other sites, lateral drilling improved well yield significantly. On basement rock for which the technique was originally designed, lateral drilling in this project has improved well yields by an average of 38 per cent. Where successful, it has the effect of reducing drawdown and increasing rate of recovery following abstraction from the well. In real terms, this improvement has converted relatively low yielding large diameter wells at site 1 and 6 into collector wells that now satisfy the required target of 15 m^3/day, and at site 3, 4 and 5, has converted satisfactory large diameter wells into relatively high yielding collector wells that could now support irrigated gardens larger than 0.5 ha.

2.1.6.4 Comparison with Alternative Well Designs

Four other well designs are currently used to abstract groundwater in Zimbabwe and elsewhere and may be compared with collector wells.

2.1.6.4.1 Traditional dug wells (TDW)

In Africa, the traditional source of groundwater is a dug well. Generally, these are dug by hand using chisels and picks and a bucket and a bucket and windlass to remove the spoil but without the aid of a dewatering pump. They are usually 0.8-1.2 m in diameter. The level of abstraction sustained is generally low but can be sufficient to supply a family with water for domestic use and small garden. Finally, depth of well is controlled primarily by the degree of weathering in the upper layers and by the water-table. The well cannot be dug much below this level without the aid of a dewatering pump and often will be deepened progressively during periods of low recharge as the water-table falls. With the occurrence of several dry years and increasing demands of a rising population, programmes of well deepening (often using explosive into the hard rock) are being undertaken in Zimbabwe. These so called ‘deep wells’ are included under this category.

2.1.6.4.2 Large diameter wells (LDW)

Like a traditional well, a large diameter well is constructed in weathered zone of basement rock. Again, it is dug by hand but with the use of hand tools that can include a pneumatic jack hammer and dewatering pump to allow construction below water table. It is lined either with rocks, bricks, concrete or steel and usually 2-3 m in diameter. Final depth is controlled by weathering and the required volume of water, digging usually stopping when fresh rock is encountered. Abstraction sustained can be higher than traditional dug well due to the greater volume of storage provided and the greater wetted surface area giving increased inflow (figure 8) at present large diameter wells are not common in Zimbabwe. But are important elsewhere, for example agro –wells in Sri Lanka (Shah, 1990). For the purpose of this comparison, large diameter wells tested at each site were those constructed prior to lateral drilling to create collector wells.

2.1.6.4.3 Conventional Deep Boreholes (CDB)

Conventional deep boreholes (or tube wells) of 150-200 mm diameter are the most widely constructed type of underground water abstraction point in southern Zimbabwe. Being mechanically drilled they are relatively easy to construct and, with modern air rigs, can be drilled in a relatively short time. Few in Zimbabwe are completed in the upper weathered zone and most penetrated the bedrock in search of fractures which yield water (figure 8). A depth of 50-70 m is typical. Where interconnected fractures occur, they allow deep conventional boreholes to draw on a higher storativity of the regolith. If major water bearing fissures are intercepted and produce locally higher transmissivity, high yields can be achieved. However, the pattern of fracturing in crystalline bedrock is highly variable and not easy to predict. Despite use of geophysical siting techniques, many boreholes do not intercept fracture systems sufficient to certify either the volume of water or longevity of supply required, and poor drilling success rates reflect this difficulty. For the purpose of this comparison, tests were conducted on existing conventional deep boreholes at each site where these existed.

2.1.6.4.4 Screened Regolith Boreholes (SRB)

Generally, the weathered zone of conventional deep boreholes is cased rather than screened. This reduces costs and eases construction but denies access to water from the upper aquifer. Where the upper aquifer is productive, screening is possible. For the purposes of this comparison, exploratory boreholes used to site collector wells were screened in the regolith and deepened into the bedrock to create a screened regolith borehole and allow the aggregate transmissivity of regolith and bedrock to be determined.

2.1.6.5 Factors that Influence Choice of Well Design

A number of factors influence choice of well design and should be considered during a groundwater development programme.

2.1.6.5.1 Geology and rainfall

Two principal physical factors influencing choice of well design are geology and rainfall. These determine the relative positions of the base of the weathered zone and the water table (Chilton and Foster, 1995). Geology and rainfall may also influence groundwater quality. Where the regolith is very thin or absent or the water table are deep, dug well construction may be virtually impossible and boreholes may be the only option. If the regolith is thick and water table shallow, dug wells are likely are likely to be visible.

2.1.6.5.2 Community preference and sense of ownership

Ensuring local involvement in designing, implementing and managing water projects bring the greatest chance of success (Carter, 1989; Underhill, 1990). Sense of ownership and responsibility for upkeep of a water point is more likely if the well is dug by the community rather than drilled by an external agency.

2.1.6.5.3 Pump capacity and easy of repair

Rural farmers owning traditional wells generally use a bucket and windlass to abstract water. Communal wells however are often fitted with a pump. The nature of the rural environment prohibits use of high technology pumps. A source of power is rarely available or within people's means and the necessary tools or skills are not available to repair such equipment. Simple hand-powered single action reciprocating pumps (eg. Zimbabwe bush pump) are thus most often used. These pumps are typically able to supply 5-10 m3/day. A borehole will only allow one such pump to be fitted. This has the added advantage of less wear and tear per unit and when one pump breaks water may still be abstracted by another. Community-based maintenance of the pumps is the ultimate goal. Irrespective of well design, this requires training of local people and provision of tools, but is made easier on a dug well than on a borehole due to the shallow depth and corresponding lighter pumping unit to be extracted.

2.1.6.5.4 Well performance and cost-effectiveness

Ultimately, the type of well-chosen must provide the volume of water required, be sustainable, and be cost-effective. It was with the aim of better understanding this aspect of well selection for use with small-scale irrigation that the following comparison of well designs was undertaken.

2.1.6.6 Well Comparison Methodology

Table 2.1.2 indicates the well types either existing or constructed that were tested at each site.

Table 2.1.1 Geology, annual rainfall (mm) and depth of well (m) either constructed or existing at each site. (Length of laterals (m) are shown where drilled)

Site	1	2	3	4	5	6	7	8
Geology	gneiss	gneiss	gneiss	granite	gneiss	gneiss	basalt	basalt
Rainfall	780	790	780	780	820	785	580	580
LDW	15.8	15	15	15	13	9.5	8.76	18
CW	15.8 (1x15, 4x30)	15 (4x30)	15 (8,9,25,27, 28)	15 (1x18, 3x30)	13(14,16,28, 30)	9.5 (2,4,8,23,30)	8.76 (none)	18 (9,19,27,27)
TDW	-	-	-	12.75	-	-	-	-
SEB	15	15	-	15	15	15	-	-
SRB	40	30	-	-	33	-	18	30
CDB	48	-	43	-	39	-	-	-

LDW=large diameter well CW= collector well

TDW= traditional dug well SEB= shallow exploratory borehole

SRB= screened regolith borehole CDB= conventional deep borehole

2.1.6.6.1Pumping test procedure

Pumping tests listed in table 2.1.2 were conducted on the large diameter well at all sites prior to lateral drilling and at seven sites on the collector well after lateral drilling. Additional tests were also completed on the best shallow exploratory borehole drilled during well siting, on the same borehole deepened to 40 m and screened in the regolith, and on the nearest conventional deep borehole and traditional dug well where present. Test results were analysed to obtain values for transimissivity and storativity at each site for both the shallow and deep aquifers. Values of available drawdown were used to model the abstraction that could be sustained by each well type at each site for a period of one dry season.

Table 2.1.2 Pumping-tests performed at each site

Well type	Pumping rate (litre/sec)	Pumping time (mins)	Site
LDW+CW low discharge	0.65	120	1 + 2
LDW + CW high discharge	4.5	120	1 + 2
	2.65	240	3 + 4
LDW + CW medium discharge	1	300	5 + 6 + 8 (no CW at 7)
SEB	0.4	60	1,2,4,5,6,7
SRB	0.6	240	1,2,5,8
CDB	0.6	60	1,3,4,5

2.1.6.6.2 Pumping test analysis

Analysis was performed using program BGSPT (Barker, 1989). The package consists of two parts: PTFIT which analyses pumping-test data; and PTSIM which simulates well drawdown using specified well and aquifer parameters and which can be used to estimate sustainable yield of the well. PTFIT accepts ranges of aquifer parameters as input and optimises the fit of the pumping tests data to the built-in conceptual model of the aquifer by verifying the parameters within these ranges. The conceptual model is that of a fully penetrating well in a confined homogeneous aquifer of infinite extent. Flow within the aquifer is assumed horizontal and within the aquitard, vertical. The water table in the aquitard was assumed to be horizontal. The well is defined by the radius of the casing Rc and the screen section of the well Rw. During this project, the program was developed to allow the radii to be varied along with other parameters. This allowed increase in effective radius, Rw, of the large diameter well due to laterals to be estimated.

Complexities of the basement aquifer

In a pumping test, drawdown during abstraction and subsequent recovery are monitored. By fitting this this response to a mathematical model, properties of the aquifer in the vicinity of the well can be estimated. The model is based on a simplification of the aquifer. Crystalline basement rocks form complex aquifer. Regolith and bedrock aquifer are different in form. The first varies both vertically and laterally. Often its composition changes from being sandy on watersheds to clayey in valley bottoms. Derived from *in situ* weathering of parent rock, permeability tends to increase with depth as the proportion of secondary clay minerals reduces. The lowest part of the regolith may be the most permeable and productive part (Wright et al., 1989; Chilton and Foster, 1995), and the collector well is targeted at this zone. Bedrock permeability in contrast depends on the connectivity of fractures. The boundary between the two aquifers is not usually sharp. The form of aquifer usually changes from site to site being dependent on lithology and geomorphological history. Even at one location lateral variation can have hydraulic significance where; for example, hard bands in the rock form barriers to ground flow.

This complex nature means that it is difficult to describe crystalline basement aquifers with a mathematical model. There may not be a single set of parameters that define the hydraulic system. To assume the aquifer to be confined in this study appeared justified at most sites, by observation that depth of first water strike was below the final rest water-level. It is unlikely, however, that the weathered aquifer remained confined throughout all tests. To assume all wells to be fully penetrating was reasonable, as they were dug to the base of the weathered zone, as digging further into unweathered rock is slow and difficult. However, the aquifers do not extend infinitely and drawdown was not always small compared to saturated thickness. Rarely also were absolute rest water level conditions possible prior to tests. The wells were dewatered to allow construction and the time required for full recovery would have affected progress overall and denied access to water by the local people. Compensation was made in the analysis where this was important. In some cases, fracturing of the aquifer in the vicinity of the well may have increased storage as the water-level did not decline as fast as expected. In other low yielding wells, a seepage face may have developed as drawdown was greater than expected.

Consequently, a number of difficulties were encountered in the analysis of the data obtained from the pumping tests. In some cases, data could not be fitted to the model without allowing input parameter ranges to be set at unrealistic values. In particular the fitting of storativity was problematic and values were obtained that were higher than would be expected even for unconfined aquifers (eg at site 6). Despite these difficulties, a level of confidence in the results was possible, based on the sum of errors of the modelling fit, consistency of aquifer properties determined by testing different wells at one site, predicted versus measured well behaviour for those wells that were monitored (eg figure 2.1.6), and the history and performance of all wells as reported by the communities. This degree of confidence was assigned a level ranging from 1 to 5, most confidence being represented by the higher values. The results presented below should be considered indicative rather than absolute, but are believed to be a true reflection of the relative performance of each well type tested and indicate the relative magnitude of the sustainable yield that might be expected from each.

2.1.6.7 Aquifer Properties at Project Sites

Table 2.1.4 shows aquifer properties estimated at each site using pumping test results for the various wells. In the shallow aquifer, transimissivity values at all sites are consistent with previous work in the regolith in southern Zimbabwe (Wright, 1992) and elsewhere in Africa (Howard and Karundu, 1992; Chilton and Foster, 1995). Sites 3-6 have somewhat higher transmissivities than site 1 and 2, and could therefore be the more favourable sites for the construction of a dug well or screened regolith borehole.

Table 2.1.4. Aquifer properties of transmissivity (T) and storativity (S) determined at each site

	SHALLOW AQUIFER					DEEP AQUIFER				
Site	Depth (m)	T (m3/d)	S	Conf	Source of data	Depth (m)	T (m3/d)	S	Conf	Source of data
Romwe	12	1.1	0.520	2	LDW	-	-	-	-	-
1	15	0.8	0.005	4	LDW	48 40	32.0 4.48	2e-6 5e-3	4 4	CDB SRB
2	15	1.4	0.008	4	LDW	30	2.4	8e-3	4	SRB
3	15	2.9	0.009	1	LDW	43	118.0	7e-3	4	CDB
4	15	2.9	0.010	3	SEB	25	0.9	1e-2	2	CDB
5	14	3.1	0.007	3	LDW	33 33	5.6 0.8	9e-3 7e-3	3 2	SRB CDB
6	10	2.5	0.077	4	LDW	-	-	-	-	-
7	9	30.2	0.565	3	LDW	18	206	2e-3	3	SRB
8	18	9.8	0.004	3	LDW	30	9.8	1e-6	2	SRB

Transmissivity in the deeper aquifer is highly variable, but the storativity is consistently smaller. Tests on the conventional deep boreholes, which are cased through the regolith, are therefore indicative of bedrock productivity. The bedrock can be highly transmissive where these boreholes intercept significant, interconnected fractures (site 1 and 3), and very high yields may be obtained. Where fractures are not encountered, the bedrock has very low transmissivity (site 2, 4 and 5) and poor productive borehole result. There is a noticeable difference at site 1 between bedrock transmissivity at the existing borehole and at the exploratory borehole, further indicating the high spatial variability of this property.

The screened regolith boreholes (Table 2.1.4) are open in both shallow and deep aquifer, and the test results provided an aggregate of the Transmissivity provided by both. At some sites, yields from the screened regolith boreholes drilled by the project as part of site exploration were higher than those existing conventional deep boreholes. This is important, as it suggests that: a) siting deep conventional borehole by shallow exploratory drilling in the regolith to identify the greatest depth of saturated weathering may be a reasonable approach for helping to overcome the spatial variability found in the deeper aquifer; and b) boreholes in this region would generally be better if screened in the regolith rather than cased. A similar conclusion that greater and more economical use could be made of regolith aquifer where sufficient saturated thickness exists was made for Malawi (Chilton and Smith-Carington, 1984).

2.1.6.8 Predicted Wells of Sustainable Abstraction

To compare performance of the alternative well types at each site, measured aquifer properties and well dimensions were input to the PTSIM program. This program uses the same mathematical model as PTFIT described earlier. The model was made to simulate a cycle of daily abstraction, pumping at 1.5 m^3/hour (an arbitrary but typical rate) for five hours in the morning (06:00 to 11:00) and for five hours in the afternoon (13:00 to 18:00) and repeated for 240 days, the average length of a dry season. As drawdown is proportional to pumping rate in PTSIM, the maximum abstraction rate was evaluated as the simulated rate (15 m^3/day) multiplied by the ratio of maximum available drawdown to simulate drawdown. Available drawdown was taken to be the inlet depth of the pump in the case of wells and two-thirds the depth of deep boreholes, a level of drawdown considered to give optimum abstraction for a borehole (Bromley, pers.comm.) Table 2.1.5 shows the results.

Table 2.1.5. Maximum sustainable yield of wells tested at each site for a simulated dry period 240 days

	LDW		CW			TWD		CBD Cased in regolith		CRB SCREENED REGOLITH		BED Cased in regolith	
Site	Q	CL	Q	CL	%Imp	Q	CL	Q	CL	Q	CL	Q	CL
Romwe10.2	10.2	3	10.8	3	6	1.6	2	-	-	-	-	-	-
1	11.5	4	18.0	4	56	-	-	200	2	43	2	35	2
2	16.7	4	17.1	4	2	-	-	NA	-	26	2	8	2
3	26.1	3	34.1	3	31	-	-	769	2	-	-	-	-
4	24.1	3	40.2	3	67	2.1	3	2	2	-	-	-	-
5	22.3	3	34.7	2	56	2.9	2	1.4	2	37	3	25	3
6	12.4	4	18.3	4	48	-	-	-	-	-	-	-	-
7	47.0	3	NA	NA	NA	-	-	-	-	407	2	347	2
8	55.3	4	62.5	4	13	-	-	-	-	31	2	<2	2

DEB = Deep exploratory boreholes (comparable to conventional deep borehole but sited by exploratory drilling

At no site did a traditional well satisfy the target of 15^3 per day. This is due in part to the smaller diameter and shallow depth typical of this well but may also be because the shallow well is dug into the less permeable, clay-rich part of the regolith weathering sequence. It must also be due in part to the method of siting employed. A communal farmer must site his well on his own land, ideally in a position convenient to house of garden. He may employ a diviner but he will still operate under these restrictions. In contrast, project wells were sited to serve the community with few restrictions on area and with the considerable benefit of exploratory drilling to locate preferred aquifer properties prior to digging. It is likely that a well of traditional diameter sited in this way would be better.

At six of the nine sites, the large diameter well is shown to supply the required volume of water. This ratio improves to eight of the nine sites with lateral drilling. Aquifer heterogeneity caused the success of lateral drilling to vary from site to site. At Romwe and Gokota improvement was minimal but at other sites it was significant, giving an average improvement on basement rock of 38%. This improvement can be thought of in two ways. Radials can increase maximum sustainable yield for the period of a dry season, or radials can increase the period for which a particular abstraction rate can be sustained. At site 1 for example, the large diameter well would sustain pumping at 11.5 m^3 per day for 240 days. The collector well could last for 300 days. This has important implications for villagers trying to water their vegetables throughout a period of low recharge or an extended dry season, and care must be taken when placing a value on this additional water. Any increase in yield made possible by lateral drilling cannot simply be divided by the additional cost of this operation.

On basement rock, the yield of large diameter wells, collector well and screened regolith bores sited by exploratory drilling, although lower than yields of conventional deep boreholes at some sites, are shown to be far more consistent and quite adequate for small –scale vegetable production. Site 7 and 8, however, highlight the hole that geology can play in well type selection. Unlike weathered basement, the shallow weathered zone of basalt is not marked. Instead, groundwater flow is concentrated in zones of horizontal sheet jointing, and these zones, both shallow and deep, can be highly trans-missive (Table 2.1.4). Both large diameter wells and exploratory boreholes gave high yields but well digging was made particularly difficult by the compact layers of rock. Consequently, although either well type would suffice, a screened regolith borehole is preferable in basalt due to the far greater ease of construction.

2.1.7 Environmental appraisal of collector well gardens

2.1.7.1 Potential of the Groundwater Resource

2.1.7.1.1 Extent of groundwater resource

Crystalline basement aquifers, which are present over much of Africa, provide a limited but extensive resource which has the potential for greater use. A review of the extent and availability of these groundwater resources in Zimbabwe and other countries of southern and eastern Africa can be found in Anon (1989). Over the last decade, interest in shallow water in the crystalline basement regolith has gained considerable momentum with the recognition that the regolith may provide a more sustainable and less costly because of rural water supplies than the underlying bedrock fractures which have traditional been exploited (Howard et al 1994). Despite increasing interest, many basic questions relating to the development and sustainability of shallow basement aquifers have been answered in part.

In principle a sustainable level of abstraction from an aquifer is one that does not exceed the recharge to groundwater. The exception of this is if the abstraction is very small compared to the storage of the aquifer, in which case the impact of the abstraction will be negligible, at least in the short term. Failure to match abstraction to recharge will result in a non-sustainable well yield and will permanently reduce ground water level with consequent effects throughout the whole ecosystem. Even abstraction that are equal to or less than recharge can have a deleterious impact; groundwater levels will not rise as much in response to seasonal rains and the flow of ephemeral rivers fed by ground water during parts of the year will be reduced in volume or become extinct. Such rivers may be important sources of water for settlements and will certainly have an ecosystem dependent upon them.

2.1.7.1.2 Groundwater recharge in south-east Zimbabwe

Houston (1989) estimated recharge for the Masvingo province of Zimbabwe using three independent methods; base flow analysis, environmental tracer (chloride) and soil moisture budget (based on a monthly recharge-run odd model). Data from 22 rain gauges and 3 class 'Á' evaporation pans were available for an area of 22000 km^2 roughly centred on Masvingo. Flow data were available from the Chiredzi, Mzero, Musokwesi and Lundi rivers for periods ranging from 8 to 17 years. Houston estimated recharge throughout the Masvingo province to be 2-5% of rainfall, a range which was shown by all three methods. Houston further suggested that recharge was very dependent on annual average rainfall with low rainfall areas likely to have a recharge of 2% of annual rainfall. Based on an area receiving 500 mm annual average rainfall and a recharge rate of 2%, Houston calculated that an area of 65 ha was required to sustain a hand pump delivering 6500 m^3/annum.

Meigh (1988) made a first attempt at estimating recharge for the Lowveld region using two independent techniques based on a soil moisture budget model and two empirical models. The soil moisture model was a simplification of the Houston (1988) model but using meteorological data from the Lowveld Research Station and the airport at Buffalo Range. The empirical approach involved relationships developed for a number of catchments between annual average rainfall and runoff and base flow index (BFI) and soil types (Bullock, 1988). A second empirical approach used relationships for annual average rainfall and runoff and annual average rainfall and BFI developed by Maigh (1987). These methods gave recharge estimates of 0.5-6%, a range that encompasses that of Houston (1988).

2.1.7.1.3 Groundwater recharge in other African basement areas

Wright (1992) reported estimates of recharge calculated by a number of methods for areas of Zimbabwe that generally have a higher rainfall than south-east Zimbabwe. Recharge calculations using base flow analysis demonstrated the importance of dambos and relief on base flow and hence on recharge. These relationships are also discussed by Farquharson and Bullock (1992). With rainfall in excess of 800 mm, ground water recharge was estimated by Wright (1992) to be typical in the range 10-20% of mean annual rainfall which he considers to be substantially in excess of the demand of rural populations (1-3 mm) estimates of recharge using chloride balance, again for an area with rainfall higher than normally experienced in south-east Zimbabwe was 9-14% of mean rainfall.

Howard and Karundu (1992) estimated recharge for basement aquifer in south-west Uganda. Annual rainfall in this area is normally in the range 750-1000 mm. using daily water balance techniques mean recharge was estimated at 30 mm (3-4% mean annual rainfall) over an eight-year period. However, median recharge for the same period was estimated at 17 mm (2% mean annual rainfall) and this value was considered to be more representative of annual recharge in an area. Howard et al (1994) used water balance calculations and information from aerial photographs and LANDSAT images to assess the influence of land use change on groundwater recharge in south-west Uganda. Recharge estimates for the period 1954 to 1961 were found to be just over half those for the period 1988-91. The difference was attributed to large-scale deforestation during the intervening 30 years which has increased recharge by causing significant reductions in rates of actual evaporation.

2.1.8. Effects of Land Management on Groundwater Recharge

Wright (1992 explains that "*the basement aquifer of this region (S.E. Zimbabwe) are distinctive in that their occurrence and characteristics are largely a consequence of the interaction of weathering processes related to recharge and groundwater through flow. A close relationship exists therefore between groundwater occurrence and relief, surface water hydrology, soil and vegetation cover. Recharge is sensitive to certain land use changes, notably those associated with desertification. Improvements in the understanding of these relationships will be fundamental to the management and planning of groundwater resources in the crystalline basement terrain.*"

The investigation of the relationships between hydrology phenomena and land use has normally been carried out within small experimental drainage basins or catchments. Results from such catchment studies, carried out in eastern and southern Africa, have been reviewed by Whitlow (1993). These catchment studies have proved invaluable particularly in providing information on the impacts of afforestation and deforestation in sparsely populated upland areas. However, the critical water and soil resource problems throughout much of Africa occur in catchments that are heavily populated by subsistence farmers and their livestock (Pereira, 1961). With few exceptions, there is a paucity of data from such catchments (Whitlow, 1983).

In response to the growing populations, an increasing area of land is being taken over for crop production throughout Africa (Whitlow, 1983). Declining productivity of existing croplands has also promoted extension of cultivation often on to more marginal terrain better suited to other less intensive forms of land use. Extension of cropland is often at the expense of grazing lands, thus placing even greater pressures on the remaining grazing land (Whitlow and Campbell, 1989). The net result is reduction in vegetation cover, land degradation and changes to the water balance. The mechanisms by which land degradation can influence different components of the water balance have been discussed by Whitlow (1983). Wallace (1994) provides an alternative framework for discussing the interactions between hydrological processes and land degradation.

In 1992 work started on a project In the Romwe Catchment in South-east Zimbabwe that is studying the influence of land use and land management on groundwater recharge (Butterworth et al., 1995). This project is currently being expanded to become a long-term assessment of the physical and socio-economic benefits of taking an integrated approach to community management of resources in semi-arid areas.

2.1.9 Ground Water Development and Dry River Flows

Investigation of the factors which determine dry season river flow reveals a dominant rainfall control, reflecting and relationship between groundwater base flow and rainfall (Farquharson and Bullock, 1992). Ephemeral rivers, such as many of the sand rivers of southern Africa, occur when the water table in local aquifers rise to the surface as a result of seasonal rains. It is clear therefore that abstraction of water from the aquifers may affect both the duration and flow rate of rivers during dry seasons. In south-east Zimbabwe, dry season river flows are very important to the ecology and biodiversity of the region.

Well abstraction levels in an area should not exceed those that will be detrimental to regional ground water levels. It is possible that widespread use of wells for increased domestic consumption and garden irrigation may reduce water tables in the dry season to levels that cannot be recovered in the wet season by anything less than exceptionally high rainfall. Thus, the local population and ecosystems that rely on the sand rivers will find that their water supply is more unreliable. The users of ground water will benefit at the expense of the river users with perhaps little net water resource benefit to the population of the whole water catchment. However, this scenario is considered unlikely given that abstraction levels are very low when compared to natural groundwater flow away from the area.

The quantification of the effects of groundwater abstraction on river flows is complicated being a combination of aquifer recharge, aquifer properties (transmissivity and storage coefficient) and location and pumping rate of the wells. This complexity requires that mathematical models of the aquifer-river system be constructed in order that these interactions are accounted for fully. Such approaches have been attempted. Oakes and Wilkinson (1972), for example, developed a model of a hypothetical aquifer-stream system and investigated the effects on stream flows for a range of abstraction regimes assuming different aquifer properties. A similar approach but based on a calibrated model of a specific catchment was undertaken by Clausen et al, (1993). Both studies showed similar results, in particular, that in the simplest case of a constant abstraction once equilibrium is reached, the river base flow is reduced by an amount equal to the pumping rate. The time to reach equilibrium depends on the properties, a small aquifer with a high conductivity and a well close to the stream will reach equilibrium quickly, whereas the other extreme set of values for these factors will result in equilibrium taking several years to be achieved. Thus, under intermittent abstraction regime the effect on the rivers is less easy to characterize simply. Oakes and Wilkinson (1972) showed that the effects of continuous abstraction could be diminished by the sequential operation of two groups of wells at different distances from the river, the nearer group being pumped in winter and the farther in summer.

It is clear that collector wells are not a universal solution when it comes to developing limited groundwater resources in semi-arid areas. In many circumstances screened regolith boreholes or large diameter wells sited by exploratory drilling are a more cost-effective option than collector wells. However, improved siting of these wells will enable more efficient and widespread development of groundwater from basement aquifer with certain characteristics.

Increased use of these wells coupled with an increase in community gardening will at some stage have a quantifiable influence on the regional hydrology of south-east Zimbabwe. It is anticipated that research being carried out in the Romwe Catchment and related modelling studies will provide information on the influence of wide spread introduction of groundwater-based gardens in basement areas. It is anticipated also that research being carried out in the Romwe Catchment will provide information on approaches to catchment management that are sustainable in terms of utilization of resources both within the catchment and in areas outside the catchment that are influenced by activities within the catchment (Butteworth et al, 1995; Macdonald et al, 1995)

2.1.10 Effects of Community Gardens on Groundwater Quality

2.1.10.1 Background

Environmental problems associated with the intensification of agriculture have been apparent in the northern hemisphere for some twenty years or so. The link between a combination of expanding cultivated areas and increased fertiliser use and nitrate concentration in groundwater has been extensively researched in both Europe and North America. Expensive measures to control nitrate pollution from agricultural sources are being implemented. More recently, pesticides originating from intensive cultivation have been identified in groundwater used for potable supply in Europe and North America, also requiring expensive measures to meet drinking water standards. While much of the work in temperate zones relates to cultivation without irrigation, nevertheless the same principles can be expected to supply to irrigation agriculture in tropical regions.

Concern has been expressed about the potential impact on groundwater quality from the intensification of agriculture in tropical regions. This impact could be felt in relation to nutrients, pesticides. However, such less quality monitoring information exists from which the overall situation with respect to the impact of agriculture on groundwater quality in tropical regions can be appraised. A recent study by BGS under the ODA TDR programme indicated that, for the most vulnerable aquifers, intensive agriculture can produce serious determination in groundwater quality, resulting principally from the leaching of nitrogen fertilisers.

Pesticide use has also produced major benefits for agriculture in reducing pest and disease attacks and weed competition. As for fertilisers, the increase in usage is now more rapid in some developing countries than in the developed world. Awareness of the leaching of nitrate has led to concern about the possible leaching of pesticides. There is now evidence of significant pesticide occurrences in groundwater in agricultural areas of Europe and North America, but hardly any routine monitoring of pesticides is carried out in developing countries. This is because of the sophisticated and costly analytical procedures, the wide range of compounds in common use the care required in sampling (Chilton et al, 1995).

2.1.10.2 Pollution vulnerability of the basement aquifer

The risk of the aquifer becoming polluted depends on the interaction between the natural vulnerability of the aquifer and the pollution load imposed on it as a result of human activity. The pollution loading can be controlled or the source of pollution removed. The vulnerability of the aquifer cannot be changed; it is determined by the intrinsic properties of the material of the aquifer and the strata above it. An assessment of pollution vulnerability thus requires knowledge of these characteristics which determine the likelihood of the aquifer being adversely affected by an imposed contaminant load. These characteristics are:

- The thickness and nature of the strata making up the unsaturated zone.
- The physico-adsorption-chemical attenuation capacity of these strata, resulting from chemical reactions, adsorption and biochemical degradation.

Aquifer and the material above them can be assessed on the basis of their overall lithology and permeability and the depth of groundwater, and many schemes have been developed for mapping aquifer vulnerability. Taking extreme cases, a fracture coral limestone aquifer with very thin soil cover on low-lying island with water table just below the ground surface would be highly vulnerable. A deep sandstone or limestone aquifer overlain by a significant thickness of confining clays would, in contrast, be relatively well protected from pollution.

Turning to the Basement regolith aquifer of southern Zimbabwe, although the water table is relatively shallow, within 5-10 m of the ground surface, both soils and aquifer material have low permeability. The almost ubiquitous rise in water level after striking water is an indication of the semi-confined nature of the regolith aquifer. Permeability is low, recharge is low, and unsaturated zone travel times are likely to be long, allowing adequate time for pollutant attenuation. The regolith aquifer is considered to have relatively low vulnerability to pollution from agricultural activities. Exceptions could occur, however, where local recharge reaches the water table more be significantly more vulnerable to pollution because of the thinner soils, different weathering patterns and greater potential for more rapid transport of pollutants to the water table through the jointed and fractured and fractured lavas.

2.1.10.3 Fertiliser and Pesticide Usage in Communal Gardens

Although nitrogen fertiliser usage is now increasing in some of the most rapidly developing countries at rates greater than in temperate countries (Chilton et al 1995), the use of such fertilisers in the communal lands in Zimbabwe remains low. This was confirmed by an informal survey of garden in Chivi Communal areas in 1991 (Lovell at al 1992). Many garden members could not afford to purchase chemical fertilisers, which were in any case not always available in rural stores. Most gardens applied animal manure if available, preferring goat manure to cattle manure and considering poultry manure to be even better. This is not surprising as, based on comparative figures for the Uk, poultry manure should have significantly higher nutrient content. Application rates of one or two 20 litres buckets of manure to each bed per crop (Lovell et al, 1992) could be equivalent to a nitrogen application of some 250 kg/ha. While this appears high, it is unlikely to be applied to all or even many of the beds in each garden. During the recent drought, for example, in the absence of cattle many garden members at Romwe used composted leaves as a substitute.

From the same survey it was clear that some cooperative gardens bought fertilisers, which were either used collectively or distributed amongst members. Thus, it was estimated (Lovell et al, 1992) that some 340 kg N/ha was applied to tomatoes and some 300 kg N /ha to cabbages. These seem somewhat high in relation to application to similar crops elsewhere. Huang et al (1994) record applications of 250 kg N/ha to tomatoes in Florida and 170 kg N/ha to cabbages in Texas. Application rates are similar in the UK for cabbages, an average of 180 kg N/ha, but perhaps half of that quoted above for tomatoes. Although these applications are made at present only to a relatively small proportion of the garden, they are nevertheless close to the well and the estimates therefore require further refinement.

Pesticides usage in communal areas gardens is likely to be limited both by local availability and by cost. Nevertheless, a range of compounds are in regular use, dominantly insecticides. A preliminary estimate of their susceptibility to leaching can be made from consideration of the physiochemical properties of the pesticides. Together with other evidence from other regions of their occurrences in groundwater, this can be used to select target compounds for monitoring, although there are few published studies for any of these compounds.

2.1.11 Climate Change and Community Gardens

In southern Zimbabwe, drought is recurrent and high inter-annual variation in the amount and distribution of rainfall occurs. Furthermore, there is evidence of cycle of rainfall above and below average that appear to occur in phases of about 11 years or so (Butterwoth et al 1995). Obviously, such changes in climate can have important impacts on both surface and groundwater resources, and upon the effectiveness and sustainability of water resource development projects.

Given our current state of knowledge, it is impossible to quantify the possible influences on regional climate of widespread introduction of groundwater-based community gardening in semi-arid areas. However, as more scientific information becomes available, it may be necessary to consider the influence on regional climate of land use and irrigation when developing natural resource management strategies. In particular, the recycling of moisture by vegetation across continents is now considered to be an important process (Savenije, 1994). This is the process by which evaporation falls as rainfall, re-evaporates and falls as rainfall many times as an air mass moves across or circulates within a continent. Current theory suggests that irrigation during the dry season does not enhance rainfall for lack of rainfall-bringing mechanisms. In the context of irrigation gardening, this would mean that if the total irrigated area became significantly large, irrigation during the wet season would have a higher environment benefit than irrigation during the dry season.

2.1.12 Conclusion

Poverty and the physical environment are linked. If people are living in areas of poor soils above difficult aquifers, they are likely to be poor and short of water. The corollary of this is that if poverty and shortage of water are used as the main criteria to select sites, prospective sites will include a high number that are technically difficult and some that are even impossible to develop. If, on the other hand, only physical and hydrogeological criteria are considered, prospective sites will include some where the communities are relatively well off.

The resources needed to develop schemes in difficult areas will be higher but the communities and environment that benefit are likely to be the most needy. Whether these costs can be met depends largely on the aims and objectives of the donor. This project has highlighted the need to identify selection criteria during the design and costing stage of a project. Any project that uses cost per well as the main criteria for site selection will either overrun on costs, implement less cost-effective well designs or end up assisting less needy rural communities.

Rainfall and geology are two principal Zimbabwe physical factors influencing choice of well design. In southern Zimbabwe (average annual rainfall 400-800 mm) collector wells have been found to be particularly well suited in areas of younger undifferentiated gneiss and younger intrusive granite, but less well suited in areas of Karoo basalt. Collector wells could not be sited in areas of older gneiss complex because either depth of saturated regolith was insufficient or depth to water table exceeded 15 m. Collector wells were also not sited in Beitbridge paragneiss in this project due to the high incidence of saline ground water encountered. However this geology underlies some of the neediest communal lands in the region, and it is recommended that the potential to use collector well laterals to skim fresh water from the upper layers of these aquifers should be tested in a small, independent R&D project.

Other African countries having similar or even more favourable, climatic and hydrogeological conditions to those of southern Zimbabwe, and countries where collector wells should be particularly well suited include Malawi, Uganda, Tanzania and parts of Ethiopia.

Basement complex aquifers in southern Zimbabwe have been found to be highly discontinuous and variable over distances of only a few meters. Consequently, when searching for water sufficient to support domestic use and community gardening in this region, it is important to identify a prospective area (4-10 km^2) rather than specific location. In southern Zimbabwe, this can usefully correspond with a Vidco (village development committee) comprising of several kraals or villages.

It is important to acknowledge the vital role of exploratory drilling in locating optimum sites for water points in this region. Present geophysical methods do not have sufficient resolution to locate optimum well sites in terrain of such high spatial variability.

At all project sites, a consistently adequate supply of water for domestic use and small-scale vegetable production has been obtained from the shallow regolith aquifer, both by large diameter wells (17.6 m^3/day, on average) and collector wells (25.9 m^3/day on average). The increase in sustainable yield by lateral drilling is variable but on average is significant at 38 per cent improvement on basement aquifers.

Some screened regolith boreholes drilled by the project gave higher yields than yields of existing conventional deep boreholes. This is important, as it suggests that: a) siting deep boreholes by shallow exploratory drilling in the regolith to identify the greatest depth of saturated weathering may be a reasonable approach for helping to overcome the spatial variability found in the deeper aquifer, and b) boreholes in this region would generally be better if screened in the regolith rather than cased.

The challenge for future groundwater resource development in the communal lands of southern Zimbabwe is suggested to lie in quick and cost-effective identification of the appropriate well design and location within a given area. Social factors should also be considered during selection of well design. The sense of ownership that a community achieves from digging their own well increases the likelihood of their assuming responsibility for maintenance and management of the water point. The ability to attach several pumps to a well is also a major advantage compared to a borehole. It provides a safety net in the case of one pump breaking, and the relatively shallow depth reduces wear and tear on pump parts and effort required both to lift water and lift the pumps during repair.

Although variable in space and time, current use of groundwater in the communal lands of south-east Zimbabwe is only about 4% of the average annual recharge, which is of the order 2-5% of annual rainfall. Increased use to develop community gardens thus appears viable if the available aquifers can be appropriately developed and managed. Land use and management have a significant influence on hydrology processes including groundwater recharge. Consideration should be given to land use and management in the vicinity of groundwater-based gardens to protect recharge and ensure physical sustainability of the schemes.

It seems likely that increased use of groundwater for community gardens will have an influence on dry-season river flows, however, the importance of the influence is difficult to quantify given data currently available. Data and modelling work from catchment studies such as Romwe should provide insight and identify methods to minimise this effect. Average abstraction recorded per scheme is 3,200 m^3/year (8.76 m^3/day or 81 gallons/hour). This is very small compared to abstraction rates of 10,000gallons per hour often quoted by commercial farmers irrigating commercial crops from high yielding boreholes in this region. Any environmental consequences of developing community gardens using groundwater should be considered in this context.

Similarly, deforestation, overgrazing and the clearing of marginal land for rainfed cropping have had, and continue to have, major detrimental effects upon catchment hydrology and dry-season river flows in the region. In contrast, the development of community gardens using groundwater has little risk of negative consequences but may have far-reaching environmental benefits by helping to alleviate some of the pressures that presently cause these problems. Concern has been expressed about the potential impact on groundwater quality of the intensification of irrigated agriculture in tropical regions. The regolith aquifer of basement rocks is considered to be relatively vulnerable to pollution because of the shallow water table, but the weathered material forming the aquifer contains a significant clay fraction and has low permeability. As yet fertiliser and pesticides usage in community gardens in southern Zimbabwe is limited, and the potential threat to groundwater quality is low. This could change if agro-chemical usage increased with time, and a simple, annual programme of groundwater quality sampling is proposed.

There remains a critical need to ensure that any new investment in water and garden facilities does not fail through failure to protect and monitor the groundwater resource. Community participation in the development of small-scale irrigation using groundwater provides the ideal springboard to other community projects, especially in health, agriculture and the environment. Future development of wells and gardens should proceed hand-in-hand with programmes of community-based catchment management and monitoring, building on the opportunity and incentive that initial water development project provides.

2.1.13 References

1. Abel and Blaikie (1989). Land degradation, stocking rates and conservation policies in the communal range lands of Botswana and Zimbabwe. Land Degradation and Rehabilitation, 1, 101-123.

2. Anon (1992). Presidents report on the Drought of 1991-92. Min. Public Information, Harare.

3. Anon. (1994). Research and development of collector well systems: Report of an evaluation study. ODA Eng Div, ODA, London. 36pp

4. Barker, J. (1989). Programs to simulate and analyse pumping tests in large diameter wells. Technical Report WD/89/24 (updated 1994), British Geological Survey, Wallingford, UK.

5. Batchelor, C.H., Foster, M.W., Murata, M., Gunston, H. and Bell. J.P. (1990). Development of small-scale irrigation using limited groundwater resources: First Interim report. IH report ODA 3/90, Institute of Hydrology, Wallingford, UK.

6. BDDA (1992). Small scale irrigation using collector wells: Pilot Project-Zimbabwe, Project Data Sheet, ODA, Lilongwe, Malawi, 30pp

7. Brown M.W. and Dube, T. (1992). Tamwa, Sihambe and Dhobhani Kraals collector well garden, 1991-92: an evaluation survey. Lowveld Research Station, P.O Box 97, Chiredzi, Zimbabwe.

8. Bullock, A. (1988). Dambo process and discharge in central Zimbabwe. PhD Thesis, University of Southampton, UK.

9. Butterworth, J.A. and Mugabe F. (1995). Romwe Catchment Study: Reducing environmental degradation and promoting sustainable agricultural development in semi-arid Zimabwe. ODA Plant Sciences Programme, Manchester, 5-6 Sept.

10. Campbell, B., du Toit, R. & Attwell, C. (1988). Relationship between the environment and basic needs and satisfaction in the Save Catchment. Univ. of Zimbabwe, Harare, pp119.

11. Carter, R. C. (ed) (1989). NGO Casebook on Small Scale Irrigation in Africa. AGL/MISC/15/89, FAO, Rome.

12. Chilton, P.J and Foster, S.S.D. (1995). Hydrogeologyical characterisation and water supply potential of basement aquifer in tropical Africa. Hydrology Journal 3 (1), 36-49.

13. Chilton, P.J. Lawrence, A.R. and Stuart, M.E. (1995). The impact of tropical agriculture on ground water quality. Chapter 10 in Groundwater Quality (eds Nash & McCall). Chapman and Hall, London pp113-122.

14. Chilton, P.J. & Talbot, J.C. (1990). Collector wells for small scale irrigation: sitting, construction, testing and operation of a collector well at the Lowveld Research Station, Chiredzi. Report WD/90/20, British Geological Survey, Wallingford, UK.

15. Chilton, P.J. & Talbot, J.C. (1992). Collector wells for small scale irrigation: construction and testing of a well at Tamwa, Sihambe, Dhobani kraals and further work at Chiredzi. Report WD/92/27, British Geological Survey, Wallingford, UK.

16. Clause, B., Young, A.R. & Gustard, A. 1993. Modelling the impact of groundwater abstraction on low river flows. In: Proc. 2nd Int. Conf. on FRIEND, Braunschweig, Germany, 11-15 October.

17. CSO (1993) Census (1992). Provincial Profile: Masvingo, Central Statistical Office, Harare.

18. Farqharson, F.A. & Bullock, A. (1992). The hydrology of basement complex regions of Africa with particular reference to southern Africa. In: The Hydrology of Crystalline basement aquifers in Africa (eds E.P. Wright & W.G. Burgess). Pp59-76. The Geological Society, London.

19. Houston, J.F.T. (1988). Rainfall –runoff-recharge relationships in the basement rocks of Zimbabwe. I. Simmers (ed), Estimation of Natural Groundwater Rechrge, pp. 349-365, Reidel Publ. Co.

20. Howard, K.W.F., & Karundu, J. (1992). Constraints on the exploitation of basement in East Africa- water balance implications and the role of the regolith. J. Hydrology 139, 183-196.

21. Howard, K.W.F., Taylor, R.G. & Salvatori, S.L. (1994). The dynamics of groundwater flow in the shallow regolith of Uganda. Proc. XXV Congress IAH, Adelaide, Australia, Vol 1 pp 435-440.

22. Lovell, C.J., Batchelor C.H., & Murata, M. (1990). Development of small-scale irrigation using limited groundwater resources: Second interim Report. ODA Report 11/90, Inst of Hydrology.

23. Lovell, C.J., Batchelor C.H., Semple, A., Murata, M., Mazhangara, E. & Brown, M (1992). Development of small-scale irrigation using limited groundwater resources: Third Interim Report. ODA Report 92/4, Institute of Hydrology, Wallingford, UK.

24. Lovell, C.J., Mazhangara, E., Mtetwa, G., Dube, T., Thompson, D.M., Macdonald, D.M.J & Batchelor, C.H. (1995). Small scale irrigation using collector wells: Pilot Project Zimbabwe. Fifth Progress Report, Oct. 1994-March 1995. ODA Report 95/6, Inst. Hydrology, pp79.

25. Macdonald, D.M.J., Thompson, D.M.E. & Herbert, R. (1995). Sustainability of yield from wells and boreholes in crystalline basement aquifers. Tech Report WC/95/50, British Geol. Survey, UK.

26. Mazhangara E., Mtetwa, G., & Dube, T. (1995). Collector well garden performance- Winter 1995. Lowveld research Stations, P.O.Box, 97, Chiredzi, Zimbabwe.

27. Meigh J. (1987). Low flow analysis of selected catchments in Malawi and Zimbabwe. Unpublished report to British Geological Survey, Wallingford, UK.

28. Meigh J. (1988). A preliminary estaimate of groundwater recharge in the Chiredzi area. Unpublished report to ODA Engineering Division, Institute of Hydrology, Wallingford, UK.

29. Murata, M., Batchelor, C.H., Lovell, C.J., Brown, M.W., Semples, A.J., Mazhangara, E., Haria, A., McGrath, S.P. & Williams, R.J. (1995). Development of small-scale irrigation using limited groundwater resources. Fourth Interim Report. Institute of Hydrology Report ODA 95/5, Institute of Hydrology, Wallingford, UK.

30. Oakes, D.B. & Wilkinson, W.B. (1972). Modelling of groundwater and surface water systems, I. Theoretical relationships between groundwater abstraction and baseflow. Water Resources Board, UK.

31. Pereira, H.C. (1961). Land use and hydrology in Africa. Proc. Inter-Africa Conf. on Hydrology, CCTA, Nairobi, pp45-50.

32. Saveniji, P.S.C. & Alley, W.M. (1994). Pesticides In: Regional groundwater quality (Alley W.M. ed.) Van Nostrand Reinhold, New Yolk, pp 345-382.

33. Scoones, (1992). Land degradation and livestock production in Zimbabwe's communal areas. Land degradation and Rehabilitation, 3, 99-113.

34. Tagwira, F. (1992). Soil erosion and conservation techniques for sustainable crop production in Zimbabwe. J. Soil and Water Conservation, 47. 370-374.tiffen, M., Mortimore, M. & Gichuki, F. (1994). More People Less Erosion: Environmental Recovery in Kenya. Wiley, Uk.

35. Underhill, H.W. (1990). Small scale irrigation in Africa –in the context of ruaral development. Cranfield Press, Bedford, UK pp90.

36. Wallace, J.S. (1994). Hydrological processes and dryland degradation. WMO Bull. 43 (1), 22-28.

37. Whitlow, J.R. (1983). Hydrological implications of land use in Africa, with particular reference to Zimbabwe. Zimbabwe Agric. J. 80, 193-212.

38. Whitlow, J.R. and Campbell, B. (1989). Factors influencing erosion in Zimbabwe. A statistical analysis. J. Environmental Management, 29, 17-29.

39. Wright, E.P. Herbert, R., Murray, K.H., Ball, D., Carruthers, R.M., McFarlane, M.J. & Kitching, R. (1989). Final Report on collector well report 1983-1989. BGS Report WD/88.3, Keyworth UK.

40. Wright, E.P. (1992). The hydrogeology of crystalline basement aquifer in Africa. In: Wright, E.P. & Burgess, W.G. (eds): The Hydrogeology of Crystalline Basement Aquifers in Africa. Geol. Soc., London, No. 1-28.

2.2 Improving Rainfed Farming with Water Harvesting Techniques in Zimbabwe

2.2.1 Assessment of the Loss of Rainwater and Techniques for its Utilization in Musikavanhu area

G. L. Silva and G. Mtetwa

Summary

On average Musikavanhu receives over 450 mm of rain per annum but the rain-fed crops grown fail frequently due to water shortages. Intra seasonal rainfall variations are so large that even in wet years disastrous mid-season droughts are only too common. With better water management, either by conservation in situ rain or by harnessing local runoff, rainfed cropping could be greatly enhanced.

Simple water conservation practices such as tied ridging could be widely applied and each farmer could benefit significantly; but extension services necessary to widely introduce such techniques are lacking. Water harvesting by comparison is more complex, its applicability is site specific and hence limited, but benefits for the individual farmers would be substantial. Under the circumstances, therefore, a limited research project to demonstrate various water harvesting techniques and to test their adaptability is recommended. The project envisages tasting those techniques appropriate for a range of sites, varying in extent from several tens of square metres to a few hectares.

The need for providing technical extension support, perhaps initially in the form of technical co-operation, to assist farmers to install water management systems and then for providing agricultural extension facilities to optimise crop production is strongly recommended.

2.2.2 Introduction

In collaboration with Lowveld Research Station (LVRS) in Chiredzi, Zimbabwe, Musikavanhu Communal Land was studied to a) determine the potential for improving rainfed farming by better management of the crop water regime and (b) identify possible development options for further support under the ODA funded, NRI managed, Agronomy and Cropping Systems Research Programme. The study was undertaken during 18November – 22 December 1992, and was funded under the same programme.

2.2.3 Project description

2.2.3.1 Background

In February 1992 Lowveld Research Station (LVRS) submitted an Outline Project Proposal entitled "Assessment of the loss of rainwater and techniques for its utilisation in Musikavanhu", for Natural Resources Institute (NRI) funding consideration under the Agronomy and Cropping Systems Programme. NRI indicated interest. In June 1992 LVRS in collaboration with AGRITEX extension officers carried "An exploratory survey on management of runoff in Musikavanhu Communal Lands". The survey, led by an agricultural economist, included interviews with 21 farmers who had been handpicked by AGRITEX extension offers as those having keen interested in water harvesting. Based on the conclusions of this study, in July 1992, LVRS prepared and submitted the project proposal "Small-Scale water harvesting in Musikavanhu Communal Lands, Southern Zimbabwe".

The exploratory survey categorised several forms of surface water runoff. These were; hill slope runoff in shallow channels, runoff originating from house compounds, village shopping premises etc, runoff from uncultivated lands and path ways, runoff in deep gullies, and flood runoff that naturally spreads into farmers' fields.

2.2.3.2 The Project Environment

The project area was part of the Musikavanhu Communal Land, and its location is shown in Map 1. It is somewhat rectangular in shape, some 30 km long and 10km wide on average and the width tapers slightly towards the southern edge. The project area was oriented roughly north to south, with its centre located at about $20^0$31'south latitude and 32^0 21' East Longitude. To the East of the southernmost two thirds of the project area are the steep west-facing slopes of the two mountain ranges, Masosote and Matsuro, which rise to over 800 m.a.s.l. to the north of these mountain ranges and to the east of the rest of the project area there is a steep west-facing escarpment. Along the break of slope runs the Chiredzi-Tanganda main road, which generally follows the 460 m contour. This road was chosen as a convenient eastern project boundary. To the south, the project boundary follows the southern boundary of Mazvire ward, while to the north the project boundary runs approximately east-west, following the Nyamuvava branch of the Nyautsa River system. To the west, the project boundary is generally formed by the Sabi River which drains from north to south.

Along the left bank of the Sabi River is a pronounced but somewhat truncated flood levee. A part of the levee, near Chibuwe, is developed under conventional irrigation. This area is excluded from the project.

2.2.3.2.1 Drainage

Outside the project area to the north east is a cluster of relatively low hills but with steep slopes. These hills total some 150 km^2 in area and are drained by a large number of heavily dissected ravines and gulches (large gullies) which combine to form the Nyautsa river. Its natural westward passage is intercepted by the Sabi flood levee, so the Nyautsa River then turns south, enters the project area and runs south for some 10km, bisecting the project area into two almost equal eastern and western halves. It then swerves sharply to the west, changes its name to Dakate River as it approaches the Dakate Business Centre, and runs some 10 km further south. The Dakate river (as it now called) then runs another 5 km south-southwest before it finally discharges into the Sabi River, near Gumira town ship. South of Gumira there is a separate water course that runs along the foot of the flood levee. This channel intercepts gully drainage from the east towards the River Sabi. It is relatively steep, less than 10 km long and meets the River Sabi near Makaza.

In contrast to the concentrated Nyautsa River runoff that enters from the north, all along its eastern boundary the project area receives runoff from a series of small, westward flowing gullies are intercepted by the Nyautsa /Dakate River before they reach the Sabi River. The small and medium sized gullies drain only the western slopes of the escarpment and the slopes of the two mountain ranges, whereas the larger gullies drain the hinterland further to the east. The great majority of the gullies are small, and they are distributed all along the eastern project boundary. The medium sized gullies are concentrated near the southern end of the project area while the larger gullies occur further north. The three groups of gullies are significantly different; hence they are discussed separately.

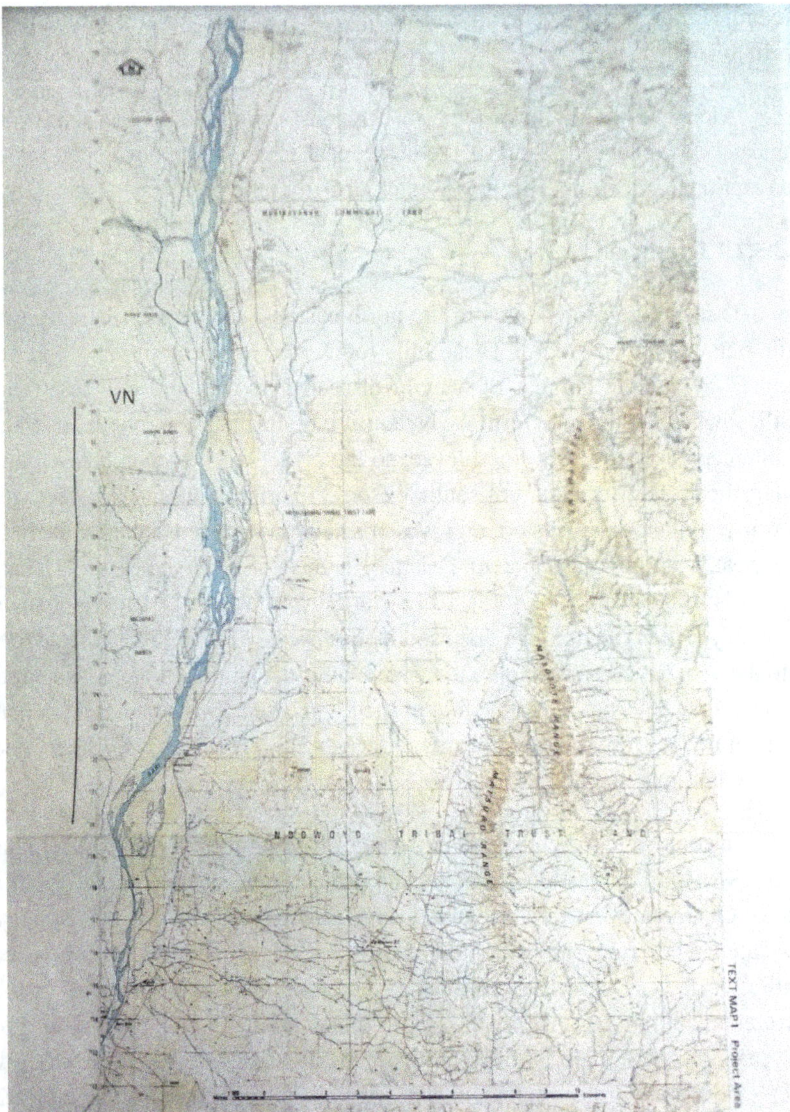

Fig 2.2.1: Map of project area in Musiskavanhu

2.2.3.2.2 The small gullies

The small gullies flowing across the road are numerous; their catchment areas are small and seldom exceed 0.5 km^2. So, the drainage density of the western slopes of the mountain ranges/escarpment is high, and runoff from the individuals gullies small. The erosivity of these small gullies is nevertheless large, due mainly to associated characteristics such as extremely steep slopes, easily erodible colluvium, high intensity storms and the lack of vegetation on serrated scarps. The high drainage density of the individual gullies is keeping erosion somewhat under control.

In the vicinity of the main road these gullies though small are in bold relief as they have dissected the mountain and escarpment foot slopes so that the land is now undulating to rolling. A few kilometres into the project area, however, these gullies gradually disappear as the flows are attenuated and/or begin to spread onto flat land. Most runoff then flows overland via a continually changing network of ill-defined fluves that meander across the flat to gently undulating terrain. Little runoff from these small gullies appears to reach the Dakate river however, as much of the water is held up as surface retention and is eventually lost by percolation and evaporation.

2.2.3.2.3 Medium gullies

Nearer Manzvire towards the southern end of the project area the several medium sized gullies that drain the eastern slopes of the Matsuro mountain range differ significantly from the above. The individual catchment areas of these southern gullies at their road crossing points are typically 3-5 km^2 (as opposed to 0.5 km^2). Furthermore, downstream of the road they converge rather than disappear, and form fewer but more formidable gullies. The resultant larger gullies then flow across the whole width of the project area before discharging into an interceptor channel running parallel to the Sabi River, about a kilometre away from it. The slope profile and the steepness of the terrain in the south are different from those, elsewhere within the project area and the flood flow regime of the gully system therein is influenced accordingly.

2.2.3.2.4 Large gullies

In addition to the small and medium sized gullies discussed above, there are also several large gullies, some big enough to be classified as gulches. These enter the project area across its eastern boundary, the Tanganda-Chiredzi main road. The largest gullies rise from the distant hills east of the escarpment and the mountain ranges, and typically drain up to about 25 km2 of rugged hill catchment which receive more frequent rain than Musikavanhu.

Like the upper tributaries of the Nyautsa River, these gulches can accordingly carry heavy discharge through the project area even at times when there is no local rain in Musikavanhu. The dynamic channel storage capacities of these gulches are large in relation to their catchment areas. Thus, along the lower reaches of these water courses, the flood peaks become attenuated and the recession flow is prolonged. On the Nyautsa and Nyamuvava Rivers for example the flood recession can last more than a day.

Given the frequent and large flood discharge they carry, most of these gulches are well defined all along their courses until they join with the Dakate all along their courses until they join with the Dakate River. However, they are not permanent; they often overflow their banks and meander across the flat terrain, changing their courses continuously. In addition to flood flows they also bring in heavy sediment loads which are deposited in their lower reaches- hence the outflanking of their banks as the flow channels becomes filled with sediment.

2.2.3.2.5 Relief

The main features that influence the topography of the project area are foot hill slopes to the east, the Sabi River with its left bank flood levee, and the Nyautsa/Dakate River. The steepest terrain is in the immediate vicinity of the main road where it slopes westwards, at gradients typically up to about 2%.

In the northern half of the project area these slopes rapidly decrease to about 0.5% a few kilometres away from the road. These gentle slopes meet the equally gentle eastward sloping sides of the Sabi flood levee along a line of intersection through which the Nyautsa River flows. This river and the numerous gullies which drain the rugged mountain terrain to the east of the project area together bring in large amounts of sediment. Smaller quantities are carried away into the River Sabi, therefore on balance that part of the Nyautsa River valley which is inside the project area accumulates sediment. This build-up of sediments accounts for the flat terrain in the middle of the project area. Further to the west, the topography is dominated by the broad truncated River Sabi flood levee. In contrast to its precipitous western slope down to the Sabi River, the eastern slope of the flood levee is gentle as a result of siltation in the Nyautsa valley.

Further to the south where the Dakate River runs closer to the Sabi River, the run of the slope away from the main road is longer, the decrease of the gradient to the west is less, and less sedimentation takes place. As the bed slope increases, the sediment carrying capacity of the Dakate River nearer its confluence with the Sabi River also increases, causing some riverine erosion. A little upstream however, there are several level pans along its course. The situation in the region to the south of the Dakate/Sabi confluence is also similar, except that the westward slope are generally somewhat steeper. While the dominant slopes are westwards, the project area also slopes southwards, following the general fall of the Sabi River. Comparatively however, this southwards slope is very slight, at a gradient of about 1 in 1000.

The micro-topography of the project area is somewhat more complex than the above described overall orthogonal east-west and north-south slopes, as there is considerable local variation. For in-field water management and local run-off harvesting, the micro-topography is more important. Nearer the eastern boundary the numerous gullies that cross the road strongly dissect the west-sloping terrain. As a result, the land nearer the road is hummocky, with relief at times exceeding several metres while the average interfluve distance is less 100 metres. The cross slopes of this land can therefore exceed 10%. These lands could thus become good runoff generating plots. Conversely, though, the prospects for them to benefit from water harvesting are minimal. Here is a good example of a land class that stands to benefit immensely from simple water conservation measures while being inappropriate for treating with run-on water.

Nearer the middle of the project area where the overall orthogonal slopes are gentler, the micro-topography is mostly flat to slightly undulating. The drainage density clearly defined gullies is low, but where there are gullies, they tend to be large and U-shaped, indicating large volumes of soil erosion in the past. This, however, is not widespread, most runoff that disperses from the gully takes place overland and their opportunities a) to extend the water spread area, b) to conserve water at macro (plot) level and c) possibly to improve infiltration by appropriate cultivation methods. Larger scale and more elaborate engineering measures would be necessary to stabilise and check gully erosion.

To the south of this relatively flat terrain there are two conspicuous kopjes, Umbwa and Zinyuni. These are some 100 m high but their slope decreases rapidly. So, the total area they occupy is no more than several square kilometres, and the runoff generated will therefore be limited, but could be harvested without too much effort and applied to nearby fields as a supplement to rainfall.

2.2.3.2.6 Rainfall

Several rainfall mechanisms give rise to rain in Zimbabwe, though a few usually dominate in a given region. For, example, throughout the year the mean surface winds in Zimbabwe have an easterly component and the warm waters of the Mozambique Channel thus form an important source of moisture. The eastern border of Zimbabwe being less than 400 km from the Indian Ocean, these easterlies are responsible for much of the rain in the eastern parts. Being induced by orographic effects, the mountain region in the east thus receives over 1200 mm of rain per annum.

In contrast, that part of the Sabi Valley which lies in the rain shadow receives not more than 500 mm. most of this rain is produced by convective processes which depend on heating of ground that takes place during the warm afternoon hours. In the hilly country though, convection usually becomes vigorous somewhat earlier, hence rain in the early afternoon is common in the region of the project area, with the incidence of rain decreasing after sundown. When other influences and rainfall processes are taking place however, rain can occur at other times. Night rain is common particularly in the hills to the east of the project area.

According to the regional statistics published by the Department of Meteorological Services, the first and the last rainy pentads in the region of the project area could be expected in late November and in mid-February respectively, constituting an agricultural useful rainy period of some 90 days. On average there are about 40 rainy days (daily rain exceeding 1.0 mm) per annum, almost all concentrated in the extended rainy season which usually begins in about late September and lasts until April.

The mean annual rainfall in the region is highly variable, with a percentage coefficient of variation in the lower thirties. Considering the country as a whole, however, the region cannot be viewed as exceptional as in more than two thirds of the country also the coefficient exceeds 30%. High variability of annual rainfall is a feature in most parts of Zimbabwe and it is not unique to Musikavanhu.

Another feature is that rainfall is organised into spells of a few days at a time when most stations in an area will record rain, followed by spells of several days when almost no stations receive rain. A rainy season characteristically composed of such alternations of rainy and dry spells in general is favourable for crop growing, and this to some extent compensates for the above-mentioned high variability of total annual rainfall.

Chiredzi Research Station provided a 31year record of daily rainfall collected from October 1952 to September 1983 at the Middle Sabi Research Station. The mean annual rainfall calculated from these records is 471 mm, with a coefficient variation equal to 32.7%. The absolute minimum and maximum annual rainfall during the period are 243 mm and 735 mm respectively. In more than half the years during this period, annual rainfall exceeded 450 mm. by subsistence rain fed farming standards therefore, total mono-modal rainfall seems to be adequate or more than adequate to grow convectional rainfed crops. Yet, crop failure and underperformances due to soil moisture deficits are widespread. Invariably these failures are caused by the mid-season droughts which may last for more than three weeks, and even those years with the heaviest annual rainfall are not necessarily immune from them. It is thought that during the 31year period of the rainfall record, crops could have totally failed in at least six years and severely underperformed in five other years in spite of annual rainfall exceeding 350 mm in all but eight years.

Given that much of the rain is produced by convective processes, the maximum instantaneous rainfall intensities can be high. But the peak intensities do not last for more than a few minutes; the complete storm lasts no more than a few hours, and total storm precipitation rarely exceeds 75 mm. during the 31 years on record, on only nine days did the total daily rainfall exceed 75 mm, and of these nine days on only two days did it exceed 100 mm. it should be noted that even during these heavy rainy days, the daily totals might have been generated by more than one storm. So, for practical purposes it could be concluded that the probability of an individual storm precipitating more than about 60 mm is small. The implication of these rainfall intensities for in situ water conservation is that if tied ridging is practised, the direct rainfall on a piece of land could rarely escape as runoff. Thus, tied ridging even in its simplest form would be extremely effective in containing almost all the direct rainfall a farm receives.

At the same time the rainfall intensities are sufficiently large to induce high runoff. Sparsely vegetated grazing on uncultivated land with steep slopes and capping soils would generate high runoff under the prevailing rainfall regime. As population pressure increases these marginal suitable lands are being made arable, but the remaining uncultivated land provides some opportunities to generate runoff for growing crops elsewhere. So, rainfall intensities in the project area are not high enough to frustrate in-field water conservation attempts employing simple tied furrow techniques nor so low that they cannot generate sufficient runoff from non-arable lands to grow crops elsewhere. Thus, the often-acclaimed high intensity rainfall in the project area is not considered by the author to be a serious problem either for conservation or for runoff harvesting.

It is suspected that no rainfall intensity data are at present being collected anywhere in or around the project area. However, there are plans to install an automatic rain gauge at the new Save Valley Research Station, and for this an instrument is already available at LVRS. Convective storms influenced by orographic effects can often give rise to localised rain causing substantial spatial variation. Only one set of rainfall data (for Middle Save Research Station) was available during the mission, and therefore the spatial variability of rainfall in the project area could not be studied.

2.2.3.2.7 Gully Erosion

There are several ravines and many gullies that enter the project area across its eastern boundary, the Tanganda-Chiredzi road. During road construction, the natural drainage regimes of many of these have inevitably been modified. The most significant changes, however, seem to be limited to some of the smaller gullies. The hydraulic regimes of the bigger water courses appear to have been left almost unaltered by providing them with large box-culverts of bridges openings at their natural road crossing points.

The more significant modifications have caused some small gullies to disappear although while others have got bigger, particularly downstream of the road. Joining two or more natural gullies together (to reduce the number of costly road culverts) cumulatively increases the flood discharge of the resultant single gully and causes erosion in it. With the upstream road embarkment acting as a runoff interceptor, cross-flows from one culvert to another could also take place above the road, especially if some culvert entrances are blocked by shifting sands. This also in effect amounts to joining or interconnecting gullies, albeit unintentionally. This concentration of flow could give rise to unanticipated heading up of water flow could give rise to unanticipated heading up of water above the road, thus generating erosive flow velocities downstream of the road culvert.

All these erosion hazards brought about by road construction have to be regarded as 'one off events" rather than as problem that would continue to aggregate indefinitely. Until a new 'hydrologic equilibrium' is attained, however, noticeable erosion and deposition changes in the vicinity of the road will continue to occur. These are unlikely to cause serious problems as the road design generously provides for drainage crossings and erosion control structures. Whether these structures would be adequately maintained in the future is another matter altogether.

The downstream ends of most culverts have been protected against scour, often with gabion-built drop structures, some as high as 3 m. basically, these structures are to protect the road infrastructure. Building similar structures to check gully erosion and protect farmers, fields from potential erosion hazards is usually not considered economically justifiable. Consideration could be given, however, to utilising some of these already built drop structures to harness gully water for supplementary irrigation, but this would require substantial technical assistance inputs including collaboration with the Road Authority.

2.2.4 Problems and intervention strategies

2.2.4.1 Rainfall amount and distribution

The rainfall regime is mono-modal with a six-month rainy season. The mean annual rainfall is about 475 mm. the annual variation is large, but in two out of three year's rainfall can be expected to exceed 300 mm, apparently sufficient to grow conventional dry land crops. But in actual practice, the Musikavanhu farmers manage to secure a decent crop in no more than two years out of five. In addition to the annual rainfall variation, a further problem is the unfavourable intra-seasonal variation.

In general, the most agriculturally productivity rains occur during the three-month crop growing period, December to February. In anticipation of these rains most farmers start planting from about mid-September and carry on until mid-December. This allows them some flexibility to benefit from any early or late rains and to spread the demand on the limited available labour and work animals.

But extended drought periods lasting several weeks and capable of destroying standing crops commonly occur during mid-season. This can happen even in the wettest years. In 1976/77 for example when the annual rainfall exceeded 730 mm, two mid-season droughts each lasting up to four weeks occurred between November and February. Middle Sabi rainfall data reveal that in one year out of every three which received more than 300 mm, such destructive droughts occurred. Considering the varied in situ moisture-holding capabilities and the effects of staggered planting dates across the project area, the combined probability of severe crop underperformance due to mid-season drought and/or inadequate total rainfall could be as high as 60%.

2.2.4.2 Water conservation

Under these circumstances better rainwater management would undoubtedly improve rainfed agriculture. In situ water conservation and supplementary irrigation using harvested runoff are considered to be feasible, affordable and effective technologies to adopt. It should be noted nevertheless that although these practices on their own would help increase crop yields, such efforts need to be accompanied by complementary improved agronomic practices to reap the full agricultural benefits. At those sites which could benefit from regular supplementary irrigation for instance, high value crops or high yielding varieties could be grown. By contrast, at less favoured sites, the cultivation of maize, particularly those varieties with low drought tolerance should be discouraged.

The effectiveness of improved water management would depend on the technology used, site topography, soil, rainfall characteristics, access to runoff water and farmer commitment, but in the long term could be positive. Agricultural benefits would accrue from increased crop yield due to improved soil moisture and also from drought risk amelioration. Many of the Musikavanhu soils are sandy with poor moisture holding capacities. During extremely unfavourable years when total crop failure is likely, in situ water conservation of local runoff harvesting would be of little value. But during the other years, when crops survive but the rainfall is insufficient to product maximum crop yields, even the marginal soil moisture improvements that can be brought about by rain water management could result in significant crop yield increases and in additional fodder for livestock. So, while tied furrows and local rainwater harvesting would not prevent total crop losses in extremely unfavourable rainfall conditions, they would nevertheless guarantee significant increases in crop yields and fodder in most other years.

2.2.4.3 Supplementary irrigation

Although mid-season droughts are quite common within the project area, it is not so in the mountain region just to the east. As a result, it is not uncommon for several of the larger gullies that drain the mountains to bring substantial quantities of runoff into the project area at times when Musikavanhu itself is experiencing extended droughts. Such timely runoff would be invaluable if it could provide supplementary irrigation to the otherwise dying crops and pastures. This would be immensely profitable, but the technology requires is somewhat demanding. Both water conservation and supplementary irrigation with harvested runoff could substantially increase crop yields during most years. With the latter technology in particular, crops could be saved from mid-season droughts if timely mountain runoff became available. The two technologies are complementary to an extent but there are also significant differences between them. These differences and various other factors relevant to the application of the different technologies have been discussed elsewhere in this report.

2.2.4.4 Conservation of high intensity rainfall

After a simple analysis of all the daily rainfall data LVRS could provide, it was possible to demonstrate that daily rainfall rarely exceeds 70 mm, no matter what the individual storm intensities may be. This means that if the farmers were to adopt simple in-field water conservation practices, such as tied furrow farming, hardly any rain falling on their fields could escape as surface runoff.

While a few farmers might have to extend significant extra resources to incorporate tied ridging into their farmer practices, some Musikavanhu farmers claim that the labour required for tied ridging is, in fact, less than that required for traditional cultivation methods. Furthermore, tied-furrows as advocated by LVRS causes crop yields to increase substantially, particularly during those years with less favourable rains. Musikavanhu farmers would there be better off by adopting tied ridging.

2.2.4.5 Concentration of drier years

LVRS research has also re-affirmed the well-establishment view that returns from water conservation measures are highest in years of poor to average rainfall and minimal in others, particularly the extremely wet and dry years. The results available so far show that the marginal increments of crop yield due to tied ridging range from a few percentage points in years with best rainfall to over 100% in years with poorer rains. The results also illustrate that in 'years with good rains' there is sufficient water to grow traditional crops reasonable rains, yield increases due to extra water management efforts are negligible. On the other hand, in drier years, the effect of tied ridging and other water conservation or water harvesting techniques, is a dramatic increase in yields.

2.2.4.6 Appropriate technology choice

Harvesting and conservation of water are two techniques that can improve crop water regimes under rainfed farming. To that extent they are complementary. But the ways in which they operate, however, are quite opposite to each other. Conservation practices such as tied ridging trap the rain directly falling on a micro unit of land and prevents surface runoff, while water harvesting brings relatively large quantities of run-on water from outside into the farm plot and encourages its run within the plot to achieve desired spatial distribution. In the case of the latter, the harnessed water and the direct rainfall the plot receives are conserved at the farm plot scale rather than at the micro basin or furrow scale. This requires easy mobility of water within the plot and any tied ridges placed in the plot will obstruct such movements. In this context the technologies become mutually exclusive alternatives which cannot be used simultaneously in the same field.

The task is to identify and recommend technologies most appropriate to a given situation, considering the many factors involved. Conducting two separately administered programmes to promote tied ridging on the one hand and water harvesting on the other is counterproductive. The requirement is to find the most appropriate combination of techniques for each field situation.

Many factors will determine the most appropriate rainwater management techniques, and they include; rainfall characteristics, access to local runoff, access to runoff generated by distant rainfall, terrain topography, plot topography plot elevation with respect that of the possible source of supplementary water, nature of soils, crops and farming practices, farmer dexterity and commitment and relevant socio-economic aspects. The challenge is how to build these considerations into the decision making help the extension workers and farmers to develop the right combination of water management techniques. In addition, the project should address the broader socio-economic and agronomic issues associated with water harvesting and conservation technologies. Of particular interest are the rights of immediate and long term access to runoff water, negotiations to secure 'right of way' for water furrows through neighbouring upstream land, group participation in constructing, managing and sharing the more elaborate water harvesting systems, farmer commitment to operate the system in real-time as the flows occur, the costs and benefits of water harvesting both in tangible terms as well as how the farmers conceive them, the availability of appropriate agronomic packages to ensure benefits are actually obtained from the improved soil moisture status, and most importantly, institutional development prospects for the wider dissemination of the technologies among the farmer population.

2.2.4.7 House compound runoff can make a small contribution

LVRS showed particular interest in harvesting runoff generated from house compounds; hence it is considered her in some detail. House compounds in the project area are generally at higher elevations than their surroundings so that the runoff generated flows away from the compound, often along a small ditch dug by the household. The amount of runoff generated is minute in the context of what is required to grow cereal crops, and if used for a few tens of square metres. In comparison with the several thousands of square metres of cereal a farmer plants on average, this is insignificant, even if the expected yield from the treated area were to double. Farmers' primary concern is to obtain sufficient grain for his livelihood, not to achieve high yields from an insignificantly small part of his land. The importance of the concept of compound runoff harvesting should not be exaggerated, but such runoff, or any other for that matter, should still not be allowed to go to waste.

A simple and practical way of exploiting such runoff for cereal growing, subject to site condition, would be to encourage farmers to cultivate the ditch itself, together with the adjoining strips of land with furrow lines slightly angled to the alignment of the ditch to spread and absorb all the runoff to grow crops. Constructing artificial channels to convey the little runoff into specially prepared and levelled strip basins located somewhat away from the compound, as LVRS had tried out at the Kuwirirana site, is not considered worth of effort. Besides, by the time water reaches the plot a considerable proportion of the compound generate runoff would have been lost by seepage through the bed of the canal. In situations where the ditch carrying the runoff naturally empties into cultivated fields, little extra intervention is needed, except perhaps to encourage farmers to retain and distribute the water efficiently along carefully aligned plough lines in relation to the natural slope of the land.

At sites where the compound runoff gets freely discharged into established gullies there is little point in trying to prevent this from happening, with the overall concept of runoff harvesting, unless further downstream the gully flows only go to waste. Other possible uses of compound runoff could be to grow tree crops-fruit eg. mango, shade, firewood, or fodder, and wet season vegetables at or near the compound itself. Marginal returns to both labour and water from such ventures could well be higher than for cereals growing.

The above illustrates the range of opportunities and possible interventions available, subject to varied site conditions of course, when developing compound runoff-the smallest 'hydrological unit' encountered in Musakavanhu in the context of runoff harvesting.

2.2.4.8 Rivulet runoff may irrigate up to one hectare.

Next on the scale is the development of rivulets or first order streams that drain undulating cultivated lands, parts of the grazing areas and built-up areas such as school compounds and business centres. Runoff from these sources should be sufficient to provide supplementary irrigation for up to a hectare of cultivated land. The peak flow rates in these rivulets are low enough to attempt total diversion using rudimentary earth bunds or dykes of height no more than 50 cm while the flows are high enough to channelize along handmade furrows and convey to the command area. Command areas are unlikely to be located any further than about 100 m from the diversion site. An example of such a development was demonstrated at Manzvire.

2.2.4.9 Small gullies may command more than one hectare

In the project area, the second order streams manifest themselves as generally well established but small gullies, typically 30 to 50 cm deep and about 1.5 m wide. At their peak flow conditions, they are likely to be erosive and their diversion should be attempted with care.

The alignment of the off-take canal needs to be carefully chosen and its regular maintenance is essential. Otherwise canal siltation would cause overflowing, leading to haphazard erosion. Gullies could be diverted by low fuse-plugs and this would reduce the risks of overburdening the offtake cannal and the gully out-flanking at or near the diversion site. With such schemes it should be possible to command areas little over one hectare in extent.

2.2.4.10 Medium gullies require permanent check structures.

To harvest water from gullies deeper than one metre, permanent check structures are needed. Farmers' vigilance to operate the structure during the time of flow cannot be guaranteed, so needle controls (as opposed to stop-logs) should be deployed. This should reduce the risks of gully erosion downstream of the structure and the overloading and siltation of the off-take canal, particularly under high flow conditions.

The hydrologic regimes of these medium size gullies are varied. Some only receive runoff locally generated within the project area while others in addition receive runoff from the mountain slopes and even possibly some runoff originating from the distant mountain areas. The area of land that can be commanded with these gully flows would be accordingly varied, ranging from a few hectares to tens of hectares. A site potentially capable of such development was identified near Manzvire.

2.2.4.11 Large gully with more potential need detailed study

Some of the larger gullies nearer the Chiredzi-Tanganda road are over 2 m deep and harvesting their runoff along these reaches would be difficult. Where gabion drop structures are already in place, built to protect road infrastructure, however, some of the gully flows could be extracted upstream of the structure. Further downstream where these gullies become shallower and flows attenuated, they fall into the category of medium sized gully, development of which is described above.

Lower ends of the major gullies and Nyamuvava River near its confluence with Nyautsa river are sites with such potential. The scale of interventions is necessarily large but so would be the benefits, particularly as these water courses bring in runoff at times when there are no local rains in the project area. Detailed field investigations, including surveying and levelling of the extended areas, hydrology, socio-political issues with regard to water rights, right of way for canals and bunds, community participation during construction and sharing of benefits etc need to be undertaken before specific construction work can begin.

When attempting to harness any form of runoff, as the case of compound runoff harvesting illustrates, the developer should not be restricted to barely thinking in terms of simply applying pre-set, stereotype solutions but be prepared to drive and adopt the most appropriate solution the site and situation demand.

2.2.5 Site development

2.2.5.1 Baera site

The site lies adjacent to a gully rising from the mountainous hinterland to the east of the project area. The gully drains some 3 km2 of the mountain catchment and 1.5 km^2 of lower escarpment slopes. The site could therefore receive gully water not only when there is local rain but also on those occasions when it rains in the distant mountains and not locally. In the latter case the flows thus derived are of great value as timely irrigation, while the value of gully flows derived from local rains is limited to providing irrigation water to supplement the direct rain the site receives anyway. According to local information, "the gully flows every time it rains'. In the absence of firm data, the exact meaning of this could not be established.

Nearer the site the gully is well defined single channel of rectangular cross-section measuring approximately 3 m wide x 0.6 m deep with a sandy bed. A few hundred metres upstream, the channel is ill-defined, and the braided channels are no more than a few centimetres deep. Here sedimentation is taking place and a thicket is flourishing. A kilometre or so further upstream the gully had developed into a gorge, measuring up to 10 m wide and over 6 m deep, cut into easily erodible silty sands. A few hundred metre downstream from the site, the channel becomes somewhat ill-defined allowing some sedimentation where a small thicket is growing. Immediately below this branch gully joins from the right and the main gully soon assumes a quasi-steady path. The mean annual flood flow near the site is thought unlikely to exceed 2 m^3/s. this is based on the assumption that the gully seldom overflows its bank at or near the site. Just upstream, however, the gully overflows its banks and some of the overflow water could well enter the site.

Sketch map 2.1.2 depicts the site and its structural developments. The developed irrigable area is arranged into four closed and two open-ended blocks, and exceeds 1.5 ha.

The areas of the individual blocks range from 0.1 ha to over 0.5 ha. Subject to the availability of water it should be possible to command this area with the built in infrastructure. Depending on performance during the coming season, the prospects for expanding the command area and extending the facilities so that the neighbouring downstream farmer might also benefit from the scheme remain to be examined.

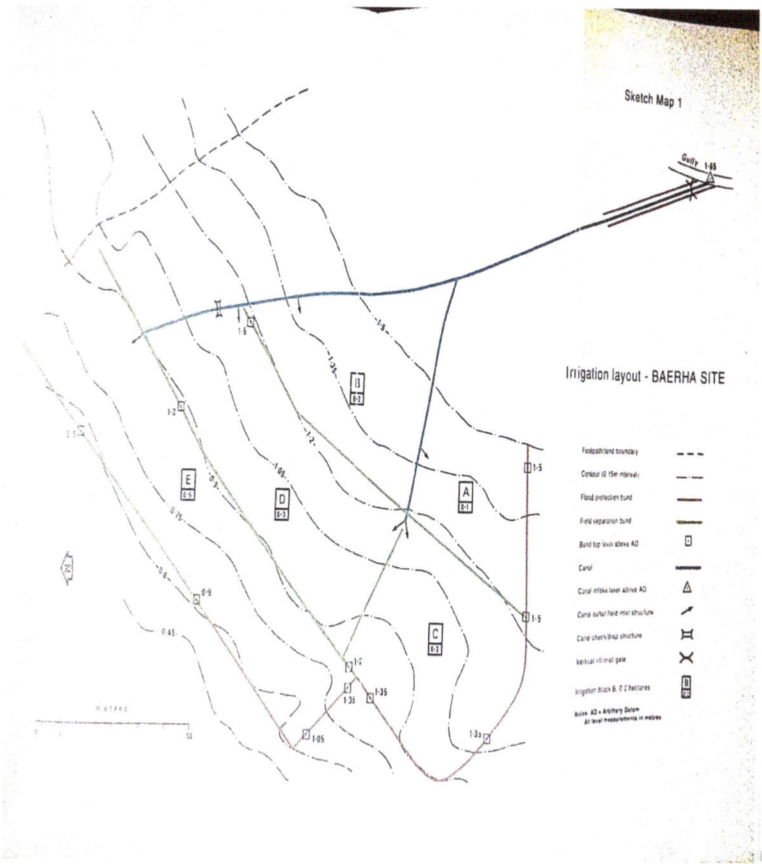

Fig. 2.2.2: Schematic map of Baerha site

Some low-lying areas of the blocks C and E occupying the extreme south-western corner of the development area are vulnerable to prolonged inundation. A flood dyke of height up to 0.4m is provided to isolate the site such natural flooding. The site is supplied with the irrigation water from the gully by way of an open cut canal through the left bank of the gully. At the inlet of the canal is of trapezoidal section 0.8 m deep, bed with 0.5 m and the banks on 1:1 slope. The inflow into the canal is by gravity, unaided by any flow diversion arrangement. The canal inlet is fitted with a vertical-left gate fabricated out of disused corrugated galvanized iron sheets previously used for roofing. According to available information, primary flooding is unlikely to happen at or near the inlet, but with the available material excavated form the canal, double embankment flood bunds 0.3 m high were built into the head-works to provide additional flood protection. The canal left bank flood bund was deliberately built to check any gully water which had out-flanked upstream from entering the canal on its overland passage to re-join the gully.

With the inlet fully opened, the conveyance capacity of the inlet canal at flood condition would be about 0.4 m^3/s, corresponding to a mean velocity of 0.7m/s. the canal is unlined but it is cut into clayey material. The 0.7m/s. the canal is unlined but it is cut into clayey material. The 0.7m/s velocity is near the threshold at which canal bed/slope erosion could begin. Under normal operational circumstances, however, with the inlet gate only partially opened, the canal would be subjected to significantly lower flow rates, and the velocities thus generated in practice would be well below the threshold limit.

Little over 100m downstream from the intake, the canal bifurcates. A few tens of metres further downstream both left bank and right bank branch canals are formed almost entirely with filled-in embankments, with little or no cut, and are thus able to command the adjacent land by gravity. The left branch canal which is some 60 m long finally discharges into the irrigation blocks C and D through a two-way outlet structure built of disused corrugated iron roofing material. Block A is also command by this branch by this branch canal.

The right branch canal has its first irrigation outlet structure located 50 m from the bifurcation to irrigate Block B. the second outlet is located some 24 m further downstream to irrigate Block D, which can also receive water from the other branch canal. Both Blocks B and D lie on the right side of the canal. Technically, some land lying to the right of the canal could also be irrigated from this branch canal but the boundary with neighbouring farmers is not far from the canal. Subject to satisfactory performance of the scheme in the coming season and if a settlement can be agreed with neighbouring farmers to share the water, the neighbouring land could be brought into the command of the scheme.

Immediately below the second outlet a check/drop structure, built of disused roofing material, is provided to maintain the canal bed slope at 0.25%. from here the right branch canal traverses a further 35 m before it freely discharges into the open-ended irrigation Block E. subject to the availability of additional gully water, consideration will be given to extending or further developing Block E and the right branch canal. This is another possible way to incorporate adjacent land belonging to the neighbouring farmer into the scheme. The irrigation blocks are separated by 0.3 m high contour bunds built approximately 15 m apart. At right angles to the bunds, the land slopes at about 1%. The irrigation inlets to these blocks are located at the respective points of highest elevation and the farmer is expected, eventually, to practice strip irrigation along plough lines set out parallel to the contour bunds. Appropriate land preparation is crucial for successful in-field irrigation and LVRS staffs are expected to assist the farmer with this. If this most appropriate land preparation is difficult to achieve during the coming season, wild flood irrigation could still be practised. In this event, some uneven water distribution is to be expected. By building smaller contour bunds, say 0.15 m high at 5 m intervals, more uniform in-field water distribution can be achieved. In total 280 m of canal, 400 m of field separation bunds and 275 m of flood bunds were constructed at this site. The total earth work volume was about 290 m^3, of which only 20% was cut.

2.2.5.2 Qumayo Site

This site located less than a kilometre away from the Baera sire could benefit from the gully which passes through it. The head reaches of this gully extends almost up to the left bank of the other medium sized gully developed to irrigate the Baera Site. The apparent catchment area of the gully, which consists almost entirely of lower escarpment slopes, is no more than 0.5 m^2, on this basis the gully receives no mountain runoff, and unlikely the Baera site, Qumayo site can benefit from gully runoff only when it is also receiving direct rainfall. The irrigation water that could be secured from the gully would therefore appear to have no timely advantage but only supplementary value. Under heavy rains, however it is suspected that overflows from medium size gully could enter this gully. Although on such occasion this gully could well carry mountain runoff, such occurrences would be comparatively less frequent.

Inspite of this limitation, Qumayo site was chosen for development because the farmer concerned in the past had attempted on his own to abstract water from the gully for crop growing, and had thereby shown commitment. Furthermore, during heavy rains the gully overflows and causes downstream flooding which adversely affects several homesteads. The dwellers therein have been made to settle on this flood path in spite of their protestations. When developing the site for irrigation the opportunity was taken to install a flood bund to provide flood protection for these homesteads.

The eidetic cross-section of the gully nearer the site is small, perhaps not more than 0.5 m^2. During the rains, however, the gully flow was said to occupy a much larger cross-section area, with substantial proportion of flow taking place as over land flow along its banks. Water abstraction from the gully is affected by building a low earth dam across the gully. The headed-up water then enters a double embankment canal through an inlet gate built of discarded corrugated roofing materials. The maximum height of the dam is 0.5m its downstream slope, which is paved with loose rip-rap is laid at a shallow gradient. Gully flows exceeding the irrigation requirement of the blocks are allowed to spill over the dam and eventually re-enter the allowed to spill over the dam and eventually re-enter the gully further downstream. Before their re-entry, however, the farmers could utilise such flows to irrigate some of the land in block A, as he had successfully done in the past. The site features and the structural development are shown in Sketch Map 2.2.3

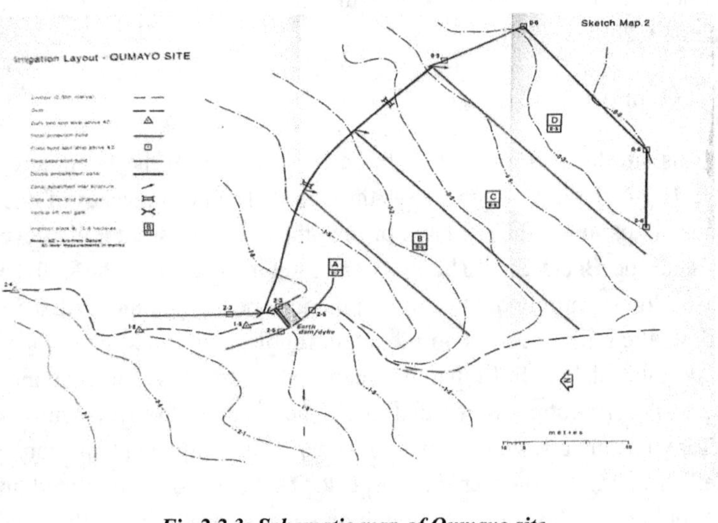

Fig 2.2.3: Schematic map of Qumayo site

Nearer the intake, flood protection works were provided. These were necessarily generous because little information on the local flood hydrology was known at the time of the system design. Furthermore, the backwater conditions created by the dam would raise the natural flood levels even higher. Accordingly, flood dykes and canal embankment were built to a maximum height of 0.7 m.

The total command area of Qumayo is 2 ha and this is divided into four irrigable blocks. Their individual areas range from 0.2 ha to 0.8 ha. The scheme comprises a single canal and this forms the uppermost boundary of each block. The first three blocks are provided with individual inlet structures to receive water from the canal. In the absence of a terminal structure, the canal discharges freely into the last block. The canal is 170 m long and trapezoidal in section. Throughout its length, the bed is 0.5 m wide with filled in banks effectively sloping at 1:1. The bed slope is 0.45%and this is controlled by two check/drop structures installed 75 m and 135 m from the inlet gate. Most of the canal is constructed in fill and very little cut was involved. The canal inlet was set at 0.3 m below the spill level of the dam and for all practical purposes this could also be the maximum possible water depth in the canal, if the gate were to be left fully opened. At this depth the canal could theoretically carry 0.25 m^3 of water per second with a mean velocity 0.9 m/s. the canal embankment are built of loose earth with little compaction (to simulate farmer construction standards) and as such they will begin to erode when the flow velocity exceeds about 0.4 m/s. the amount of water allowed to enter the canal and the velocity it thus generates therefore have to be controlled by the inlet gate.

Each irrigation block is separated by earth bunds less than 0.12 m high. The bunds are set out at an angle to the general direction. Parallel to the bunds, the land slopes at about 0.05% away from the canal and back towards the gully from which the irrigation water is abstracted. At this site the irrigation layout allows the farmer to practise furrow irrigation along furrows 150 m long, constructed during land preparation. Alternatively, the blocks could be further sub-divided into basins by constructing lateral bunds, and basin to basin wild flood irrigation could be attempted. Another possibility is to build a secondary canal system and practice basin irrigation.

2.2.6 Construction work, Standards and Costs

Construction works primarily consisted of earth work and erecting hydraulic control structures. The earthwork accounted for more than 75% of the total direct cost.

2.2.6.1 Earth work

In total 450 m of canals, 500 m of flood bunds and 380 m of field separation bunds were constructed. For most of their lengths, the canals were formed with filled-in double embankments are 0.35 m high, 0,4 m wide at the top and with sides sloping at 1:1. The total earthwork volume of these canals' excavation was small, amounting only to some 70m^3. At their deepest cut sections, the canals were 0.7 m deep. Throughout, canal bed width was maintained at 0.5 m and side slopes were cut at 1:1. Where a canal section was formed with both cut and fill, 0.15 m wide berms were allowed. Where necessary, flood bunds were installed to provide protection against floods. In land surface depressions, the maximum height of the flood bunds constructed was 0.6 m. the bund top width in general was maintained at 0.5 m with the sides sloping at 1:1. The earth volume of the flood bunds was 165 m^3.

Filed separation bunds were designed and constructed to two different standards. Where they were built on the contour, eg at Baera Site, they were 0.3 m high, allowing basin irrigation by ponding, if necessary. At the Qumayo site where furrow irrigation is advocated, similar bunds 0.2 m high were provided. The bund top width of the larger contour bunds was 0.5 m and that of the smaller bunds was 0.3 m. in both cases the side slopes were 1:1. The total earthwork volume of the field separation bunds was 160 m^3.

2.2.6.2 Earthwork standards

All the canal embankment, flood bunds and field separation bunds were essentially built of top soil, which was predominately silty. Purely from a technical view point this is not good engineering. Under the project circumstances however, with regard to its scope and objectives in particular, this form of construction was thought appropriate. An important consideration when setting up these trial sites was to demonstrate to the individual farmers that the amount of work involved in establishing water harvesting schemes is not larger. With construction material available locally, for example, the required canals and bunds can easily and quickly be built. If high engineering standards were stipulated for these essentially self-help initiatives, farmers would have to confront several practical problems. These include: farmer comprehension of the specified earth-fill material, gaining access to such material (invariably located in privately owned land), huge labour requirements to haul this material from its source to the site, and refilling of the borrow pits to ensure that disease vector breeding sites are not proliferated. Particularly at this very early stage of the project, the aim is to reassure farmers that the work involved is well within their individual capabilities: this may then go on to promote self-help initiatives.

The need to keep the job simple outweighs any technocratic desire to adhere to engineering standards. Besides, it was felt that the effect of the technical compromise thus adopted is unlikely to be serious. The depth of water the bunds will have to retain is shallow, and seldom exceeds 0.2 m over prolonged periods. Similar bunds locally made in the past seem to have withstood the test of time, helped by continuing compaction as a result of repeated wetting (rain) and drying. It is also expected that some vegetation, natural or otherwise, would soon establish and stabilise the bund slopes.

2.2.6.3 Hydraulic control structures

The hydraulic control structures built into the two schemes were; two vertical-left canal inlet gates, three check/drop structures and eight canal outlets. Conventionally these structures would be built in concrete or stone masonry with necessary steel or timber fittings. Within the scope of the project, this form of construction was judged inappropriate because it cannot be undertaken by individual farmers on self-help basis. No doubt these structures could be built locally by the village craftsmen, but the project emphasis at this stage is to use only those technologies that could be adopted by individual farmers without having to rely too much on ex-farm resources. On this count the use of timber would have been acceptable except for the potential ant problems: evidence is overwhelming that ant attacks on timber structures would be severe. So, structures built of untreated timber would not last long under the Musikavanhu soil environment.

Locally investigations revealed that without too much difficulty disused corrugated galvanised iron roofing sheets could be obtained locally at nominal prices, albeit from an informal market. These sheets, cut into appropriate shapes and sizes, were assembled to form vertical-lift gates, check/drop structures and canal outlets. Before assembling at site the sheets were wire brushed and coated with bitumen to give them extra protection against corrosion. The available sheets are of standard width, 0.7 m, and this was a major constraint in the design of the structures. When choosing system design parameters such as canal bed width, width of gate openings, bund toe widths and canal drops, care was taken to limit their dimensions to accommodate the width limitation of the available roofing sheets.

Being of improvised design, the hydraulic, structural and operational performance and durability of these structures are very much on test during the forthcoming season. During the life of the project, opportunity was taken to monitor their performance, identify the scope to modify the design and carry out further improvements.

2.2.6.4 Construction method

In order to complete the schemes within the short period of time available, all construction work had to be carried out intensively. The labour that could be provided solely by the beneficiary farmer was extremely small and totally inadequate for the timely completion of the task in hand. Almost all the construction works were therefore planned and carried out with hired casual labour. A total of 22 casual labour was employed. Some days the work was carried out in two staggered labour shifts which extended up to 11 hours of construction. Each labourer, however, was effectively employed only for 8 hours. The total number of 8-hour man days deployed for on-site construction was 360.

2.2.6.5 Tools

The tools used were rudimentary and limited to hoes, spades, pick axes, wheel-barrows and hand (push) carts. Except for the wheel-barrow and hand carts which were borrowed from local farmers, all the other tools were supplied by LVRS.

2.2.6.6 Labour productivity

About 95% of the construction labour procured was deployed for carrying out earthwork. Earthwork labour productivity amounted to only about 1.6 m^3/man/day, but properly motivated, organised and supervised casual work gangs should perform at about 4m^3/man/day. The labour efficiency was therefore poor.

2.2.6.7 Costs

Labour accounted for most of the direct cost incurred. Hired casual labour was paid at the government approved ate of Z$6.88 per 8-hour day. In total the 425 man-days cost Z$2900.00. Some 80% of this was incurred in the construction of earthworks while 15% and 5% respectively were incurred in carrying out site surveying and in fabricating and installing the hydraulic control structures. Six disused corrugated iron sheets were used to fabricate the hydraulic control structures. At a nominal price of Z$35 a sheet, this amounted to Z$210. The cost of miscellaneous items such as bitumen paint, wire brushes, material for setting out works etc was Z$340.

Accordingly, the total direct cost (direct labour and materials) incurred in establishing the two-water harvesting/irrigation schemes was Z$3025. This, when spread over the total developed area (3.5 ha) amounts to Z$864 (£86) per hectare development cost. Although these schemes can only provide opportunistic irrigation on a supplementary basis compared to the irrigation on demand provided by the formal irrigation scheme, their unit development cost thus incurred is only a minute fraction of those of conventional schemes.

2.2.6.8 Affordable Technology

With the overall scope of the programme, most of the earth work would be carried out less intensively by family labour during off-season, and more efficiently. So the opportunity cost of productive labour, in practice, would be substantially less than the financial cost paid to the hired labour. On this basis, therefore, the true economic unit cost of the water harvesting developments could be as low as Z$250 (£25) per hectare. At such low developments costs the local farmers could afford to build water harvesting/irrigation schemes like those demonstrated. Also, the technology adopted is simple and the resources mobilised within easy grasp of most peasant farmers in the communal areas. Farm family labour required to develop a hectare of land is likely to be less than 75-man days, an input the family can well afford during agricultural off season.

2.2.6.9 Problems and Prospects

Some problems encountered during the visit are set out below with the hope that action can be taken to ameliorate them. Some however are inherent in the programme because of the very nature of the funding and institutional arrangements under which the project was set up.

PART 3: Science and Innovation

MS.ID-IJPBS-2596

International Journal of Pharmacy and Biological Sciences-IJPBSTM (2025) 15 (4): 72-91

Online ISSN: 2230-7605, Print ISSN: 2321-3272

Research Article | Biological Sciences | OA Journal | MCI Approved | Index Copernicus

Received: 29 May 2025/ Accepted: 2 Jun 2025 / Published online: 1 Oct 2025

**Corresponding Author Email: mtetwaanswer@gmail.com*

3.0 Title: Preparation of antimicrobial soap using *Kigelia africana* (sausage tree)-extracted phytochemicals.

Authors:

Answer Mtetwa, Department of Applied Chemistry, National University of Science and Technology (NUST), Bulawayo, Zimbabwe

Dr. Audacity Maringa, Department of Applied Chemistry, National University of Science and Technology (NUST), Bulawayo, Zimbabwe

Dorcas Nyama, Department of Applied Chemistry, National University of Science and Technology (NUST), Bulawayo, Zimbabwe

Corresponding Author: *Answer Mtetwa: mtetwaanswer@gmail.com*

: https://www.linkedin.com/in/answer-mtetwa-a125b2269

3.1 Abstract

Bacterial skin infections are prevalent, and the use of synthetic antibacterial agents in commercial soaps has raised concerns about health and environmental impacts. This study explores *Kigelia africana*-extracted phytochemicals, specifically flavonoid rich extracts, as natural antibacterial agents in soap production. Phytochemical screening, extraction, and antimicrobial susceptibility testing were conducted on leaves, bark and fruit samples collected from Bulawayo, Zimbabwe. The bark sample showed superior effectiveness against *Escherichia coli* and *Staphylococcus aureus*. Methanol was used as the preferred extraction solvent due to its superior efficacy in extracting phytochemicals with high antimicrobial activity against *E. coli* and *S. aureus*, compared to distilled water and ethanol. The flavonoid rich extracts, extracted with methanol exhibited the highest antimicrobial activity amongst other extracted phytochemicals such as tannins, saponins, and alkaloids, therefore the flavonoid-rich extracts were used to manufacture antimicrobial soap. The manufactured soap not only demonstrated superior antimicrobial activity but also exhibited good skin compatibility, with no irritation observed in a 2-week study involving human volunteers. The manufactured soap outperformed commercial soaps (containing Triclosan and Active silver as synthetic antimicrobial reagents) in all assessed aspects, demonstrating superior antimicrobial activity, pH, foam height, foam retention, and percentage-free alkali. The Single factor ANOVA and Tukey's Post Hock data analysis results validated the manufactured soap's antimicrobial activity. This research contributes to developing natural, effective, and safer alternatives to synthetic antimicrobial agents, reducing health and environmental risks. The findings suggest *Kigelia africana* flavonoid crude extracts are a promising source of natural antibacterial agents for soap production, offering a potential solution to the drawbacks of synthetic agents. This study's results have implications for the development of sustainable and eco-friendly antimicrobial products.

Keywords

1. *Kigelia africana*
2. Phytochemical extraction
3. Phytochemicals (flavonoids, saponins, alkaloids and tannins)
4. Phytochemical Qualitative analysis
5. Percentage yield
6. Antimicrobial activity
7. Kirby Bauer disc diffusion
8. Antimicrobial soap
9. Natural antimicrobial agents
10. Synthetic antimicrobial agents
11. Soap production
12. Health consequences
13. Environmental impact
14. *Staphylococcus aureus*
15. *Escherichia coli*

3.1.1 Introduction

3.1.1.1Background of the study

Pathogenic bacteria cause a wide range of ailments, especially when conditions are suitable for their growth (Asumang *et al.*, 2021). According to the study by Asumang et al., (2021), infectious diseases continue to be a major cause of illness and death in many countries around the world, particularly in developing countries. Other diseases, such as wounds and the infections that develop from them, can risk people's lives (Asumang *et al.*, 2021).

Antimicrobial formulated soaps are recognized as a solution to lower the number of bacteria on the surface of the skin and other distinct materials used in the home (Teniola *et al.*, 2019). *Staphylococcus aureus* and *Escherichia coli* are significant sources of infection in humans and animals (Bachir and Abouni, 2015). They cause infections of the skin and soft tissues, the surgical site, and the bones and joints (Bachir and Abouni, 2015). *Staphylococcus aureus* is a leading cause of hospital-acquired bacteremia and has been associated to respiratory tract infections (Bachir and Abouni, 2015). *Escherichia coli* is the most common cause of UTIs, as well as enteric and systemic infections (Bachir and Abouni, 2015).

According to the study by Chaudhari (2016), an antibacterial soap can remove 65% to 85% of bacteria from human skin. Antibacterial soaps have been experimentally proven to kill bacteria at a specified dose. They also have antimicrobial action and can limit the development of bacteria (Chaudhari, 2016). Triclosan (TCS) and Triclocarban (TCC) are antimicrobials that are present in most consumer items, where they are routinely added to home soaps, detergents, disinfectants, cosmetics, and medical disinfectants to inhibit bacteria (Bakare and Adeyinka, 2022). Triclosan is a "down the drain" pollutant that is conveyed in home sewage to municipal WWTPs (waste water treatment plants), eventually ending up in wastewater effluents and, lastly, in the aquatic environment, most commonly in surface water, groundwater, or partitioned into the soil or sediment (Bakare and Adeyinka, 2022). In this context, aqueous environmental factors such as temperature, photolysis, and humidity, as well as living microorganisms (microalgae, fungi, and micro bacteria), are capable of degrading parent Triclosan into more harmful daughter by-products such as chlorinated phenoxy-phenols, chlorinated phenols, and trihalomethanes (Bakare and Adeyinka, 2022).

The negative effects of AgNPs antimicrobials are linked to free silver ions, but they may also penetrate the blood-brain barrier and impair both short- and long-term memory (Talapko *et al.*, 2020). Smaller silver nanoparticles have a bigger surface area and are more reactive, they have more serious adverse effects than larger ones (Talapko *et al.*, 2020). Depending on the kind of silver molecule found in the environment, silver can have harmful bio-toxic consequences (Talapko *et al.*, 2020).

According to the study by Kaushik et al., (2020), natural products are an important part of the human health care system since there is widespread concern about the toxicity and side effects of modern medications. Since ancient times, *Kigelia africana* has been utilized to treat human illnesses (Nabatanzi *et al.*, 2020). It is also the sole species in the genus *Kigelia* and a member of the Bignoniaceae family (Nabatanzi *et al.*, 2020). According to the study by Nabatanzi et al., (2020), the traditional uses of *Kigelia africana*, which include treating gynecological issues, cancer, and skin ailments, have been documented by ethno botanists. The study notes that scientists have been interested in this and have looked at the bioactivity of plant components of *Kigelia africana*.

Herbal soaps are manufactured by incorporating dried herbs, flowers, and stems into the soap base (Chaudhari, 2016). One very telling quote from Chaudhari's research is that, "herbs are the natural products could be found in the treatment of almost all diseases and skin problems owing to their high medicinal value, cost effectiveness, availability and compatibility, hence it can be used in soap base". The soap's properties include tenderness on the skin, rich foam, protection against skin diseases (including rashes, eczema, and scabies), treatment of infections that affect the skin (such as ringworm), protection of even skin tone, and smoothness of the skin (Chaudhari, 2016).

This research project will conduct a thorough analysis of *Kigelia africana*, including phytochemical screening, antimicrobial susceptibility testing, and antimicrobial soap manufacture. The study makes use of leaf, bark, and fruit samples gathered from Bulawayo, Zimbabwe. Phytochemicals are extracted using cold maceration and soxhlet extraction procedures, utilizing solvents such as ethanol, methanol, and distilled water. The most effective extraction process is identified via phytochemical screening. The extracts with high antimicrobial susceptibility will be further processed to separate the class of phytochemicals, with the class with the highest antimicrobial activity being utilized to make antibacterial soap. This experiment is unique in that it focuses on using a class of phytochemicals to manufacture antimicrobial soap, whereas earlier studies have often employed crude phytochemical extracts for this process.

3.1.1.2 Classification

Kigelia africana

- **Kingdom**: Plantae-Plants
- **Shona name**: Mubvee
- **Subkingdom**: Tracheobionta-Vascular plants
- **Class**: Magnoliopsida-Dicotyledons
- **Order**: Scrophulariales
- **Family**: Bignoniaceae - Trumpet-creeper family
- **Species**: *Kigelia africana* (Lam.) Benth.- Sausage Tree

3.1.1.3 Statement of the problem

The extensive use of synthetic antibacterial agents in commercial soaps, including silver nanoparticles, Triclocarban, and Triclosan, has aroused serious concerns about their negative health consequences and environmental impact. These compounds may provide health hazards such as antibiotic resistance, hormone disruption, and carcinogenicity, while they may also pollute water and destroy aquatic life. Notably, the FDA issued a Final Rule effective September 6, 2017, declaring that 19 active ingredients including Triclosan and Triclocarban used in over-the-counter (OTC) consumer antiseptic wash products are not Generally Recognized as Safe and Effective (GRAS/GRAE). The negative effects of AgNPs antimicrobials are linked to free silver ions, but they may also penetrate the blood-brain barrier and impair both short- and long-term memory (Talapko *et al.*, 2020). As a result, it is critical to discover alternative antibacterial agents that are both environmentally friendly and safe. The purpose of this study is to look into the potential of *Kigelia africana* (sausage tree) extracted phytochemicals as a natural and effective source of antibacterial agents in soap manufacture, offering a long-term alternative to reduce the environmental and health effects of synthetic antibacterial agents.

Aim

To investigate the potential of *Kigelia africana* extracted phytochemicals as a natural and effective source of antibacterial agents in soap production, evaluating their antimicrobial activity, physicochemical properties (pH, foam height, foam retention, percentage free alkali), and skin compatibility.

Research Objectives

1. To screen the phytochemicals present in the leaves, bark, and fruit of *Kigelia africana.*
2. To extract the phytochemicals from *Kigelia africana* using cold maceration and soxhlet extraction methods.
3. To evaluate the antimicrobial susceptibility of the extracted phytochemicals against selected microorganisms.
4. To identify the isolated class of phytochemical with the highest antimicrobial activity.
5. To manufacture the antimicrobial soap from the class of phytochemical that exhibited the highest antimicrobial activity.
6. To determine the antimicrobial susceptibility of the manufactured antimicrobial soap and that of available commercial antimicrobial soaps.
7. To carry out soap evaluation tests (pH, foam height, foam retention, percentage free alkali, and skin compatibility).

Research Questions

i. Did the extracted phytochemicals of *Kigelia africana* have antimicrobial properties?
ii. Which isolated class of phytochemical exhibited high antimicrobial susceptibility?
iii. Did the manufactured antimicrobial soap exhibit better antibacterial susceptibility as compared to commercially used antibacterial soaps?
iv. Did the manufactured antimicrobial soap pass the evaluation tests?

Hypothesis of the study

Null hypothesis– The manufactured antimicrobial soap does not exhibit antimicrobial activity.

Alternative hypothesis – The manufactured antimicrobial soap exhibits antimicrobial activity.

3.1.1.4 Significance of the study

This study aims to develop a natural, effective, and safer alternative to synthetic antimicrobial agents, which can reduce the risk of adverse health effects and environmental pollution. By providing a natural antimicrobial soap, this study contributes to the reduction of antimicrobial resistance, promotes sustainable practices, empowers communities to adopt natural personal hygiene practices, and paves the way for further research and development of natural antimicrobial agents from African plants, ultimately improving public health and well-being.

3.1.2 Materials and Methods

3.1.2.1 Chemicals and reagents used

Acquired reagents

Petroleum ether (Sigma Aldrich, Catalog No. 232-453-7), Ammonium hydroxide (Sigma Aldrich, Catalog No. 215-647-6), Picric acid (Sigma Aldrich, Catalog No. 201-865-9), Copper Sulphate (Sigma Aldrich, Catalog No. 231-847-6), Ferric Chloride (Sigma Aldrich, Catalog No. 231-729-4), Palm oil (Sigma Aldrich, Catalog No. 232-316-1), Coconut oil (Sigma Aldrich, 232-282-8), Mueller Hinton agar (Sigma Aldrich, Catalog No. 1272/2008), 99.9% ethanol (Associated Chemical Enterprise, Catalog No. 200-578-6), 99.5% Methanol (Associated Chemical Enterprise, Catalog No. 200-659-6), n-Hexane (Associated Chemical Enterprise, Catalog No. 203-777-6), diethyl ether (Associated Chemical Enterprise, Catalog No. 200-467-2), 32% HCL (Associated Chemical Enterprise, Catalog No. 231-595-7), Sodium sulphate anhydrous (Associated Chemical Enterprise, Catalog No. 231-820-9), Iodine solution (Associated Chemical Enterprise, Catalog No. 231-442-4), Lead acetate (Associated Chemical Enterprise, Catalog No. 206-104-4), Sodium hydroxide pellets (Associated Chemical Enterprise, Catalog No. 215-185-5).

3.1.2.2 Prepared reagents

20% ethanol was prepared by diluting 20 ml of 99.9% ethanol with distilled water to the mark in a 100 ml volumetric flask. 75% methanol was prepared by diluting 150.75 ml of 99.5% methanol with distilled water to make up 200 ml in a beaker. 5% NaCl was prepared by dissolving 5 g of NaCl in 100 ml of distilled water. Hager's reagent was prepared by dissolving 1 g of picric acid in 100 ml of distilled water. Wagner's reagent was prepared by dissolving 2 g of iodine and 6 g of KI in 100 ml of distilled water. 10% NaOH and 20% NaOH were prepared by dissolving 10 g and 20 g of NaOH in 100 ml of distilled water respectively. 5% ferric chloride was prepared by dissolving 5 g of ferric chloride in 100 ml of distilled water. 1% lead acetate was prepared by dissolving 1 g of lead acetate in 100 ml of distilled water. 1.5% v/v HCl was prepared by 4.69 ml of 32% HCl with 95.31 ml of distilled water. 25% ammonium hydroxide was used as provided.

3.1.3 Plant collection

For this study, a comprehensive plant sampling was conducted in Bulawayo, Zimbabwe, during the summer season when temperatures reached 29°C. From a single plant specimen, samples were collected from various parts, including the leaves, bark and fruit. The plan was growing in red clay soils. This sampling approach allowed for an in-depth examination of the plant's characteristics in its natural environment. The national herbarium and botanic gardens plant identification report is shown in the image below:

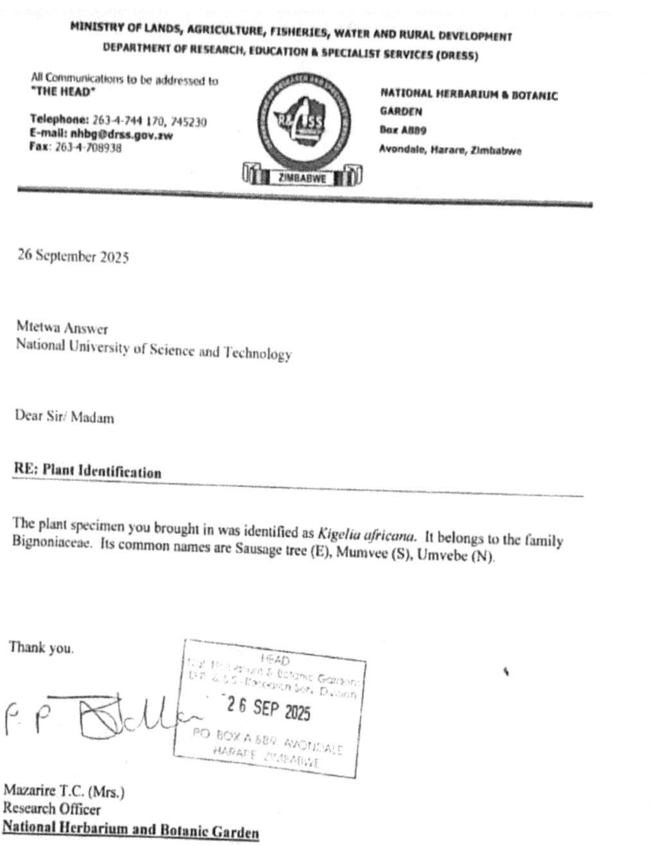
MINISTRY OF LANDS, AGRICULTURE, FISHERIES, WATER AND RURAL DEVELOPMENT
DEPARTMENT OF RESEARCH, EDUCATION & SPECIALIST SERVICES (DRESS)

All Communications to be addressed to
"THE HEAD"

Telephone: 263-4-744 170, 745230
E-mail: nhbg@drss.gov.zw
Fax: 263-4-708938

ZIMBABWE

NATIONAL HERBARIUM & BOTANIC GARDEN
Box A889
Avondale, Harare, Zimbabwe

26 September 2025

Mtetwa Answer
National University of Science and Technology

Dear Sir/ Madam

RE: Plant Identification

The plant specimen you brought in was identified as *Kigelia africana*. It belongs to the family Bignoniaceae. Its common names are Sausage tree (E), Mumvee (S), Umvebe (N).

Thank you.

HEAD
Nat. Herbarium & Botanic Gardens
DR & SS - Extension Serv. Division
26 SEP 2025
PO BOX A 889 AVONDALE
HARARE ZIMBABWE

Mazarire T.C. (Mrs.)
Research Officer
National Herbarium and Botanic Garden

Figure 1: National Herbarium & Botanic gardens plant identification report

3.1.3.1 Preparation of samples

The leaf, bark and fruit samples were washed carefully with distilled water to remove dust and foreign matter. The different plant organs were left to dry in sunlight for 15 days. The dry samples were ground into powder form using pestle and mortar for extraction purposes.

3.1.4 Extraction

3.1.4.1 Soxhlet extraction procedure

10 g of dry, ground *Kigelia africana* bark, fruit and leaf were added to different 100 ml beaker respectively. The plant material was transferred into a

thimble and placed into a soxhlet extractor. Extraction was carried out separately with 150 ml of distilled water, ethanol and methanol according to their polarities for 4 hours. Each plant material was extracted with the different solvents. The extracts were concentrated by using a rotary evaporator at 40 ºC to get the crude extract (Swarnalatha, Kishore Babu and Babu, 2019).

3.1.4.2 Cold maceration extraction

a) Ethanol, Methanol and Distilled water extracts:

10 g of plant powder was added to 150 mL of the extraction solvent in a conical flask and was covered with aluminum foil to avoid evaporation of ethanol. After 2 days, the extract was filtered with Whatman (No 1) filter paper, in order to obtain an aqueous extract and solvent was removed completely under reduced pressure, according to the study by Hussain *et al.*, (2016). Filtered extract was collected in a conical flask. The ethanol extract was evaporated by rotary evaporator at 40 ºC to get the crude extract (Hussain *et al.*, 2016).

3.1.4.3 Extraction of crude alkaloids

Powdered plant material (20 g) was wetted with 30 ml of NH_4OH (25%, m/m) and room temperature solvent extraction was performed with 300 ml of ethyl acetate for 72 hours. The extract was filtered and the solvent was evaporated in a rotary evaporator under reduced pressure at 40°C. The residue dissolved in 40 ml of distilled water and acidified with H_2S0_4 to pH 3-4, was extracted with petroleum ether and diethyl ether to remove lipophilic, acidic and neutral material. After basifying the aqueous solution to pH 9-10 with NH_4OH (25%, m/m), it was extracted with chloroform, the extract washed with distilled water to neutral pH, dried with Na_2SO_4 and concentrated to dryness under reduced pressure to obtain crude alkaloids (Djilani *et al.*, 2006).

3.1.4.4 Extraction of crude flavonoids

Powdered plant material (30 g) was defatted with n-hexane (250 ml) using Soxhlet extraction. The extraction was carried out for 4 hours at a temperature between 65-70 ºC. The defatted marc was further extracted with methanol (250 ml) at 60 °C and the solvent in the extract was evaporated on a rotary evaporator (Mohammed *et al.*, 2023).

3.1.4.5 Extraction of crude tannins

Crude tannins were extracted by placing 200 ml of 75% methanol into conical flask containing 20 g of the sample, then put into a beaker glass that was filled with water. The conical flask was placed in an ultrasonic water bath and extracted for 20 min at room temperature. The sample was centrifuged at 4°C for 10 min. The procedure was repeated twice and the supernatants combined. The solvent in the extract was evaporated using a rotary evaporator (Yuliana *et al.*, 2014)

3.1.4.6 Extraction of crude saponins

Crude saponins were extracted by heating the sample (20 g) for 4 hours in a water bath at 55 °C with 100 ml of 20% ethanol. The extract was filtered and the residue was re-extracted with 200 ml of 20% ethanol. The extract was concentrated on a water bath till the volume reduced to 40 ml, which was mixed with 20 ml of diethyl ether in a separating funnel. The mixture was shaken and then the separating funnel was fixed in a stand till the development of aqueous and diethyl layer. Aqueous portion was collected while the diethyl ether was discarded. To the aqueous layer, n-butanol (60 ml) was added and properly mixed by shaking. The n- butanol extract was treated with 10 ml of 5% NaCl solution. The resultant solution was concentrated on a water bath and the crude saponins were dried in an oven (Yuliana *et al.*, 2014).

Percentage yield

The percentage yield after rotary evaporation, was calculated using the following formula:

$$Percentage\ yield\ (\%) = \frac{\mathrm{m1-m2}}{m3} * 100\ \% \qquad (1)$$

Where:

m1 = mass of round bottom flask + dry mass of extracts

m2 = mass of empty round bottom flask

m3 = dry mass of sample

3.1.5 Qualitative phytochemical screening

3.1.5.1 Preparation of analytical reagents

Hager's reagent: 1 g of picric acid was dissolved in 100 ml of distilled water (Fatima and Lokare, 2019)

Wagner's reagent: 2 g of iodine and 6 g of KI was dissolved in 100 ml of distilled water (Fatima and Lokare, 2019)

Test for alkaloids

a) **Hager's test:**

2 mg of the test extract was taken in a test tube; a few drops of Hager's reagent was added. Formation of a yellow precipitate indicated the presence of alkaloids (Fredrick *et al.*, 2014).

b) **Wagner's test:**

2 mg of extract was acidified with 1.5% v/v of hydrochloric acid and a few drops of Wagner's reagent were added. Formation of a reddish brown precipitate indicated presence of alkaloids (Fredrick *et al.*, 2014).

3.1.5.2 Test for flavonoids

10 ml of ethyl acetate was added to 0.2 g of the extract and heated on a water bath for 3 minutes. The mixture was cooled, filtered and the filtrate was used for the following test:

a) **10% NaOH test:** 1 ml of 10% NaOH was added to 2 ml of the extract. Formation of a yellow solution indicated the presence of flavonoids (Fredrick *et al.*, 2014).

3.1.5.3 Test for saponins

To 20 ml of distilled water, 0.5 g of the extract was added and boiled on a hot water bath for 2 minutes and filtered. The filtrate was allowed to cool and was used for the following test:

a) **Foaming test**: 5 ml of the filtrate was diluted with 15 ml of distilled water and shaken vigorously. Formation of stable Foam indicated the presence of saponins (Fredrick *et al.*, 2014).

3.1.5.4 Test for tannins

To 0.5 g of the extract, 20 ml of water was added, boiled and, filtered and used for the following test:

a) **Lead acetate test:**

To 3 ml of the filtrate, 1% lead acetate solution was added. Formation of a precipitate indicated the presence of tannins (Fredrick *et al.*, 2014).

3.1.5.5 Test for phenols

a) **Ferric chloride test:** 2 ml of 5% ferric chloride solution was added to 1 ml of extract. Formation of a blue-black solution indicated the presence of phenols (Swarnalatha, Kishore Babu and Babu, 2019).

3.1.6 Antibacterial activity assay:

Two bacterial strains were collected from the environment using aseptic techniques. Necessary materials such as sterile loops and containers were prepared. The hands were disinfected by washing with soap and water. Alcohol was applied to he hands and allowed them to dry. The area around collection site was cleaned and disinfected and any debris was removed. A sterile loop was used to collect bacteria and the loop was placed into a sterile container and the samples were labelled and transported to the Microbiology lab at National University of Science and Technology, Zimbabwe. The bacteria were propagated in Microbiology lab at National University of Science and Technology, Zimbabwe. These strains included *Staphylococcus aureus* and *Escherichia coli*. The Gram stain procedure was used to identify the bacteria, with *Staphylococcus aureus* staining purple because it is a Gram-positive bacteria and *Escherichia coli* staining pink because it is a Gram-negative bacterium. The minimum inhibitory concentration of the various extracts against the bacteria strains (*E. coli* and *S. aureus*) was determined by the Kirby Bauer disc diffusion technique as described below:

3.1.6.1 Kirby Bauer disc diffusion method

The diameter of the zone of clearance of the various extracts against the bacteria strains (*E. coli* and *S. aureus*) was determined by using a ruler (empirically) in the Kirby Bauer disc diffusion technique. Nutrient broth medium was prepared by suspending 13 g of the nutrient broth in 1 liter of distilled water. It was sterilized in an autoclave at 121°C for 15 minutes. After autoclaving, the bacteria (*E. coli* and *S. aureus*) were inoculated into the nutrient broth and incubated for 24 hours at 37 °C to allow for bacteria growth. The discs to be soaked into the extracts were autoclaved at 121°C for 15 minutes. The discs were soaked into the extracts for 4 hours. Bacterial isolate was spread on sterilized Mueller Hinton agar plates respectively with the aid of a glass spreader. The soaked discs of each extract were then placed on its designated area on the petri dishes containing the respective bacteria strains. The inoculated plates with the extract were allowed to stay on the bench for about one hour; this is to enable the extract to diffuse on the agar. The plates were then incubated at 37°C for 24 hours. At the end of incubation period, the plates were observed for any evidence of inhibition which will appear as a clear zone that was completely devoid of growth around the discs (zone of inhibition). The diameter of zones was measured using transparent ruler calibrated in millimeter and the result were recorded (Kaushik *et al.*, 2020).

3.1.7 Soap making

3.1.7.1 Solid soap

A weight of 12 g of coconut oil and 54 g of palm oil was mixed and stirred well at room temperature for 15 minutes. Thereafter, 51 mL of 20 % NaOH solution was added and the soap mixture was well stirred approximately for 30 minutes until formation of soap was visible. 9.3190 g of the active *Kigelia africana* flavonoid rich extract was added to the measured weight of the above soap mixture and mixed well to prepare a solid soap containing 7% (w/w) extract. The concentration of 7% (w/w) was determined to be ideal after evaluating soap formulations with 1% (w/w), 4% (w/w), and 7% (w/w) extract concentrations. The mixture was poured into moulds, covered with a plastic sheet, and allowed to solidify for 12 hours. The resultant solid soaps were taken out of the moulds and kept in open air for 7 days for drying. Control soap was prepared using the same procedure without the addition of the plant extract (Wmank and Bgk, 2016).

3.1.8 Soap evaluation tests

3.1.8.1 pH:

1 g of soap sample was dissolved in 10 ml of water and a pH meter was used to measure the pH range of the sample (Ahmed *et al.*, 2021).

3.1.8.2 Foam retention:

1% of soap solution i.e., 25 ml taken in 100 ml measuring cylinder, was covered with hand and shaken for 10 minutes, and left to stand. The foam retention time for a period of 10 minutes was recorded (Ahmed *et al.*, 2021).

3.1.8.3 Foam height:

0.5 g of sample was dissolved in 25 ml of distilled water. The solution was transferred to a 100 ml measuring cylinder; the volume was made up to 50 ml with water and shook by 25 strokes. It was kept aside for 2 minutes, the foam height above the aqueous volume was measured (Ahmed *et al.*, 2021).

3.1.8.4 Determination of percentage free alkali:

2 g of sample was dissolved in 20 ml of distilled water and boiled for 10 minutes and titrated with 0.1 N HCl while using phenolphthalein indicator (Ahmed *et al.*, 2021).

Percentage free alkali is calculated using the following formulas:

$$NaOH + HCL \rightarrow NaCl + H_2O \qquad (2)$$

$$\frac{\mathrm{M1\,V1}}{n1} = \frac{M2\,V2}{n2} \qquad (3)$$

Where: M1 = molarity of Base

M2 = molarity of Acid

V1 = volume of mixture

V2 = titre value

n1 = moles of base

n2 = moles of acid

$$Normality = Molarity * the\ number\ of\ proton\ present\ in\ acid \quad (4)$$

$$strength = \frac{amount\ of\ alkali}{cubic\ decimeters} = molarity\ of\ base * mol.wt \quad (5)$$

$$alkali\ in\ the\ volume\ of\ mixture = \frac{amount\ of\ alkali}{1000} * volume\ of\ distilled\ water \quad (6)$$

$$Percentage\ free\ alkali = \frac{amount\ of\ alkali}{amount\ of\ dissolved\ soap} * 100\% \quad (7)$$

3.1.8.5 Skin Compatibility test

A controlled use study was conducted to evaluate the skin compatibility of the antimicrobial soap under controlled conditions. Five volunteers were recruited and provided with the soap. The study protocol involved using the soap on the elbow and neck areas twice daily for a period of two weeks. These areas were chosen for their relatively sensitive skin. Participants were instructed to rinse thoroughly after each application (Leroy, 2017).

3.1.9 Soap anti-microbial test

The prepared soap was subjected to antimicrobial screening by disc diffusion method. Organisms used were *Escherichia coli* (*E. coli*) and *Staphylococcus aureus* (*S. aureus*):

3.1.9.1 Preparation and dilution of soap samples extract

With the help of sterile sharp knife, soaps were scraped atone side. 10-500 mg of each soap sample was weighed and dissolved in 1 ml of sterile distilled water separately and used for the preparation of discs. 10 – 500 mg concentration of each soap was used because it's the concentration range that gives the solution which is not too viscous thereby not causing problems in antimicrobial susceptible testing for the given study as in accordance to other studies (Chaudhari, 2016).

3.1.9.2 Preparation of sterile discs

Filter paper discs were prepared from Whatman's No.1 filter paper. Discs of 6 mm size were prepared in Petri plate and sterilize in an autoclave at 121 ℃ for 15 minutes. Each sterile disc was incorporated individually with 10-500 mg of soap solution. Paper discs soaked in the soap solution were allowed to stand for a period of four hours to ensure full saturation of the soap preparations (Chaudhari, 2016).

3.1.10 Assay of antimicrobial activity

Overnight cultures were kept ready for anti-microbial activity. Assay of the antimicrobial activity of soaps were done by disc diffusion method (Chaudhari, 2016):

3.1.10.1 Kirby Bauer disc diffusion assay

Agar disc diffusion method was used to detect antimicrobial assay. The standardized 0.1 ml saline suspension of test organisms were inoculated on the surface of sterile Mueller-Hinton agar plates. Sterile filter paper discs prepared from different concentrations of the various soap samples were aseptically transferred directly into the surface of plates with the help of a sterile forceps. All plates were incubated at 37°C for 24 hours and then were examined for zone of inhibition around the disk. The zone of inhibition was determined by measuring the diameter in millimeters of zone to which the soap inhibited the growth of the organism by using a ruler. Triplicate measurements were done for each concentration to overcome any technical errors that might have occurred during a single attempt (Chaudhari, 2016).

3.1.11 Data analysis method

Single factor ANOVA was performed for the inhibition zones (mm) of different soaps (Commercial soap containing Active silver, Commercial soap containing Triclosan and Manufactured antimicrobial soap) against *E. coli* and *S. aureus* at the concentration of soap at which the inhibition of the bacteria started to show, thereby determining the significance in the differences in inhibition zones of the soaps. After Single factor ANOVA method, the Tukey's post-hoc test was done to identify which specific group means were significantly different from each other (Lakshmi *et al.*, 2019).

3.1.12 Results

Phytochemical screening

Table 3.1.1 Qualitative phytochemical screening of Soxhlet and Cold maceration, Bulawayo leaf extracts

		Bulawayo, leaves (Soxhlet)	Bulawayo, leaves (Cold maceration)
S. No.	Screening tests	Methanol extract	Methanol extract
1.	Alkaloids:		
a)	Wagner's reagent	Present	Present
b)	Hager's reagent	Present	Present
2.	Flavonoids:		
a)	10% NaOH	Absent	Present
3.	Saponins:		
a)	Foam	Present	Present
4.	Tannins:		
a)	1% lead acetate	Absent	Present
5	Phenols:		
a)	5% ferric chloride	Absent	Present

Table 3.1.2: Qualitative phytochemical screening of cold maceration extracts from bark sample obtained in Bulawayo

		Bulawayo, bark sample		
S. No.	Screening tests	Methanol extract	Ethanol extract	Distilled water extracts
1.	Alkaloids:			
a)	Wagner's reagent	Present	Present	Present
b)	Hager's reagent	Present	Present	Present
2.	Flavonoids:			
a)	10% NaOH	Present	Present	Present
3.	Saponins:			
a)	Foam	Present	Present	Present
4.	Tannins:			
a)	1% Lead acetate	Present	Present	Present
5	Phenols:			
a)	5% ferric chloride	Present	Present	Present

Table 3.1.3: Qualitative phytochemical screening of cold maceration extracts from leaf sample obtained in Bulawayo

S. No.	Screening tests	Bulawayo, leaf sample		
		Methanol extract	Ethanol extract	Distilled water extracts
1.	Alkaloids:			
a)	Wagner's reagent	Present	Present	Present
b)	Hager's reagent	Present	Present	Present
2.	Flavonoids:			
a)	10% NaOH	Absent	Present	Present
3.	Saponins:			
a)	Foam	Present	Present	Present
4.	Tannins:			
a)	1% Lead acetate	Present	Absent	Absent
5	Phenols:			
a)	5% ferric chloride	Present	Absent	Present

Table 3.1.6: Qualitative phytochemical screening of cold maceration extracts from fruit sample obtained in Bulawayo

S. No.	Screening tests	Bulawayo, fruit sample		
		Methanol extract	Ethanol extract	Distilled water extracts
1.	Alkaloids:			
a)	Wagner's reagent	Present	Present	Present
b)	Hager's reagent	Present	Present	Present
2.	Flavonoids:			
a)	10% NaOH	Present	Present	Absent
3.	Saponins:			
a)	Foam	Present	Present	Present
4.	Tannins:			
a)	1% Lead acetate	Present	Present	Present
5	Phenols:			
a)	5% ferric chloride	Present	Present	Present

Extraction and percentage yield of the dry class of phytochemicals from Bulawayo, bark sample

Flavonoid rich extracts

$$Percentage\ yield\ (\%) = \frac{108.1673 - 97.4250}{30.0275} * 100\ \% = 35.77\%$$

Tannin rich extracts

$$Percentage\ yield\ (\%) = \frac{101.7172 - 97.4250}{40.0350} * 100\ \% = 10.72\%$$

Saponin rich extracts

$$Percentage\ yield\ (\%) = \frac{146.5325 - 140.3502}{20.0250} * 100\ \% = 30.87\%$$

Alkaloid rich extracts

$$Percentage\ yield\ (\%) = \frac{97.9518 - 97.4250}{20.0000} * 100\ \% = 2.63\%$$

3.1.13 Antimicrobial susceptibility test

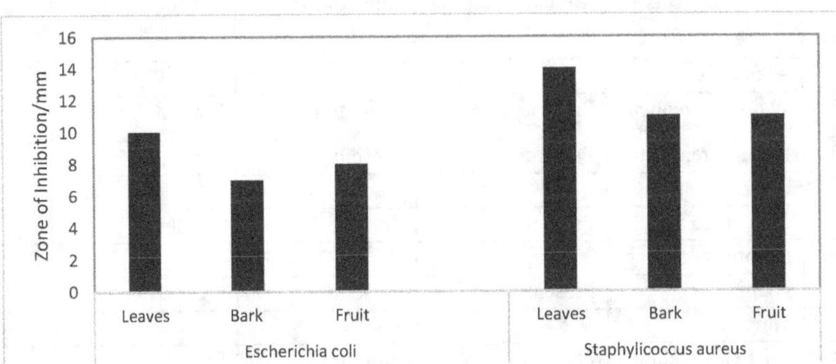

Figure 3.1.1: Antimicrobial susceptibility results of ethanol extracts from plant samples collected in Bulawayo

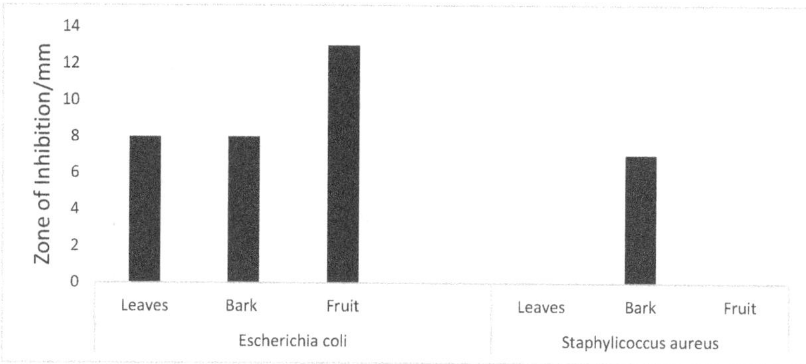

Figure 3.1.2: Antimicrobial susceptibility results of methanol extracts from plant samples collected in Bulawayo

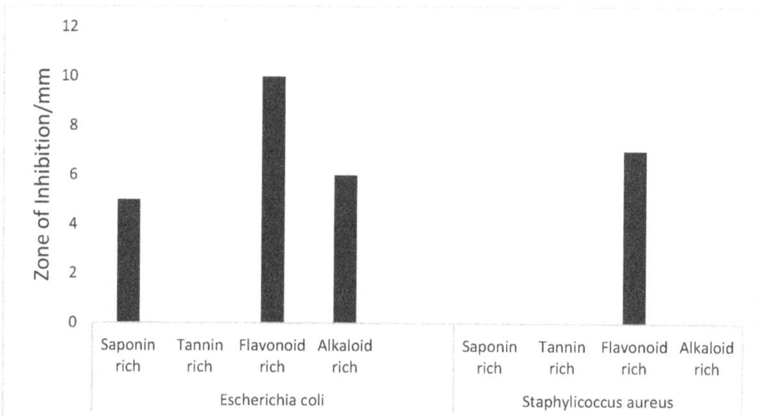

Figure 3.1.3: Antimicrobial susceptibility results of phytochemicals extracted from bark sample collected in Bulawayo

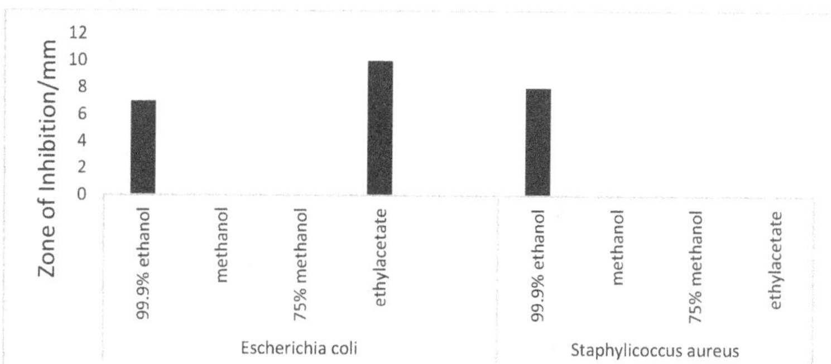

Figure 3.1.4: Antimicrobial susceptibility results of Negative controls (extraction solvents)

Table 3.1.5: Antimicrobial susceptibility results for the different soaps and their different zone of inhibition

		Zone of inhibition of the different soaps/mm			
Bacteria	**Concentration of soaps (mg/ml)**	**Commercial soap containing Active silver**	**Commercial soap containing Triclosan**	**Manufactured antimicrobial soap**	**Control soap**
E. coli	20	no clear zone of inhibition	no clear zone of inhibition	no clear zone of inhibition	no clear zone of inhibition
	60	no clear zone of inhibition	no clear zone of inhibition	no clear zone of inhibition	no clear zone of inhibition
	100	no clear zone of inhibition	no clear zone of inhibition	no clear zone of inhibition	no clear zone of inhibition
	200	10 mm	12 mm	11 mm	no clear zone of inhibition
		8 mm	10 mm	13 mm	no clear zone of inhibition
		9 mm	11 mm	12 mm	no clear zone of inhibition
	500	16 mm	16 mm	20 mm	no clear zone of inhibition
		15 mm	17 mm	21 mm	no clear zone of inhibition
		17 mm	15 mm	19 mm	no clear zone of inhibition
S. aureus	20	no clear zone of inhibition	no clear zone of inhibition	no clear zone of inhibition	no clear zone of inhibition
	60	no clear zone of inhibition	no clear zone of inhibition	no clear zone of inhibition	no clear zone of inhibition
	100	no clear zone of inhibition	no clear zone of inhibition	no clear zone of inhibition	no clear zone of inhibition
	200	no clear zone of inhibition	no clear zone of inhibition	10 mm	no clear zone of inhibition

		no clear zone of inhibition	no clear zone of inhibition	12 mm	no clear zone of inhibition
		no clear zone of inhibition	no clear zone of inhibition	11 mm	no clear zone of inhibition
	500	18 mm	16 mm	22 mm	no clear zone of inhibition
		17 mm	14 mm	20 mm	no clear zone of inhibition
		19 mm	15 mm	21 mm	no clear zone of inhibition

3.1.14 Soap production

Mass of soap mixture formed during agitation = 133.1230 g

Mass of crude flavonoid rich extract added to soap = 9.3190 g

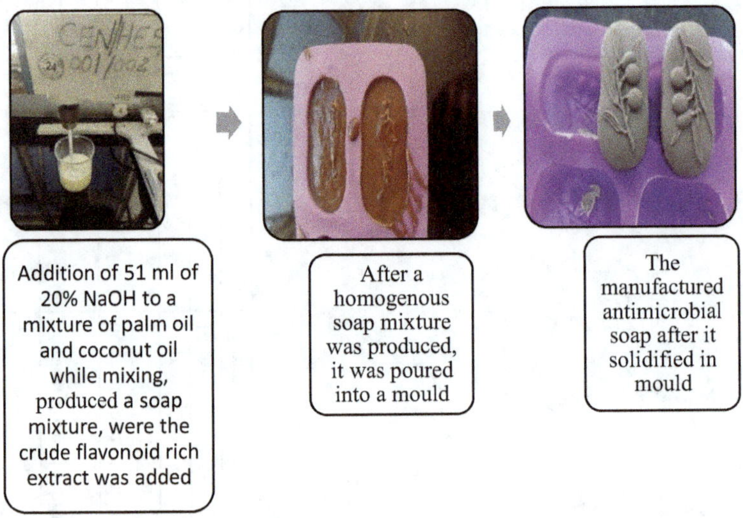

Figure 3.1.5: The process flow diagram for production of the antimicrobial soap

3.1.15 Soap evaluation tests

Table 3.1.7: Free alkali titration results for the soaps

Soap type	Manufactured antimicrobial soap	Commercial soap containing Active silver	Commercial soap containing Triclosan	Control soap
Titre value/cm^3	2.30	3.35	3.20	7.55

Table 3.1.7: Soap evaluation results

Soap Type	pH	Foam Height (cm)	Foam Retention	% Free Alkali	Skin compatibility
Commercial Soap containing Triclosan	10.50	8	Maintained a stable for 10 minutes	0.64	No irritation observed
Commercial Soap containing Active Silver	10.86	10	Maintained a stable for 10 minutes	0.67	No irritation observed
Manufactured Antimicrobial Soap	9.90	14	Maintained a stable for 10 minutes	0.46	No irritation observed
Control Soap	11.05	12	Maintained a stable for 10 minutes	1.51	No irritation observed

3.1.16 Data analysis

Table 3.1.8: Single factor ANOVA to compare the zones of inhibition values (mm) between different soaps (Commercial soap containing Active silver, Commercial soap containing Triclosan and Manufactured antimicrobial soap) against E. coli at 200 mg/ml soap concentration

Source of Variation	*SS*	*df*	*MS*	*F*	*P-value*	*F crit*
Between Groups	14	2	7	7	0.027	5.14325285
Within Groups	6	6	1			
Total	20	8				

Table 3.1.9: Single factor ANOVA to compare the zones of inhibition values (mm) between different soaps (Commercial soap containing Active silver, Commercial soap containing Triclosan and Manufactured antimicrobial soap) against S. aureus at 200 mg/ml soap concentration

Source of Variation	*SS*	*df*	*MS*	*F*	*P-value*	*F crit*
Between Groups	242	2	121	363	5.51E-07	5.143253
Within Groups	2	6	0.333333			
Total	244	8				

Table 3.1.10: Single factor ANOVA to compare the zones of inhibition values (mm) between different soaps (Commercial soap containing Active silver, Commercial soap containing Triclosan and Manufactured antimicrobial soap) against E. coli at 500 mg/ml soap concentration

Source of Variation	*SS*	*df*	*MS*	*F*	*P-value*	*F crit*
Between Groups	32	2	16	16	0.003936	5.143253
Within Groups	6	6	1			
Total	38	8				

Table 3.1.11: Single factor ANOVA to compare the zones of inhibition values (mm) between different soaps (Commercial soap containing Active silver, Commercial soap containing Triclosan and Manufactured antimicrobial soap) against S. aureus at 500 mg/ml soap concentration

Source of Variation	*SS*	*df*	*MS*	*F*	*P-value*	*F crit*
Between Groups	54	2	27	27	0.001	5.143253
Within Groups	6	6	1			
Total	60	8				

Table 3.1.12: Tukey's Post hock test results at 200 mg/ml against S. aureus (HSD = 1.45)

Soap Comparison	Mean Difference	Significant?
Manufactured vs. Active Silver	10	Yes
Manufactured vs. Triclosan	10	Yes
Active Silver vs. Triclosan	0	No

Table 3.1.13: Tukey's Post hock results at 500 mg/ml against S. aureus (HSD = 2.51)

Soap Comparison	Mean Difference	Significant?
Manufactured vs. Active Silver	4	Yes
Manufactured vs. Triclosan	6	Yes
Active Silver vs. Triclosan	2	No

Table 3.1.14: Tukey's Post hock results at 200 mg/ml against E. coli (HSD = 2.51)

Soap Comparison	Mean Difference	Significant?
Manufactured vs. Active Silver	3	Yes
Manufactured vs. Triclosan	1	No
Active Silver vs. Triclosan	2	No

Table 3.1.15: Tukey's Post hock results at 500 mg/ml against E. coli (HSD = 2.51)

Soap Comparison	Mean Difference	Significant?
Manufactured vs. Active Silver	4	Yes
Manufactured vs. Triclosan	4	Yes
Active Silver vs. Triclosan	0	No

3.1.17 Discussion

3.1.17.1 Qualitative phytochemical screening

Phytochemical screening revealed that cold maceration was the preferred method, as it yielded a broader range of phytochemicals (alkaloids, flavonoids, saponins, tannins, and phenols) compared to soxhlet extraction, which yielded alkaloids, saponins, and tannins as shown in Table 3.1.1. Cold maceration yielded a broad range of phytochemicals, which would aid in antimicrobial susceptibility due to the synergistic effect of the extract's combined phytochemicals, helping to combat antimicrobials (Vaou *et al.*, 2022). Tables 3.1.2-3.1.4 show the qualitative results for phytochemicals in the methanol, ethanol, and distilled water extracts of the Bulawayo-based bark, leaf, and fruit samples, respectively. The phytochemical screening results for the Bulawayo methanol cold maceration extracts of the leaves shown in Table 1 agreed with the study by Said et al., (2022), which showed presence of flavonoids, alkaloids, tannins, and saponins whereas the Bulawayo methanol soxhlet extracts of the leaves did not agree to the study by Said et al., (2022), as there were absence of flavonoids and phenols. According to Rodr et al., (2020), the absence of flavonoids from soxhlet extracts was due to degradation caused by the prolonged exposure of flavonoids to the high temperatures of the extraction process. Generally, during the extraction processes, the flavonoids are solubilized in organic solvents, and due to their weak acid character, the necessary conditions are created in the chemical medium, for an alcohol dehydration to occur by a mechanism of elimination, in this case, by the polyphenolic structure of the flavonoids. During the extraction process, those flavonoids exposed to high temperatures and having greater affinity to the solvent will degrade more easily. For this reason, avoiding the degradation of flavonoids ensures the obtention of extracts with greater flavonoid richness (Rodr *et al.*, 2020).

The phytochemical screening results for the Bulawayo methanol, ethanol, and distilled water extracts of the bark sample shown in Table 3.1.2 agreed with the study by Badgujar and Mistry (2017), which showed the presence of alkaloids, flavonoids, saponins, tannins, and phenols.

The phytochemical screening of methanol leaf extracts by Said et al., (2022), showed the presence of flavonoids, alkaloids, tannins, and saponins. In contrast, the phytochemical screening results for the Bulawayo methanol leaf extracts shown in Table 3.1.3, showed the presence of alkaloids, saponins, and phenols.

The findings of phytochemical screening of Bulawayo ethanol leaf extracts shown in Table 3 showed the presence of alkaloids, flavonoids, and saponins and was not per the study by Fredrick et al., (2014). The phytochemical screening of the methanol fruit extracts done by Oseni (2018), showed the presence of alkaloids, flavonoids, tannins, and saponins and this was per the findings of phytochemical screening of the Bulawayo methanol and ethanol fruit extracts as shown in Table 3.1.4.

The phytochemical screening of the distilled water fruit extracts done by Ojediran et al., (2024), showed the presence of alkaloids, flavonoids, tannins, saponins, and phenols, and the findings of the phytochemical screening of the Bulawayo distilled water fruit extracts shown in Table 3.1.4 gave an absence in flavonoids.

3.1.17.2 Extraction

Phytochemicals were extracted from *Kigelia africana* leaves, bark, and fruit samples collected from Bulawayo. Various solvents (ethanol, methanol, and distilled water) were used to determine the most suitable solvent for extracting phytochemicals, considering the sample texture and phytochemical nature. Two extraction methods, soxhlet extraction and cold maceration, were employed. Phytochemical screening revealed that cold maceration was the preferred method, as it yielded a broader range of phytochemicals (alkaloids, flavonoids, saponins, tannins, and phenols) compared to soxhlet extraction, which yielded alkaloids, saponins, and tannins as shown in Table 3.1.1. Cold maceration was then used to extract phytochemicals for antimicrobial susceptibility testing. The extraction was done under the same conditions for all samples, ensuring consistency in the extraction process and allowing for more accurate comparisons between samples. The classes of phytochemicals were extracted from the bark sample collected in Bulawayo because it was effective against *E. coli* and *S. aureus* as compared to the other samples. The bark sample collected in Bulawayo yielding a diverse composition of bioactive compounds. The percentage yields of the various phytochemicals extracted from the bark sample collected in Bulawayo are as follows: flavonoid rich extract (35.77%), saponin rich extract (30.87%), tannin rich extract (10.72%), and alkaloid rich extract (2.67%). These findings suggest that the bark sample is rich in flavonoids and saponins, which may contribute to its potential therapeutic applications.

3.1.17.3 Antimicrobial susceptibility test

For the antimicrobial susceptibility test, Mueller Hinton agar was used. This nutrient-rich medium is free from inhibitors that could interfere with bacterial growth, allowing almost all organisms plated on it to grow (Hudzicki, 2012). Additionally, it facilitates better diffusion of antibiotics, leading to a more accurate zone of inhibition (Hudzicki, 2012). The test bacteria used were *Escherichia coli* (*E. coli*) and *Staphylococcus aureus* (*S. aureus*), with *E. coli* being a gram-negative bacterium and *S. aureus* being a gram-positive bacterium. The bacteria were grown in nutrient broth at 37 °C in an incubator for 24 hours. Before use, discs for antimicrobial tests were sterilized to ensure they were free from microorganisms that could interfere with test results or grow alongside the target bacteria, leading to inaccurate interpretations.

The Bulawayo ethanol leaf, bark, and fruit extracts were determined for their antimicrobial susceptibility against *E. coli* and *S. aureus* as shown in Figure 3.1.1. The ethanol leaf extracts showed a 10 mm and 14 mm zone of inhibition against *E. coli* and *S. aureus*, respectively, in accordance with the study done by Hussain et al., (2016), which gave a 12 mm zone of inhibition against both *E. coli* and *S. aureus*. The ethanol bark extracts showed a 7 mm and 11 mm zone of inhibition against *E. coli* and *S. aureus*, respectively, whereas the study done by

Hussain et al., (2016) showed a 10 mm and 1 mm zone of inhibition against *E. coli* and *S. aureus*, respectively. The ethanol fruit extracts showed an 8 mm and 11 mm zone of inhibition against *E. coli* and *S. aureus*, respectively, whereas the study done by Hussain et al., (2016) showed that *E. coli* was resistant, with a 2 mm zone of inhibition against *S. aureus*. From the results, it was shown that the ethanol extracts may have a significant effect against *E. coli* and *S. aureus*. However, the negative control, which was the impact of the 99.9% ethanol used for extraction, showed a negative effect, with zones of inhibition of 7 mm and 8 mm against *E. coli* and *S. aureus*, respectively.

The Bulawayo methanol leaf, bark, and fruit extracts were determined for their antimicrobial susceptibility against *E. coli* and *S. aureus* as shown in Figure 3.1.2. The methanol leaf extracts showed an 8 mm zone of inhibition against *E. coli* and no clear zone of inhibition against *S. aureus*, respectively. This is consistent with the study by Said et al., (2022), which reported zones of inhibition of 6 mm and 8.70 mm against *E. coli* and *S. aureus*, respectively. The methanol bark extracts showed zones of inhibition of 8 mm and 7 mm against *E. coli* and *S. aureus*, respectively, whereas the study by Said et al., (2022) reported zones of inhibition of 9 mm and 7 mm against *E. coli* and *S. aureus*, respectively. The methanol fruit extracts showed a 13 mm zone of inhibition against *E. coli* and no clear zone of inhibition against *S. aureus*, respectively. The results indicate that the methanol bark extracts had a significant effect against *E. coli* and *S. aureus*, whereas the negative control (the impact of the methanol used for extraction) showed no effect as shown in Figure 3.1.4, with no clear zones of inhibition against *E. coli* and *S. aureus*. The methanol leaf and fruit extracts showed no clear zones of inhibition against *S. aureus*, suggesting that *S. aureus* was resistant to the extracts, likely due to the low concentration of available phytochemicals.

The Bulawayo distilled water leaf, bark, and fruit extracts were tested for antimicrobial susceptibility against *E. coli* and *S. aureus*. However, no zones of inhibition were observed, indicating that both bacteria were resistant, likely due to the low concentration of phytochemicals in the extracts. The study by Hussain et al., (2016) showed that *E. coli* was resistant to the distilled water leaf extracts, while a 5 mm zone of inhibition was observed against *S. aureus*. Additionally, the study found that the distilled water extracts inhibited both *E. coli* and *S. aureus* with a 3 mm zone of inhibition. The distilled water bark extract showed a 15 mm zone of inhibition against *S. aureus*, while *E. coli* was termed resistant (Hussain et al., 2016).

The antimicrobial susceptibility of the classes of phytochemicals extracted from the Bulawayo bark, the: alkaloid rich, saponin rich, flavonoid rich, and tannin rich extracts - was determined as shown in Figure 3.1.3. The saponin rich extract showed a 5 mm zone of inhibition against *E. coli* and no clear zone of inhibition against *S. aureus* as shown in Figure 3.1.3, indicating moderate sensitivity of *E. coli* and resistance of *S. aureus* to saponins. The 75% methanol extract of tannins showed no clear zone of inhibition against both *S. aureus* and *E. coli*, shown by Figure 3.1.3, indicating resistance of both bacteria to tannin extracts. The 75% methanol solvent, used as a negative control, produced no clear zones of inhibition against *S. aureus* and *E. coli*, showing no negative impact on the bacteria. In contrast, the methanol flavonoid rich extracts showed zones of inhibition of 10 mm and 7 mm against *E. coli* and *S. aureus*, respectively which is shown in Figure 3.1.3, indicating sensitivity of both bacteria to methanol flavonoid extracts. Figure 3.1.4 shows that the methanol solvent used to extract flavonoids had no clear zone of inhibition on both *E. coli* and *S. aureus*. In Figure 3.1.3, the ethyl acetate alkaloid rich extracts showed a 6 mm zone of inhibition against *E. coli* and no clear zone of inhibition against *S. aureus*. However, the ethyl acetate solvent used as a negative control showed a significant zone of inhibition against *E. coli* and no clear zone of inhibition against *S. aureus* as shown in Figure 3.1.4, indicating that the 6 mm zone of inhibition observed in ethyl acetate alkaloid rich extract against *E. coli* is due to the negative effect of the solvent. *S. aureus* was resistant to both the ethyl acetate alkaloid extracts and the ethyl acetate solvent. The methanol flavonoid rich extracts, which showed significant zones of inhibition against *E. coli* and *S. aureus*, were used to make antimicrobial soap.

Flavonoids have been found as polyphenolic chemicals that can show antibacterial activity via various methods (Shamsudin *et al.*, 2022). Numerous investigations have shown that flavonoids can inhibit nucleic acid synthesis, cytoplasmic membrane function, and energy metabolism (Shamsudin *et al.*, 2022). Flavonoids have also been shown to diminish adhesion, biofilm development, porin on the cell membrane, membrane permeability, and pathogenicity, all necessary for bacterial growth (Shamsudin *et al.*, 2022). As a result, developing and deploying flavonoid-based treatments might be a viable strategy for antimicrobial resistance (Shamsudin *et al.*, 2022).

For the soap antimicrobial susceptibility test, commercial soaps containing Active silver and Triclosan were compared to the Manufactured antimicrobial soap and control soap at concentrations of 20, 60, 100, 200 and 500 mg/ml as shown in Table 3.1.5. Commercial soap containing Active silver showed no clear zone of inhibition against *E. coli* and *S. aureus* at concentrations of 20, 60, 100 and 200 mg/ml, indicating that *E. coli* and *S. aureus* were resistant to the active silver synthetic antimicrobial reagent available at these concentrations. Commercial soap containing Active silver showed a significant zone of inhibition of 18 mm against *S. aureus* at 500 mg/ml, this indicated that the *S. aureus* was sensitive to the active silver antimicrobial reagent available at 500 mg/ml concentration of the soap as shown in Table 3.1.5. Commercial soap containing Active silver showed average significant zones of inhibition of 9 mm and 16 mm at 200 and 500 mg/ml concentrations of Commercial soap containing Active silver respectively against *E. coli.* This indicated that the active silver antimicrobial reagent available at these concentrations was effective against *E. coli.*

In Table 3.1.5, Commercial soap containing Triclosan soap also showed no clear zones of inhibition against *E. coli* and *S. aureus* at concentrations of 20, 60, 100 and 200 mg/ml, indicating that the bacteria *E. coli* and *S. aureus* were resistant to the concentration of the Triclosan antimicrobial reagent available at these concentrations. Commercial soap containing Triclosan showed an average significant zone of inhibition of 15 mm against *S. aureus* at 500 mg/ml, this indicated that the *S. aureus* was sensitive to the Triclosan antimicrobial reagent available at 500 mg/ml concentration of the soap. Commercial soap containing Triclosan soap showed average significant zones of inhibition of 11 mm and 16 mm at 200 and 500 mg/ml concentrations of Commercial soap containing Triclosan soap respectively against *E. coli.* This indicated that the Triclosan antimicrobial reagent available at these concentrations was effective against *E. coli.*

The control soap showed no clear zones of inhibition against *E. coli* and *S. aureus* at the concentrations of 20, 60, 100, 200 and 500 mg/ml as shown in Table 3.1.5, clearly indicating that the control soap had no antimicrobial capabilities.

The Manufactured antimicrobial soap also showed no clear zones of inhibition against *E. coli* and *S. aureus* at concentrations of 20, 60 and 100 mg/ml, indicating that the bacteria *E. coli* and *S. aureus* were resistant to the concentration of the flavonoid antimicrobial agent available at these concentrations. The Manufactured antimicrobial soap showed average significant zones of inhibition of 11 mm and 21 mm at 200 and 500 mg/ml respectively against *S. aureus*, this indicated that *S. aureus* was sensitive to the flavonoid antimicrobial agent available at these concentrations. The Manufactured antimicrobial soap showed significant zones of inhibition of 12 mm and 20 mm at 200 and 500 mg/ml concentrations of Commercial soap containing Active silver soap respectively against *E. coli*. This indicated that the flavonoid antimicrobial agent available at these concentrations was effective against *E. coli*.

S. aureus was resistant to some extracts, potentially due to mechanisms such as biofilm formation or efflux pump activity, although these were not explored in this study. The antimicrobial results differ from those reported in other studies, likely due to variations in bacterial strains sourced from different environments, which may have distinct inherent resistance properties. Additionally, differences in experimental conditions, including temperature, pH, and media used, may have also influenced the growth and resistance of *S. aureus* and *E. coli*, contributing to the observed discrepancies.

3.1.17.4 Soap production

The production of the antimicrobial soap was a cold process. It involved the saponification reaction between a mixture of palm and coconut oil with 20% NaOH. The process does not involve heat and it takes place at room temperature. The weighed amount of palm and coconut oil was mixed with an agitator to attain homogeneity. A measured volume of NaOH was poured into the oil mixture and stirred by an agitator until the formation of a soap was seen. Then the weight of the soap was measured to be 133.1230 g. The crude Flavonoid rich extract of 9.3190 g was added to make a 7% (wt/wt) soap. The concentration of 7% (w/w) was determined to be ideal after evaluating soap formulations with 1% (w/w), 4% (w/w), and 7% (w/w) extract concentrations. After the soap mixture was homogenous, it was poured into a mold and left to solidify over seven days. The Manufactured antimicrobial soap is shown in Figure 3.1.5.

The soap evaluation tests for the manufactured antimicrobial soap, commercial soap containing Active silver, commercial soap containing Triclosan, and control soap were performed as shown in Table 3.1.6-3.1.7. The soap evaluation tests included pH, foam height, foam retention, and percentage-free alkali. The pH of commercial soap containing Triclosan, commercial soap containing Active silver, the manufactured antimicrobial soap, and the control soap were 10.50, 10.86, 9.90, and 11.05, respectively as shown in Table 3.1.7. This showed that the manufactured antimicrobial soap had a moderate alkalinity compared to the commercial soaps. A pH of 9.90 is acceptable since the pH level of bar soaps typically ranges between 9 and 10, making them slightly alkaline (Ecocraft, 2023). The alkalinity of bar soaps helps to break down oils and remove dirt effectively, thereby making them suitable for individuals with oily or acne-prone skin, as the higher pH can help balance the excess oil production (Ecocraft, 2023). According to the study by Tarun et al., (2014), the majority of bar soaps used had a pH ranging between 9.01 and 11.

It is crucial to remember that, while pH is an important feature of a skincare product, it does not directly correlate with how harsh or gentle a product is (Twincraft, 2019). The acid mantle, a skinny, slightly acidic coating on the skin's surface, is a barrier to bacteria, viruses, environmental pollutants, and other possible contaminants (Twincraft, 2019). The acid mantle's pH ranges from 4.5 to 6.2, depending on age, gender, ethnicity, environment, and body part (Twincraft, 2019). Twincraft's study indicates that short-term contact with a mildly acidic or alkaline substance, such as bar soap, does not affect the acid mantle. Healthy skin can regulate its acid mantle in only a few minutes and long-term or chronic contact with a high pH substance, such as a lotion or other leave-on treatment, can cause damage and negatively impact the acid mantle (Twincraft, 2019).

The foam height of commercial soap containing Triclosan, commercial soap containing Active silver, the manufactured antimicrobial soap, and the control soap were 8 cm, 10 cm, 14 cm, and 12 cm, respectively shown in Table 3.1.7. The foam retention of commercial soap containing Triclosan, commercial soap containing Active silver, the manufactured antimicrobial soap, and the control soap maintained a stable foam for the stipulated 10 minutes. The percentage of free alkali in commercial soap containing Triclosan, commercial soap containing Active silver, the manufactured antimicrobial soap, and the control soap were 0.64%, 0.67%, 0.46%, and 1.51%, respectively as shown in Table 3.1.7. These results show that the manufactured antimicrobial soap had a lower percentage of free alkali compared to the commercial soaps (commercial soap containing Active silver and commercial soap containing Triclosan). Healthy skin has a pH of 5.4 to 5.9 (Sivani *et al.*, 2021). The low percentage of free alkali (0.46%) shown in Table 3.1.6 suggests a soap which is good for health and the environment (Betsy *et al.*, 2021). According to Bureau of Indian Standards (BIS), good quality soaps must have less than 5% of alkali content whereas according to ISO specification, soaps should have only below 2% of alkali content (Betsy *et al.*, 2021). The manufactured antimicrobial soap outperformed commercial soap containing Active silver and commercial soap containing Triclosan in all aspects of pH, foam height, foam retention, and percentage of free alkali.

The results of the skin compatibility test showed that the five volunteers who used the manufactured antimicrobial soap for two weeks did not experience any irritation. This is likely due to the soap's formulation, which resulted in a pH of 9.90, within the acceptable range for bar soaps. Although this pH is slightly alkaline, it is still within the range that is considered suitable for skin, particularly for individuals with oily or acne-prone skin.

Moreover, the manufactured antimicrobial soap had a lower percentage of free alkali (0.46%) compared to the commercial soaps tested. This low free alkali content suggests that the soap is gentle on the skin and less likely to cause irritation, which aligns with the observations made during the study.

The absence of irritation in the volunteers can be attributed to the skin's natural ability to regulate its acid mantle. According to Twincraft (2019), short-term exposure to mildly alkaline substances like bar soap does not significantly affect the skin's acid mantle, allowing healthy skin to recover quickly.

Overall, the manufactured antimicrobial soap's formulation appears to be gentle on the skin, making it a promising option for individuals seeking effective and skin-compatible cleansing products.

3.1.17.5 Data analysis

The results of the Single factor ANOVA test for the zones of inhibition of the different soaps (Commercial soap containing Active silver, Commercial soap containing Triclosan, and Manufactured antimicrobial soap) showed that there were significant differences between the zones of inhibition of the soaps against *E. coli* at a soap concentration of 200 mg/ml (p-value = 0.027 < alpha = 0.05) as shown in Table 3.1.8. Similarly, there were significant differences between the inhibition zones of the soaps against *E. coli* at a soap concentration of 500 mg/ml (p-value = 0.003936 < alpha = 0.05) as shown in Table 3.1.9. The same was observed for *S. aureus* at both soap concentrations: 200 mg/ml (p-value = 5.51×10^{-7} < alpha = 0.05) and 500 mg/ml (p-value = 0.001 < alpha = 0.05) as shown in Table 3.1.10 and Table 3.1.11 respectively. Since the p-values for both soap concentrations (200 mg/ml and 500 mg/ml) were less than the alpha level of 0.05, we reject the null hypothesis and accept the alternative hypothesis, indicating that the manufactured antimicrobial soap exhibited antimicrobial activity.

The Tukey's post-hoc test results revealed significant differences in antimicrobial activity between the manufactured soap and commercial soaps containing Active Silver and Triclosan. At 200 mg/ml against *S. aureus*, the manufactured soap showed significantly higher antimicrobial activity compared to both Active Silver and Triclosan soaps, with mean differences of 10 mm in both cases as shown in Table 3.1.12. This suggests that the manufactured soap has a potent antimicrobial effect against *S. aureus* at this concentration.

However, at 500 mg/ml against *S. aureus*, the manufactured soap showed significant differences in antimicrobial activity compared to both Triclosan and Active Silver soaps, with mean differences of 6 mm and 4 mm, respectively as shown in Table 3.1.13. This indicates that the manufactured soap has superior antimicrobial activity at this higher concentration.

Against *E. coli*, at 200 mg/ml, the manufactured soap showed significantly higher antimicrobial activity compared to Active Silver soap, with a mean difference of 3 mm, but not compared to Triclosan soap as shown in Table 3.1.14. At 500 mg/ml, the manufactured soap showed significantly higher antimicrobial activity compared to both Triclosan and Active Silver soaps, with mean differences of 4 mm in both cases as shown in Table 3.1.15.

Overall, the results indicate that the manufactured soap had superior or comparable antimicrobial activity to commercial soaps containing Active Silver and Triclosan, depending on the concentration and microorganism tested. These findings support the potential use of the manufactured soap as an effective antimicrobial agent.

Acknowledgments

I extend my sincere gratitude to Dr. A Maringa for his guidance and mentorship, and Mrs. D Nyama for her laboratory supervision. I'm also thankful to the Applied Biology and Biochemistry laboratory staff, particularly Jefferson Mathe and Belinda Chiangwa, for their technical assistance. Special thanks to my parents for their support and prayers. This project wouldn't have been possible without their contributions.

Declaration of Interest statement

The author(s) declares that there are no conflicts of interest associated with this research, including financial, personal, or organizational interests that could influence the outcome or interpretation of this study.

Declaration of Funding

The author(s) declare that no funding was received for this research project.

Data Availability statement

The data that support the findings of this study are available from the corresponding author upon reasonable request

3.1.18 References

[1] Ahmed L., Sarma D., Hazarika M., Formulation and evaluation of an ayurvedic bath soap containing extracts of three ayurvedic herbs, Journal of Medicinal Plants Studies, 9 (2): 115–117, (2021).

[2] Asumang P., Boakye Y.D., Agana T.A., Yakubu J., Entsie P., Akanwariwiak N.G., Adu F., Agyare C., Antimicrobial, antioxidant and wound healing activities of methanol leaf extract of Bridelia micrantha (Hochst.) Baill., Scientific African, 14: e00980, (2021).

[3] Bachir G., Abouni B., in: Méndez-Vilas A. (Ed.), *Escherichia coli* and *Staphylococcus aureus* most common source of infection, The Battle Against Microbial Pathogens: Basic Science, Technological Advances and Educational Programs, Formatex Research Center, Spain, 2015, pp. 637–648.

[4] Badgujar N.V., Mistry K.N., In-vitro antioxidative, cytotoxic activity and phytochemical screening from stem bark of *Kigelia pinnata*, International Journal of Green Pharmacy, 13 (1): 11–18, (2017).

[5] Bakare B.F., Adeyinka G.C., Occurrence and Fate of Triclosan and Triclocarban in Selected Wastewater Systems across Durban Metropolis, KwaZulu-Natal, South Africa, International Journal of Environmental Research and Public Health, 19 (11): 6769, (2022).

[6] Betsy K.J., Fathima R., Jilu M., Varkey J.T., Determination of Alkali Content & Total Fatty Matter in Cleansing Agents, Asian Journal of Science and Applied Technology, 2 (1): 8–12, (2021).

[7] Chaudhari V.M., Studies on antimicrobial activity of antiseptic soaps and herbal soaps against selected human pathogens, International Journal of Current Microbiology and Applied Sciences, 5 (6): 201–204, (2016).

[8] Djilani A., Legseir B., Soulimani R., Dicko A., Younos C., New extraction technique for alkaloids, Journal of the Brazilian Chemical Society, 17 (3): 518–520, (2006).

[9] Ecocraft., The importance of pH level for soap: Effects on skin, cleansing ability, and moisture, Ecocraftyliving.com, 2023. Available at: https://ecocraftyliving.com/ph-level-for-soap/ (Accessed 17 June 2024).

[10] Fatima S., Lokare P.D., Pharmacognostic Study of Leaves of *Kigelia pinnata* (Lam) Benth, Journal of Drug Delivery and Therapeutics, 9 (4): 594–597, (2019).

[11] Fluhr J.W., Darlenski R., Surber C., Glycerol and the skin: Holistic approach to its origin and functions, British Journal of Dermatology, 159 (1): 23–34, (2008).

[12] Fredrick A.C., Ebele O.P., Chioma O., Nedosa U., Analgesic, Phytochemical and Toxicological Investigations Of Ethanol Extract Of The Leaves Of *Kigelia africana* (Lam.) Benth (family Bignoniaceae)-Sausage Tree, Journal of Pharmaceutical and Biomedical Sciences, 4 (7): 588–595, (2014).

[13] Górniak I., Bartoszewski R., Króliczewski J., Comprehensive review of antimicrobial activities of plant flavonoids, Phytochemistry Reviews, 18 (1): 241–272, (2019).

[14] Hudzicki J., Kirby-Bauer Disk Diffusion Susceptibility Test Protocol, American Society For Microbiology, Washington D.C., 2012.

[15] Hussain T., Fatima I., Rafay M., Shabir S., Akram M., Bano S., Evaluation of antibacterial and antioxidant activity of leaves, fruit and bark of Kigelia africana, Pakistan Journal of Botany, 48 (1): 277–283, (2016).

[16] Kaushik P., Shrivastava V., Mehata P., Rai J.P., Screening of Aerial Parts of *Kigelia Pinnata* for its Antioxidant, Wound Healing and Antimicrobial Activity, Journal of Drug Delivery and Therapeutics, 10 (5): 86–93, (2020).

[17] Lakshmi A., Vishnurekha C., Baghkomeh P.N., Effect of theobromine in antimicrobial activity: An in vitro study, Dental Research Journal, 16 (2): 76–80, (2019).

[18] Leroy F., The Global Publication of the International Federation, Meat Science, 20 (3): 253–255, (2017).

[19] Mohammed Y.A., Kabiru A.Y., Hamzah R.U., Lukman H.Y., Aliu T.B., Alqarni M., Batiha G.E.S., De waard M., Lawal B., Rutin-Rich Flavonoid Subfraction of Annona senegalensis Mitigates Trypanosoma

brucei brucei Infection and Hematobiochemical Changes in Infected Mice, BioMed Research International, 2023: 6820338, (2023).

[20] Nabatanzi A., Nkadimeng S.M., Lall N., Kabasa J.D., McGaw L.J., Ethnobotany, phytochemistry and pharmacological activity of *Kigelia africana* (Lam.) benth. (bignoniaceae), Plants, 9 (6): 753, (2020).

[21] Ojediran T.K., Alagbe O.J., Victor D., Adewale E., Analysis of *Kigelia africana* (Lam.) Benth fruit powder's antioxidant and phytochemical properties, Brazilian Journal of Science, 3 (7): 38–49, (2024).

[22] Oseni O.A., In-vitro compositional investigations of antioxidant, phytochemicals, nutritional and minerals in the fruit of *Kigelia africana* (Lam.) Benth, The FASEB Journal, 32 (S1): 20259–20268, (2018).

[23] Rodr S.L., Oomah B.D., Campos-Vega R., Environmetally Friendly Methods for Flavonoid Extraction from Plant Material: Impact of Their Operating Conditions on Yield and Antioxidant Properties, Scientia Agropecuaria, 11 (4): 439–454, (2020).

[24] Sivani D.V., Sarath G., Anusha D., Formulation and Evaluation of Antimicrobial Herbal Soap, International Journal of Pharmaceutical Sciences and Research, 71 (19): 122–125, (2021).

[25] Swarnalatha K., Kishore Babu C.V., Babu B.H., Phytochemical screening, anti-diabetic and antioxidant activities of *Kigelia africana* (LAM.) and Sterculia foetida L., Rasayan Journal of Chemistry, 12 (2): 907–914, (2019).

[26] Talapko J., Matijević T., Juzbašić M., Antolović-Požgain A., Škrlec I., Antibacterial activity of silver and its application in dentistry, cardiology and dermatology, Microorganisms, 8 (9): 1400, (2020).

[27] Tarun J., Susan J., Suria J., Susan J.V., Criton S., Evaluation of pH of bathing soaps and shampoos for skin and hair care, Indian Journal of Dermatology, 59 (5): 442–444, (2014).

[28] Teniola O.D., Folounsho V.T., Ogunlusi P.S., Aderounmu A.E., Omemu A.M., Antimicrobial Activities of Different Soaps on Selected Human Skin Pathogens, Journal of Advances in Microbiology, 17 (2): 1–10, (2019).

[29] Twincraft., Understanding pH and its Impact on Soap, Twincraft.com, 2019. Available at: https://www.twincraft.com/twincraft-skincare-blog/understanding-ph-bar-soap (Accessed 11 June 2024).

[30] Vaou N., Stravropoulou E., Voidarou C., Tsakris Z., Rozos G., Tsigalou C., Bezirtzoglou E., Interactions between Medical Plant-Derived

Bioactive Compounds: Focus on Antimicrobial Combination Effects, Antibiotics, 11 (8): 1017, (2022).

[31] Wmank W., Bgk P., Preparation of Medicinal Soap Products Using the Leaf Extracts of Punica Granatum (Pomegranate), International Journal of Pharmacy and Biological Sciences, 6 (2): 7–16, (2016).

[32] Yuliana P., Laconi E.B., Wina E., Jayanegara A., Extraction of tannins and saponins from plant sources and their effects on in vitro methanogenesis and rumen fermentation, Journal of the Indonesian Tropical Animal Agriculture, 39 (2): 91–97, (2014).

3.2. Fabrication of Fe-W Oxide Alginate Composite Beads For Gold (III) Ions Recovery

EVIDENCE MTETWA evidence.mtetwa@gmail.com

Submitted in partial fulfilment of the requirements for the degree of Bachelor of Science Honours in Chemical Technology Department of Chemical Technology in the Faculty of Science and Technology at the Midlands State University.

Supervisor: Mr T. Nharingo

Co-supervisor: Dr U. Guyo

November 2016

Dedication

This thesis is dedicated to the Mtetwa family for their unconditional love.

Acknowledgements

I would like to give glory to the almighty God for giving me good health from the onset of the degree programme up to this level. Furthermore, I would like to extend my gratitude to my father and mother and my siblings Beloved, Desire and Answer for their prayers and support and also Kudzai Jinjika my colleague for his help and support. I would also like to express my gratitude to my supervisors Mr Nharingo and Dr. U Guyo for their assistance and support in pursuing this work.

Declaration

I Evidence Mtetwa hereby declare that this dissertation entitled fabrication of Fe-W oxide alginate composite beads for gold (III) ions recovery is my own unaided work. It is being submitted in partial fulfilment for the degree Bachelor of Science Honours Degree in Chemical Technology in the department of Chemical Technology Faculty of Science and Technology Midlands State University. It has not been submitted before for any degree.

Abstract

Gold (Au) recovery from aqueous solutions has always been a subject of great interest due to ever increasing demand of this precious metal. An attempt had been made to recover gold (III) ions on synthesized Fe-W alginate composite beads (FWB). The Fourier Transform Infrared Spectroscopy (FTIR) spectra confirmed that the Fe-W oxide was successfully entrapped in the alginate beads. The concentration of gold (III) in solution was measured by FAAS. The experimental results showed that the pH of Point Zero Charge was at 5.278 and the optimum pH was at 8.13 and the optimum dosage was at 0.2 g. The equilibrium data suited best the Dubinin Radushkevich (D-R) and the Psuedo-second-order model with R2 values of 0.969 and 0.9996 respectively. The thermodynamic studies showed that the adsorption of gold was endothermic with the adsorption enthalpy of ΔH0 29.4249 kJ/mol, ΔS0 of 11.103 J/(mol-1K) and ΔG0 of -2687.1 kJ/mol. Therefore, an increase in temperature favoured the removal of gold.

3.2.1 Introduction

3.2.1.1 Background

Mining activities have long roots in history worldwide [1]. Abidin et al. [2] states that among countless minerals gold is one of the treasurable metals due to its exceptional conductivity of electricity and its tremendously resistance to corrosion. Gold uses are categorized in three sections which are industry, investment and jewellery. The various uses include electrical and electronic industries, medicine, jewellery and as catalysts in many chemical processes [3,4]. The diminishing accessibility of gold and its growing demand has increased annually worldwide hence the importance of its recovery [2,3].

Several technologies have been employed in the recovery of gold from aqueous solutions and these include co-precipitation [5], ion exchange [6], solvent extraction [7] and adsorption onto solid-phase material [8]. Amongst the former technologies, adsorption on solid support is an auspicious technology for the recovery of metal ions [9].

Various sorbents have been studied in many researches for the recovery of gold and these include coconut activated carbon [10], carbon surface of C/Co nanoparticles [8], L-Arginine silica modified [11], oxidized tungsten [12], calcium alginate beads [13] and Sargassum biomass [14]. Nanomaterials are dominating in this era and are increasingly being applied as adsorbents. Nanomaterials are defined as materials that are smaller than 100 nm in at least one dimension [15]. The adsorbents necessary for adsorption of metals require properties such as higher surface area which leads to higher adsorption capacity and these features are observed when the particle size is reduced to nanoscale [16].

Nanomaterials are the main focus in the area of adsorption in this era because of their exclusive properties such as small particle sizes, large surface to volume ratio and their ability to be fixed onto the solid matrices for enhanced treatment of water, waste water, and gaseous process streams. Several bimetallic nanoparticles have also been shown to be useful in decontaminating water containing toxic heavy metals [17–19].

Metal oxide nanoparticles have been reported in literature for their capability to remove heavy metals [20,21]. The use of bimetallic oxide metal nanoparticles has showed to exhibit a new bond between metal elements through an oxygen atom which increase the amount of hydroxyl group on the adsorbent surface and adsorption capacities of the adsorbents [22,23]. Apart from the advantages laid out there are also drawbacks in the use of nanoparticles. Nanoparticles have a tendency of dispersing and being mobile in water, they also agglomerate if they are present in higher concentrations thus forming larger particles which tend to precipitate. Failure to control the mobility and particles' dispersion behavior leads to difficulties in using them in situations where a stationary medium is required [24,25].

Harikumar and Joseph [17], Finotelli et al. [5] and Bezbaruah et al [25] reported that contamination of the environment by free nanoparticles can be escaped by the immobilization of the nanoparticles using calcium alginate. Ca-alginate is well-known for its entrapping characteristics in various sectors in industries. The porosity in Ca-alginate allows solutes to diffuse into the beads and come in contact with the entrapped material. Furthermore, it is nontoxic, biodegradable, and non-immunogenic, and produces thermally irreversible and water insoluble gels [17,25].

3.2.1.2 Limitations

There were several limitations which did not make the project a success and this included the unavailability of instrumentation which included the SEM, TEM and XPS.

Aims:

- To synthesise Fe-W bimetallic oxide/hydroxides beads using co-precipitation method using Ca-alginate as a scaffold

- To investigate the effectiveness of the synthesized beads as an adsorbent for gold (III) ion recovery from aqueous solution and its comparison with commercial activated carbon.

3.2.1.3 Objectives:

- To synthesise Fe-W bimetallic oxide using the co-precipitation method
- To synthesis ca-alginate beads incorporated with the synthesized Fe-W oxide nanoparticles (FWB)
- To characterize the synthesized beads before and after adsorption using FTIR
- To determine the pH at Point Zero Charge (PZC)
- To investigate the effects of pH, dosage, contact time, temperature, ionic strength and the effect of initial concentration of the beads on gold adsorption
- To compare the maximum adsorption capacity obtained by the beads with those of commercial activated carbon.
- To determine the kinetic, thermodynamic and isotherm studies sing the equilibrium data obtained

3.2.1.4 Problem statement

The diminishing accessibility of gold and its growing demand has increased worldwide and this has led to researchers eyeing for more techniques for the recovery of this exquisite metal [2,3]. Gajigo et al. [4] highlights that the broad-based development of gold has underperformed due to highly capital-intensity in the current mining sector this also leads to limited employment generation thus the economy declines. Coconut activated carbon has been widely used all over the world for adsorption of gold for many years, however the problem arises with its high cost of production and its unavailability in some countries hence it has to be exported [26,27]. Carbon in a Carbon in Pulp or carbon in Leach circuit is exposed to vigorous agitation thus leading to abrasion and inevitably, the production of carbon fine hence the loss of gold on the escaping carbon fine to the tailings dam. As the carbon is lost, more carbon is required to replenish the total carbon inventory and also the lost gold is lost forever. This directly affects the production cost of gold

[10,28]. Activated carbon also has a relatively poor selectivity for the pollutants being removed and sometimes causes ineffective recovery of precious metal in practice, furthermore, the adsorption of many unmodified activated carbons is not very auspicious compared to other novel sorbents [29].

3.2.1.5 Justification

Gold is one of the most precious metals that is used in a wide variety of fields and because of this the country's wealth is dependent on it. The shrinking availability of gold and its increasing demand highlight the need for its recovery. Nanomaterials have been reported as efficient and cost-effective over the existing adsorbents [17,30]. According to Gajigo et al. [4] the use of cheaper material in the gold sector leads to creation of employment hence the increase in Gross Domestic Product. The introduction of inorganic supports to the alginate intensifies the mechanical strength in order to avoid the damage or mass loss during adsorption processes hence there are no gold losses due to fines in the adsorption tanks [31].

The metal oxides nanomaterials are considered as the auspicious sorbents for heavy metals removal from aqueous systems. This is because of their large surface areas and high activities caused by the size quantization effect [32]. In other researches it is reported that coconut carbon is less efficient in the removal of heavy metals compared to nanoparticles [33]. Hence the motive of this research is to discover adsorption effect of the studied sorbent.

Of late, nanomaterials have been opted as environmentally friendly substituting the existing sorbents, from the perspectives of both resource conservation and environmental remediation [34]. Literature highlights that metal oxide nanoparticles and alginate are non-toxic and can be safely used in waste treatment and recovery of precious metals [34,35]. The potential use of FWB might be possible since pure alginate beads and bimetallic oxide nanoparticles entrapped in alginate have been used in the recovery of precious metals and other pollutant metals like arsenic. Alginate and iron are also reported to be inexpensive since they are readily available in the earth crust and from seaweeds respectively. Alginate has –COO- which are the main metal chelators hence making it suitable for this research [36,37]. Tungsten oxide nanoparticles have been reported by several researchers on its effective adsorption characteristics and these include the adsorption of organic pollutants and other metal [12,38]. Amongst these researches there is no literature which has the information about the adsorption behavior of tungsten oxide combined with iron oxide in alginate beads hence the importance of this project.

3.2.2 Literature Review

3.2.2.0 Introduction

This chapter looks at the chemistry of gold, its different methods of recovery, description of sorbents used in research and their synthesis, classification of adsorption and the adsorption isotherms.

3.2.2.1 Background of Gold

Gold is precious metals found naturally in the earth. It is found in about 3.5 ppb in the earth's crust and about 75-90% pure native gold has been released by the weathering of sediments and igneous rocks [7]. It is the only metal that is not affected in air or water by oxygen and sulfur and its durability under most corrosive conditions resulted in its famous use in coinage and jewellery throughout time.

Being the only metal which is durable it is also the only metal found in metallic form in nature in the telluride (AuTe2) and stibnite (AuSb2) form. This famous metal reacts with all halogens; it reacts with bromine more vigorously which undergoes an exothermic reaction with gold powder at room temperature with a product of Au2Br6.

Gold complexes exist in either of the two oxidation states which are aurous (+1) or auric (+3) and all the gold complexes of hydrometallurgy fall in these two states. According to thermodymics both auric and aurous cations are not stable in aqueous solution, they will be reduced by water to metallic gold. Stability of these ions is accomplished by introducing a complexing ligand and to employ a suitable oxidizing agent [39].

3.2.2.2 Methods for gold recovery

3.2.2.2.1 Zinc Precipitation

The zinc cementation process was introduced in the 1890's and became part of the cyanidation process. The main reactions that take part in this process are the deposition of gold on the zinc particle surfaces as shown by the following reactions:

Primary equations

$$Au(CN)_{2\,(aq)}^{-} + e^{-} \rightarrow Au_{(s)} + 2CN_{(aq)}^{-} \;\ldots\ldots\ldots (1)$$

$$Zn_{(s)} + 4CN_{(aq)}^{-} \rightarrow Zn(CN)_{(aq)}^{2-} + 2e^{-} \ldots (2)$$

In the first half equation gold cyanide loses electron to form Au and CN- ion. Zinc then reacts with cyanide ion where it is oxidized to form zinc cyanide.

Secondary equations

$$2H_2O_{(l)} + 2e^{-} \rightarrow OH_{(aq)}^{-} + H_{2\,(g)} \;\ldots\ldots\ldots\ldots\ldots\ldots (3)$$

$$O_{2\,(g)} + H_2O_{(l)} + 4e^{-} \rightarrow OH^{-} \;\ldots\ldots\ldots\ldots\ldots\ldots\ldots\ldots\ldots (4)$$

The above reactions are the reaction of water with two electrons from the reaction of zinc with cyanide. Gold is recovered then in the pre-coat filters which are smelted and shaped into blocks for marketing [40].

3.2.2.2.2 Sorption by activated charcoal

Activated carbon has been used extensively in gold recovery, which is known as the Carbon In Pulp (CIP) process. It involves the use of granular activated carbon for extraction of dissolved gold directly from pulps or slimes. The activated carbon can be produced from a variety of organic products such as nut shells, coconut shells, coal and petroleum coke. Literature provide evidence that the gold adsorbs on the carbon surface is dependent on the porosity of the carbon which show higher gold retention rate and indicating that the equilibrium gold retention was mainly a function of both pore distribution and chemical properties of activated carbon surfaces [7,41].

$$4HAuCl_{4\,(aq)} + 3C_{(s)} + 6H_2O_4 \rightarrow Au_{(s)} + 16HCl_{(l)} + 3CO_{2\,(g)} \;\ldots\ldots\ldots\ldots\ldots\ldots\ldots\ldots\ldots (5)$$

3.2.2.2.3 Ion-exchange

Ionic exchange involves the recovery of the metal from a pregnant solution by ion-exchange resins using resin-in-column (RIC) technique. The process is identical to CIC with only one difference which involves the operation of the elution (stripping) stage which does not require elevated temperatures and pressures for removal of precious metals from loaded resins. This technique is scarcely used in the commercial extraction of gold hydrometallurgical solutions despite the various advantages over activated carbon which are higher loading rate and easier gold elution [42].

3.2.2.2.4 Electrowinning

Electrowinning is also named electroextraction. This is the electrodeposition of metals from their ores that have been liquefied. A current is passed from an inert anode through a liquid leach solution containing the metal so that the metal is extracted as it is deposited in an electroplating process onto the cathode [43].

3.2.2.3.0 Nanomaterials and their synthesis

Nanomaterials are the main focus in this era due to their numerous realistic applications and also their improved physical and chemical properties. Richard Feynman is referred to as the father of nanotechnology owing to his speech titled “There’s plenty of room at the bottom” in 1959, but the term “nanotechnology” was first used by Norio Taniguchi in 1974. Nanotechnology is best described as the manipulation of materials on the molecular level and the size of the particles are usually between 1 and 100 nm (1 nm = 10-9 m) in each spiral dimension. The technique focuses on both reducing the size and increasing the efficiency of materials. Nanomaterials are synthesized in different forms which include colloids, clusters, powders, tubes, rods, wires and thin films Mostly metal nanoparticles are prepared by many different methods for example solution-phase chemical reduction, UV Photolysis, metal vapor deposition, solvo thermal method [44], sol-gel method [45], electrochemical techniques [46], Laser ablation, sputtering, Microwave irradiation and ball milling/ mechanical milling. However, these methods require specific apparatus, vacuum conditions and chemical costs [47]. Co-precipitation as a chemical pathway to obtain nanoparticles is a prominent method reported to exhibit many advantages compared to the conventional methods. It is said to be the simplest, straightforward and most effective method which produces well-dispersed nanoparticles in water and there is no need for phase transfer [48].

3.2.2.4.0 Description of sorbents

3.2.2.4.1 Iron oxide nanoparticles

Iron oxides are widespread in the environment and can be readily synthesized in the laboratory. Magnetic iron oxide nanoparticles have many technological applications like targeted drug delivery, magnetic resonance imaging (MRI), magnetic hyperthermia and thermoablation, bio separation, and biosensing [49]. Fe_3O_4 is ferromagnetic in nature, spinel structure of the magnetite has the oxygen anions forming face center closed packing queue with iron located at the interstitial and octahedral sites. The presence of Fe (III) ions provides the magnetic moment of 4μB [50].

3.2.2.4.2 Synthesis by coprecipitation

The coprecipitation method is one of the simplest most efficient routes for the synthesis of magnetic particles. It has an advantage of synthesizing a large number of products. Fe_2+ or γFe_2O_3 are mostly prepared by an aging stoichiometric mixture of ferrous and ferric salts in aqueous medium. The formation of Fe_3O_4 is shown below:

$$Fe_{(aq)}^{2+} + 2Fe_{(aq)}^{3+} + 8OH_{(aq)}^{-} \rightarrow Fe_3O_{4(s)} + 4H_2O_{(l)} \quad \ldots\ldots (6)$$

The complete precipitation of Fe3O4 according to the thermodynamics of the above equation happens at pH between 8 and 14 with the stoichiometry of 2:1 (Fe_3+/ Fe_2+) in a non-oxidizing environment. Nevertheless, Fe3O4 is not stable and is sensitive to oxidation hence transformed into maghemite (γFe_2O_3) in an oxygen environment.

$$Fe_3O_{4(s)} + 2H_{(aq)}^{+} \rightarrow \gamma Fe_2O_{3(s)} + Fe^{2+} + H_2O_{(l)} \quad \ldots\ldots (7)$$

Oxidation in the atmosphere is the only route for the conversion of magnetite to maghemite since many electron or ion transfers depending on the pH of the suspension are involved as shown in the latter equation. Fe2+ ions are desorbed as hexa-aqua complexes in acidic and anaerobic conditions. While in basic conditions, oxidation of magnetite involves oxidation-reduction of the surface of the magnetite. The oxidation of ferrous ions is always linked with migration of cations through the lattice framework, hence producing cationic vacancies to maintain the charge balance thus explaining the structure of maghemite.

In maghemite, iron ions are distributed in the octahedral (Oh) and tetrahedral (Td) sites of the spinel structure. Maghemite is different from magnetite due to the presence of cationic vacancies within the octahedral site. The vacancies ordering scheme is closely correlated to the sample preparation method and fallout in symmetry lowering and possibly superstructures. The vacancies can be completely random or partly or entirely ordered. The structure of magnetite and maghemite are shown below [51,52]:

Fe_3O_4: $[Fe_3+]Td[Fe_3+Fe_2+]OhO_4$

γFe_2O_3 : 0.75[Fe3+]Td[〖Fe〗 _(5/3)^(3+)V1/3]OhO_4

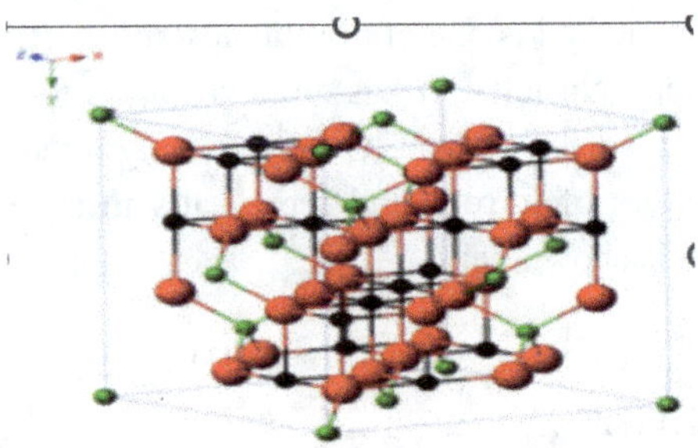

Figure 3.2.1a: Magnetite cubic, Fd3m

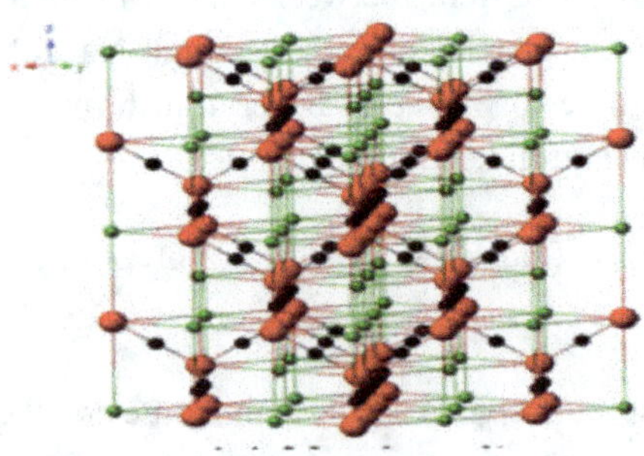

Figure 3.2.1b: Maghemite cubic, P4332/tetragonal, P41212

Figure 3.2.2: Crystal structure and crystallographic data of the a) magnetite and b) maghemite (the black ball is Fe_2+, the green ball is Fe_3+ and the red ball is O_2−) [49].

In the coprecipitation method, two stages are involved and these are a short burst of nucleation when the concentration of the species reaches critical supersaturation and a slow growth of the nuclei by fusion of the solutes to the surface of the crystal. Hence in the monodisperse iron oxide nanoparticles the two stages should be separated, that is, nucleation should be avoid during the period of growth [5,53].

The development of several iron composites has been reported in literature for the adsorption of a various pollutants and metals. These include Al–Fe hydr(oxides), Fe–Mn binary oxide, Fe–Zr binary oxide, mesoporous spheres containing iron and aluminum oxide, hydroxyl-iron-lanthanum doped activated carbon fiber and Fe–Al–Mn trimetal oxide and Fe-Cu binary metals [22,54].

3.2.2.4.3 Tungsten Oxide (WO3) nanoparticles

Tungsten oxide, WO_3, is classified as an n-type semiconductor metal oxide which has optical gap of approximately 2.5 eV. It has unique properties which include photocatalytic, electrochromic and photochromic which are mostly used in environmentally friendly and energy-renewal applications such as solar-powered water splitting, smart-windows, gas-sensing, and dye photodegradation. The tungsten oxide surface adsorption is the determent of optimal performance of these technologies is dependent on tungsten oxide surface adsorption [46,55].

Several methods have been reported for the synthesis of W03 nanoparticles and this include sol gel method, acidification method, chemical vapour deposition, hydrotherm, electrodeposition, thermal evaporation, electro spinning and pulsed laser deposition [46]. Some of the mentioned methods do not offer the advantage of controlled size which is important in the adsorption performance of tungsten oxide [55].

3.2.2.4.4 Ca-alginate beads

Alginate, which naturally occurs in different species in brown seaweeds, is among the most widely used biopolymer for the elimination of anionic and cationic pollutants from aqueous solutions. It is the ammonium or alkali salt of alginic acid. This biopolymer has been reported in literature to exhibit numerous characteristics such as being nontoxic, biodegradable, hydrophilic, biocompatible, non-immunogenic and inexpensive [24,56].

Sodium alginate comprises of a linear, unbranched chains of 1,4 linked β – D – mannuronic and α – L – guluronic acid residues. The two uronic acids transpire in homopolymeric M-blocks, G-blocks, and heteropolymeric random MG-blocks. The length, proportion and sequence of the acid units present in the biopolymer vary with the source from which it is extracted. It converts to a negatively charged anionic polymer on acidification with the formation of –COO- groups and these –COO- groups are the main metal chelators [56,57].

Divalent metal ions such as calcium, barium and stannous ions can be reacted to alginate to yield insoluble hydrogels in an aqueous medium. Tous et al. [58] illustrates that when the metal ions diffuse into an alginate solution, the rapid ion binding and formation of a polymeric network produces an inwardly moving gelling zone. The alginate moves from the gel core towards this gelling zone, leading to the deletion of alginate within the core. This scenario takes place by the chelating of ions with carboxylate group on the alginate hence making crosslinks between chains. This phenomenon takes place at the junctions in the G-G sequence rich chain region, known as "egg box junctions" [36,58].

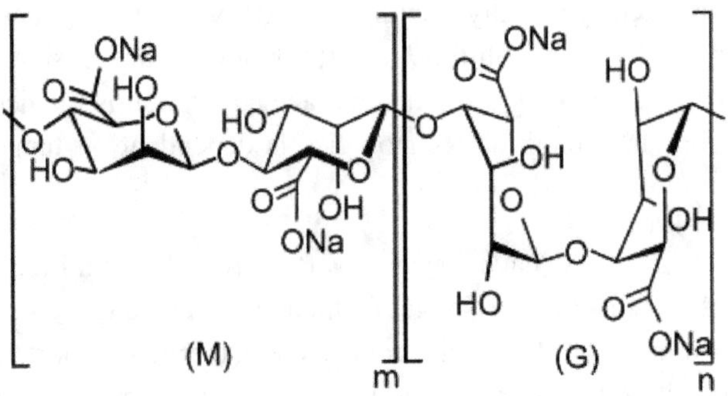

Figure 3.2.2: Idealized chemical structure of sodium alginate [59]

Calcium (Ca)–alginate is a porous material which allow contaminants to diffuse through it and interact with the entrapped materials [24]. It famously used for the entrapment of microbial cells used in the food and beverage industries, water treatment [24] and delivery of low solubility or macromolecular drugs [58]. Literature highlights that this material has been used in the encapsulation of metals and their oxides [17,36], and other sorbents such as Penicillium sp. [36] in adsorption purposes.

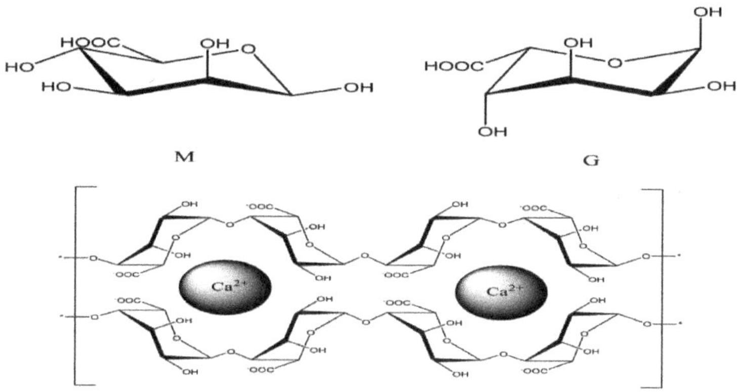

Figure 3.2.3: Molecular and egg-box structure in Ca-alginate gels [60].

3.2.2.5 Adsorption

Adsorption is a phenomenon where atoms, ions or molecules adhere to or accumulate on a suitable interface from an effluent, it uses activated sites of sorbents as medium. This technique involves the separation of substances from one phase following their accumulation at the surface of another. Particles may bind on to the surface of the sorbent by chemical precipitation, physical adsorption, chelation and electrostatic interaction, hence the importance of characterization of surface properties of a sorbent before using it for adsorption [35,61].

3.2.2.6.0 Classification of adsorption

Adsorption may take place either physically or chemically and these are termed as physisorption or chemisorption.

3.2.2.6.1a Physisorption

Physisorption is the most common form of adsorption where van der Waals forces, hydrogen, and dipole dipole bonds attract the molecules and attach on the surface of the sorbent. The molecules have a low heat of adsorption that is only slightly greater than heat of sublimation of the adsorbate. There is absence of chemical bonds but some distortions are possible. The binding energy is dependent on the polarizability and on the number of atoms involved, it also varies between few light gases and several large organic molecules. These molecules can be removed easily since the interaction is weak. This form does not affect the structure or texture of the sorbent [62].

3.2.2.6.1b Chemisorption

In this form of adsorption, a covalent chemical reaction is involved and chemical bonds are formed with the substrate. The target molecule binds to specific sites on the adsorbent with a much greater heat of adsorption approximately equal to the heat of reaction. The energies are mostly that of several large organic molecules. In the course of chemisorption, the adsorbing gas or vapour molecules splits into atoms, radicals, or ions that form chemical bonds with the active site of adsorbent. There is sharing of electrons between gas and the solid surface and may be considered as the formation of surface compound [62].

3.2.6.1b Chemisorption

In this form of adsorption, a covalent chemical reaction is involved and chemical bonds are formed with the substrate. The target molecule binds to specific sites on the adsorbent with a much greater heat of adsorption approximately equal to the heat of reaction. The energies are mostly that of several large organic molecules. In the course of chemisorption, the adsorbing gas or vapour molecules splits into atoms, radicals, or ions that form chemical bonds with the active site of adsorbent. There is sharing of electrons between gas and the solid surface and may be considered as the formation of surface

3.2.2.6.2 Factors affecting adsorption

Adsorption is affected by different parameters hence their importance for experimental designs. The following factors will be discussed on the effect of adsorption.

3.2.2.6.2a Effect of pH

Adsorption of alginate is affected by pH according to the sorbent used. An increase in pH results in the decrease of Au(III) uptake. Carboxylic acid dissociation constant (pKa) of mannuronic (M) and guluronic (G) acids are 3.38 and 3.65 respectively hence when pH exceeds pKa the carboxylic group gradually become less protonated or entirely deprotonated depending on the prevailing pH. This results in the repulsive forces between these groups and the [AuCl4] complex [37,63,64].

Adsorption of other heavy metals on iron oxide loaded alginate beads such as As (III) was reported to be almost independent of pH in the range of 4-10 with a little higher adsorption in the acidic pH [65,66].

3.2.2.6.2b Effect of contact time

The sorption kinetics is very important in adsorption processes since it provides valuable insight of reaction pathways and mechanisms of the adsorption reaction and equilibrium time is also equally important in the design of industrial effluent treatment systems since it influences the size of the reactor hence the economics of the plant. Some literature show that Stenotrophomonas sp. has an equilibrium time of around 150 min for biosorption of Au(III) while fungal strain around has 72 h for complete removal of gold ions with an initial concentration of 50 mg/L and Rhodobacter capsulatus showing an equilibrium time of 30 min for Au(III) biosorption [14].

3.2.2.6.2c Effect of adsorbent dosage

Literature reveals that an increase in alginate beads concentrations leads to a decrease in specific gold uptake. This is also supported by similar results for gold adsorption using other biosorbents. This is so because the results produced with 0.1g of alginate per liter are similar to those with a concentration of 10 folds [14,37].

3.2.2.6.2d Effect of initial concentration

The effect of initial gold concentration which is higher than 150 mg/L resulted in a massive decrease in sorption capacity according to the work which was by others hence giving the conclusion that the mechanism leading to gold adsorption is very complex. The maximum amount of AuCl-4 adsorbed by Sargassum biomass for the first 3, 6, 9, 12, 15min for an initial AuCl-4 concentration of 100 mg/L, was at an average adsorption rate of 8.13, 6.63, 1.97, 1.77, and 1.1 mg per g per min, respectively. After that, the adsorption rate declined and proceeded at an average adsorption rate of 0.13 mg per g per min until 60 min. Hence this concludes that the adsorption of AuCl4- ions by Sargassum biomass is a very rapid process, since most of the adsorption completed within the initial 6 min and it then steadily decreased to reach equilibrium. This phenomenon of initial rapid adsorption is assumed to be because of the lavish availability of vacant sites on the adsorbent which aided easy diffusion of Au(III) ions into the pores of the particles. Some literature illustrates that the speedy sorption process might have been as a result of strong attractive forces between the adsorbate molecules and the adsorbent, leading to rapid diffusion into the intra-particle matrix to attain rapid equilibrium [14,37].

Literature specifies that adsorption efficiency decreases as the initial concentration increases. Furthermore, the rate of this adsorptive reaction in the optimized period of contact varies directly with the concentration of the adsorbate [67].

3.2.2.7 Adsorption Isotherm

Adsorption isotherms describe the adsorption phenomenon at a constant temperature and pH; and the mathematical correlation which is depicted by the modeling analysis is important for operational design and applicable practice of the adsorption systems. The performance of an adsorbent can be studied by adsorption isotherm data, which can be obtained by a series of experimental tests in labs. Modeling the adsorption isotherm data is an essential way for predicting and comparing the adsorption performance, which is critical for optimization of the adsorption mechanism pathways, expression of the adsorbents capacities, and effective design of the adsorption systems [68].

3.2.2.7.1 Langmuir isotherm model

The Langmuir adsorption model assumes that the sorption sites are identical and energetically equivalent and also maximum adsorption corresponds to a saturated monolayer of solute molecules on the adsorbent surface [9,69]. Langmuir isotherm model has four equations of which the one listed below is the most popularly used by researchers due to minimum deviations from the fitted equation.

The equation is as follows:

$$C_e/C_{ads} = 1/bq_m + C_e/q_m \quad \text{.......................................} \quad (8)$$

where, qm is the monolayer adsorption capacity of the adsorbent (mg/g), b is the Langmuir isotherm constant (L mol-1) that is related to the adsorption energy, Ce is the equilibrium concentration (mg L-1), and Cads is the amount of adsorbate retained per mass unit adsorbent [9].

3.2.2.7.2 Freundlich model

The Freundlich isotherm model describes adsorption through a multilayer adsorption mechanism onto a heterogeneous surface that has different sorption energies [9,23]. It states that the presence of different functional groups and various adsorbent–adsorbate interactions make the adsorption heterogeneous.

The equation is as follows:

$$\text{Log } C_{ads} = \text{Log } K_F + \frac{1}{n} \text{ Log } C_e \quad \ldots\ldots (9)$$

where, KF is the Freundlich isotherm constant, 1/n is adsorption intensity, Cads is the concentration of the retained adsorbate ions onto the adsorbent per unit mass (mg g-1) at equilibrium and Ce is the concentration of adsorbate ions in the aqueous solution (mg L-1) [9].

3.2.2.7.3 Temkin model

The Temkin isotherm model considers the adsorbent-adsorbate interactions and also the change of heat of adsorption during adsorption on the surface of the adsorbent. The model assumes that there is a linear decrease in heat of adsorption with increase monolayer surface coverage due to the adsorbent-adsorbate interactions and also that adsorption is characterized by a uniform distribution of binding energies up to some binding energy the equation is illustrated as [33,70]:

$$q_e = \frac{RT}{b_T} \text{ In } K_T C_e + \frac{RT}{b_T} \text{ In } C_e \quad \ldots\ldots (10)$$

The above reaction can be linearized as follows

$$qe = B_T \text{ In } K_T + B_T \text{ In } C_e \quad \ldots\ldots (11)$$

where BT = RT/bT, T is the absolute temperature in degree K, R the universal gas constant, 8.314 J·mol−1·K−1, KT is the equilibrium binding constant (Lmg−1), BT is related to the heat of adsorption and bT is Temkin constant (J·mol−1) [71].

3.2.2.7.4 Dubinin-Radushkevich model

The Dubinin-Radushkevich isotherm model is one of the empirical models that can be used to determine the adsorption mechanism. It describes the adsorption process with respect to pore filling mechanism. It is usually applied to both homogeneous and heterogeneous surfaces to show the adsorption process that is taking place. The equation is as follows:

$$\text{In}q_e = \text{In}q_e - (\beta\varepsilon)^2 \quad \text{...} (12)$$

Where $\varepsilon = RT\text{In}(1 + 1/C_e)$, qe is the amount of sorbate adsorbed per unit weight of adsorbent (mg/g), qm is the adsorption capacity (mg/g), Ce is the equilibrium concentration of adsorbate in solution (mg/L), β is the constant related to the adsorption energy, R is the gas constant (kJ/K mol) and T is the temperature in Kelvin (K) [72].

3.2.2.8 Adsorption kinetics

Adsorption kinetics are studied so as to understand the dynamics of adsorption reactions. It does not only describe the adsorption mechanism but describes also the adsorption rate which controls the residence time of the sorbate t the solid-liquid interface [61,73].

3.2.2.8.1 Pseudo-First-Order Kinetics Model

This model is used for the adsorption kinetics of liquids on the solid phase. It is harmonized with the adsorption rate based on adsorption capacity. The equation is shown below

3.2.2.8 Adsorption kinetics

Adsorption kinetics are studied so as to understand the dynamics of adsorption reactions. It does not only describe the adsorption mechanism but describes also the adsorption rate which controls the residence time of the sorbate t the solid-liquid interface [61,73].

3.2.2.8.1 Pseudo-First-Order Kinetics Model

Model is used for the adsorption kinetics of liquids on the solid phase. It is harmonized with the adsorption rate based on adsorption capacity. The equation is shown below

$$\log(q_e - q_t) = \log(q_e) - t.k_1/(2.3) \quad \text{...............} (13)$$

where qe is the amount of gold adsorption (mg/g), qt is the adsorption dosage at time of t (mg/g), K1 is pseudo first-order kinetic constant (1/min), and t is time (min). The equation constants can be attained by drawing changes of log (qe – qt) [61].

3.2.2.8.2 Pseudo-Second-Order Kinetics Model

The model is used to give the kinetics of absorption based on absorption capacity of the solution; hence this means that the rate of adsorption is dependent on the availability of the adsorption sites rather than the concentration of the adsorbate in the bulk solution. . It is expressed as

$$t/q_t = t/q_e + 1/(k_2 (q_e)^2) \quad \text{.................................. (14)}$$

where qe is the amount of cerium adsorption (mg/g), qt is adsorption dosage at time of t (mg/g), K2 is pseudo second-order kinetic constant (1/min), and t is time (min). The constants of the equation can be attained by drawing changes of t/q to t [61].

3.2.2.8.3 Adsorption thermodynamics

Thermodynamic studies give a description of adsorption behavior related to temperature. The thermodynamic parameters include free energy (ΔG0), enthalpy (ΔH0), and entropy (ΔS0) and they are calculated the following equations (15)-(17) [74]:

$$K_d = q_e/C_e \quad \text{...(15)}$$

$$\Delta G^o = -RT \ln(K_d) \quad \text{... (16)}$$

$$\Delta G^o = \Delta H^o - T\Delta S^o \quad \text{.. (17)}$$

$$\ln(K_d) = (\Delta S^o)/R + (\Delta H^o)/RT \quad \text{... (18)}$$

where Kd is the distribution coefficient, R is the gas constant (8.314 J/(mol-1 K), and T (K) is the tem- perature in Kelvin, qe is the concentration of adsorbate adsorbed at equilibrium (mg/L) and Ce is the equilibrium concentration of adsorbate in the liquid phase (mg/L). A plot of the Van't Hoff of lnKd versus 1/T can be used to calculate values of ΔH° and ΔS° from the slope and intercept respectively.

3.2.3. Methodology

3.2.3.1 Introduction

This chapter gives an insight on how the research experiment was carried out and this includes sorbent synthesis, characterization of the sorbent and batch experiments.

3.2.3.1.1 Preparation of FWB

The synthesis of Fe-W binary oxides was carried out with the alteration of previous literature [5]. Ferric chloride hexahydrate (5.4050g) and sodium tangstate (9.9001g) were dissolved in 200mL distilled water separately and then mixed thoroughly using a magnetic stirrer in a beaker. Under vigorous magnetic stirring at 1100 rpm sodium carbonate (2%) was added drop wise to raise the pH to 7.5. The formed suspension was stirred continuously for 1h, then aged for 24h [5]. The mixture was centrifuged at 400rpm for 10 minutes then washed with distilled water and the centrifuging repeated again for 30 minutes. A mass of 10g of the synthesized nanomaterials were added in 400mL of distilled water and sodium alginate (2%) was sprinkled in the solution and continuously stirred at 1100 rpm with the aid of spatula until a homogenous mixture was obtained. The mixture was then added to 400mL calcium chloride (2%) drop wise to form beads [24]. The beads were aged for 48 h in the calcium chloride solution was filtered with a Whatsman filter paper, washed with distilled water and dried for 3 days in an oven at 300C [17,27].

3.2.3.2 Characterization of FWB

FTIR spectra were collected on a Nicolet 6700 Company, USA FTIR spectrophotometer using a transmission model in the range of 4000 to 400 cm-1. Samples for FTIR determination were ground with spectral grade KBr in a pestle and mortar. IR spectra of the binary oxides as beads (before and after adsorption), as pure nanoparticles and Ca-alginate were obtained as dry samples in KBr pellets corresponding to 1mg of sample in approximately 80 mg of spectral grade KBr. All IR measurements were carried out at room temperature [75,76].

3.2.3.3.0 Preparation of solutions used during the experiment

3.2.3.3.1 Preparation of 10 mg/L gold (III) chloride adsorbate

A volume of 20 mL of analytical grade gold (III) chloride was measured using a measuring cylinder and added into a 2000 mL volumetric flask. Distilled water was then added up to the mark. The flask was closed using a stopper and the contents mixed by inverting it several times.

3.2.3.4.0 Experimental conditions

The experiments were carried out with 50 mL of 10 mg/L gold (III) chloride in 250 cm3 conical flasks and a mass of 0.1g of Fe-W oxide nanoparticle entrapped alginate beads (FWB). The water bath shaker (ZWY-110X30) was set to perform at 300rpm at 25 oC (±2). The experiments were carried out for 24 h except for contact time. The pH was adjusted using HCl (0.1M) and NaOH (0.1 M). After each experiment the beads were filtered off using a Whatsman filter paper and the concentration of gold in the filtrate was determined by FAAS. Characterization for pure alginate, FWB before and after adsorption was done using a Nicolet 6700 Fourier transform infrared spectroscopy (FTIR) spectrophotometer.

3.2.3.4.1 Determination of Point Zero Charge

NaCl (10 mL 0.1 M) solution was added in different conical flasks and pH was adjusted to 3, 5, 7, 11 in different conical flasks. A known mass of adsorbent was then added to each of the conical flasks and placed in a water bath for 24 h, then pH and potential were recorded after adsorption [23,77].

3.2.3.5 Batch experiments

3.2.3.5.1 Effect of pH on gold adsorption

NaOH (0.1M) and HCl (0.1) were used to adjust the pH of gold (III) chloride solution in different conical flasks. The pH was adjusted to 1,3,5,7,9,11 in respective conical flasks. A known mass of FWB was then weighed and added in the flasks, placed into the water bath shaker [78] and the gold adsorbed was determined after the period of adsorption ended.

3.2.3.5.2 Effect of dosage on gold adsorption

Gold solution was placed in 7 different conical flasks and the pH was adjusted to the optimum pH 8 obtained from the pH experiment. The masses of 0.05, 0.2, 0.4, 0.6, 0.8, 1 and 2g were added in the conical flasks respectively and placed in the water bath shaker [79]. The gold concentration was then determined.

3.2.3.5.3 Effect of contact time on gold adsorption

The pH of gold solution in 12 different conical flasks was adjusted to pH 8. A mass of 0.1g was added in each flask and put in a water bath shaker with respect to time which was noted using a stop watch. The time intervals used in this experiment were 2, 4, 6, 8, 10, 12, 20, 40, 40, 120, 180 and 360 min [23]. The conical flasks contents were filtered and the supernatant liquid was analysed for gold concentration

3.2.3.5.4 Effect of temperature on gold adsorption

The gold solution was measured, placed in 4 different conical flasks and its pH adjusted to 8. The adsorbent was added to the flasks and placed in the water bath at temperatures of 25, 30, 50 and 70°C each [71]. After adsorption the gold concentration was determined.

3.2.3.5.5 Effect of ionic strength on gold adsorption

The pH of the gold solution was adjusted to 8 after being added in 5 different conical flasks. Concentrations of 0, 0.001, 0.01, 0.1 and 1 M of NaCl was added in each flask and the sorbent was added in each flask and placed in a water bath shaker [33]. The sorbent was then filtered off and the solution was analysis for gold concentration.

3.2.3.5.6 Effect of initial concentration on gold adsorption

The initial concentrations of 3, 6, 9, 12, 15, 20, 25, 30, 50 and 100mg/L were prepared and 50 mL of each solution was added in the respective conical flasks. The pH was adjusted to 8 and the sorbent was added into each conical flask [80]. The flasks were put in a water bath shaker and the mixture was filtered. The filtrate was taken and analyzed for gold concentration. The flasks were put in a water bath shaker and the mixture was filtered. The filtrate was taken and analyzed for gold concentration.

3.2.4.0 Results and Discussion

3.2.4.1 Introduction

This chapter mainly centers on results presentation and their discussion. It includes FTIR spectra for pure alginate beads and FWB before and after adsorption, point zero charge and the batch experiments carried out. It also covers the isotherm, kinetics and thermodynamic studies and compares the experimental qe values of CAC and FWB.

3.2.4.2 FTIR Analysis of adsorbent

Characterization of pure alginate, FBW before and after adsorption was done using FTIR.

3.2.4.2.1 FTIR spectra for pure alginate beads

The characterization of pure alginate beads showed several functional groups. The peak at 1608.40 cm-1 is consigned to the strong asymmetry stretching vibration band (CO) in the carboxyl group and the band at 1407.11 cm-1 is assigned to be a weak symmetrical stretching vibration band (CO) in carboxyl group (figure 3.2.4). The peak at 1096.54 cm-1 corresponds to C-O stretching of the ether group. The band at 821.69 cm-1 can be attributed to C-O stretching vibration of pyranosyl ring and the C-O stretching with contributions from C-C-H and C-O-H deformation [81].

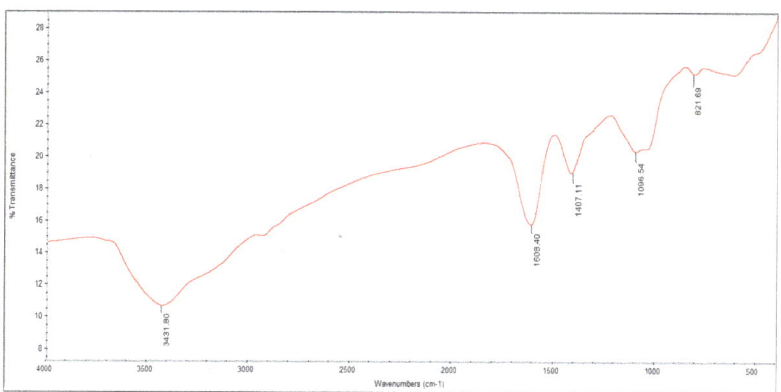

Figure 3.2.2.1: FTIR spectra of pure alginate beads

3.2.4.2.2 Spectra for FWB before and after sorption

FWB before adsorption showed new peaks, the peak at 470.60 cm-1 represented the metal oxygen stretching vibration which can be consigned to the Fe-O stretching vibration. And the peak of 694.67 and 618.36 cm-1 are in the range of 600-700 cm-1 which represent the superposition of metal oxygen stretching vibrations of M-O [27]. Wave number at 842.96 cm- is attributed by M-O-M. The stretching vibration of WO3 nanoparticles is represented by a peak at 931.77 cm-1 (figure 3.2.5). The peak at 1625.90 cm-1 is attributed to the deformation vibration of coordinated interstitial iron tungstate water molecules and indicated the presence of physisorbed water on the oxides which are reported to be responsible for ion exchange process. A peak at 2361.41 cm-1 is due to Intercalated water molecules (M-OH...H2O) [44]. A peak at1384.77 cm-1 might be due to alginate which shifted from1407.11 cm-1 when the metal oxides were introduced. This shows that there was an interaction between the alginate and the metal oxides [38].

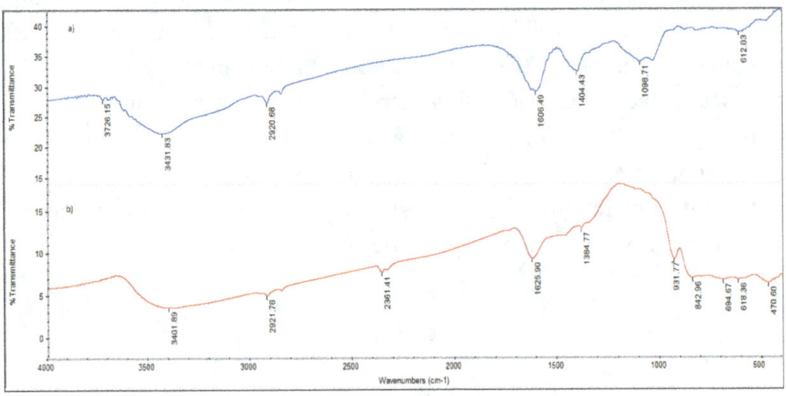

Figure 3.2.4: FTIR spectra of a) FWB after gold adsorption and b) FWB before gold adsorption

In the characterization of FWB after adsorption, there were new peaks which immerged. Lim et al [63], states that a peak at 3726.15cm-1 is due to the stretching vibrations of function group -OH which is due to the gold ions adsorbed on the surface of the sorbent. The shifting of 2361.41 to 2920.68 cm-1 to higher frequencies is said to be as a result high electron density induced by sorption of gold on the adjacent hydroxyl group. The shifting of band 1625.90 to 1606.49 cm-1 to lower frequencies is attributed to the association of the hydroxyl group with the gold (III) ions. Torres et al. [37] highlights that the peak at 1404.43 cm-1 is due to the adsorption of gold on the C-OH. This group shifted from 1384.77 cm-1 which was observed to have been affected by metal oxide nanoparticles during metal entrapment. The adsorption of gold also took place on the ether group of alginates, this is shown by the shifting of the 931.77 cm-1 to 1098.71 cm-1 which was also affected by metal oxide nanoparticles during gelation. The band at low frequency zone of 500-700 cm-1 is attributed by the M-O bond, this peak shifted from 694.67 to 612.83 cm-1 hence this is evident that gold (III) was adsorbed on the metal oxide [70].

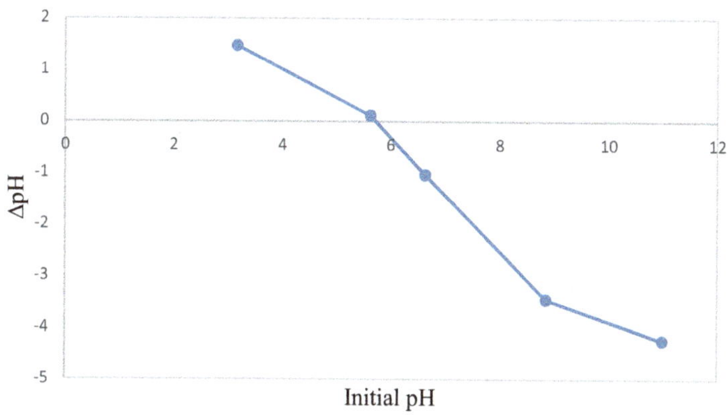

Figure 3.2.5: Point zero charge in adsorption of [AuCl4]- complex

3.2.4.3 Point Zero Charge

The point zero charge of the FBW was found to be 5.278 as illustrated in Fig 3.2.5. Many researchers like Moreno-Barbosa et al. [82] point out that a sorbent is negatively charged at pH above the point zero charge and below the point it will be positive (figure 3.2.6). They further explain this characteristic to be vital since it determines the capacity of the sorbent for ion exchange and liquid phase adsorption of ionic species that have electrostatic interaction as the main adsorption mechanism. Zhou et al. [83] highlights that when the pH values are above the point zero charge, the surface of the sorbent will be negative hence active groups such as carboxyls and hydroxyls are available for interacting with the metal ions. The [AuCl4]- complex is negatively charged hence tis shows that the interaction between the sorbent and sorbate is not based on electrostatic attraction thus other mechanisms can be used to explain the adsorption process of [AuCl4]- ions on the FBW. The point zero charge also helps in describing the features of the sorbent, this is done by comparing the point zero charge of individual components of the sorbent with combined one [84]. In the case of FBW it was 5.278 while Laurent et al. [51] states that the point zero charge of iron oxide is around 6.8. This might mean that the FWB system was not a physical mixture of iron oxide, tungsten oxide and alginate.

3.2.4.4 Batch Experiments

3.2.4.4.1 Effect of pH

The figure 3.2.4 shows the results of Ph on the equilibrium adsorption capacity (qe). The graph shows a gradual increase up to the optimum pH of 8 then a steady decrease of AuCl4- uptake. This might have been due to the gold hydroxychloride complexes (AuCl2(OH)2- and AuCl(OH)-3) which are the most strongly adsorbed species and dominant at pH 8 and not exclusively electrostatic in nature [85]. Torres et al. [37] highlights that the carboxylic acid dissociation constants (pKa) of mannuronic (M) and guluronic(G) acids are 3.38 and 3.65 respectively hence when the pH surpasses these pKa values the carboxylic groups on the alginate become less or completely deprotonated resulting in repulsive forces between these groups and the [AuCl4]- complex. This explanation shows that the interaction of the alginate and that of the complex was not electrostatic. This is also supported by Soni et al. [27] who reports that the surface of the metal oxides entrapped in alginate beads are positively charged in alkaline solutions.

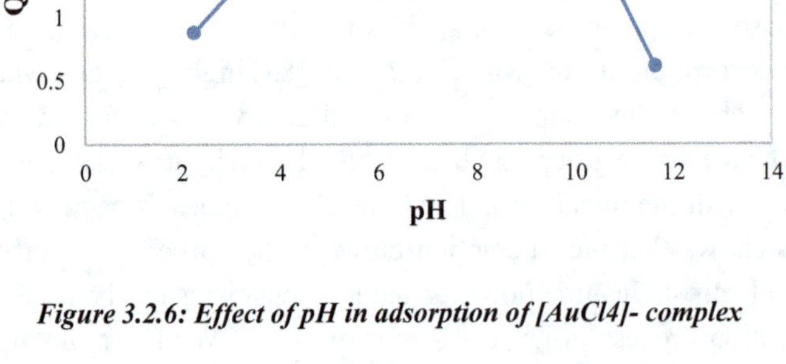

Figure 3.2.6: Effect of pH in adsorption of [AuCl4]- complex

3.2.4.4.2 Effect of sorbent dosage

The effect of adsorbent dose on the removal of gold show that as the adsorbent dosage increases the percentage removal of gold decreases (figure 3.2.7). Torres et al [37] highlights that many researchers in biosorption have produced similar results and concluded that an increase in alginate concentration reduces the amount of metal removal per dry weight unit of sorbent used.

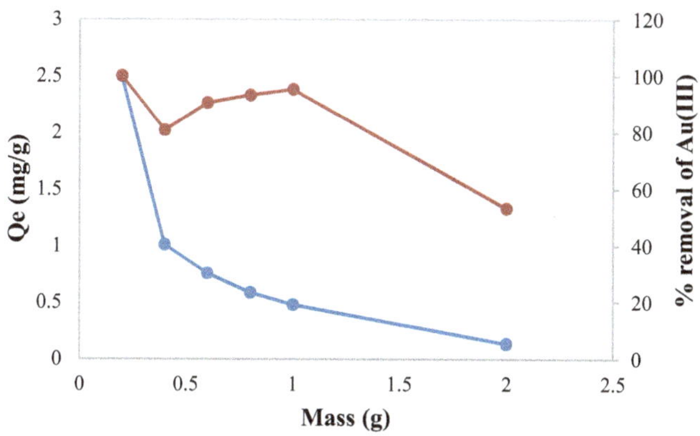

Figure 3.2.7: Effect of sorbent dosage in adsorption of [AuCl4]- complex

Hence this proposes that a higher sorbent concentration is not necessary to produce the highest percentage removal (99.97 %) and sorption yield since the results obtained with 0.2 g of sorbent are better than those obtained by higher masses of sorbent.

3.2.4.4.3 Effect contact time

The results from figure 3.2.8 show that there was a rapid removal of gold for FWB from solution at 2 min follows by a sharp decrease of the removal. The removal of the metal increases again at 40-80 mins then deceases gradually thereafter. The initial rapid phase of adsorption as due to the abundant availability of vacant sites on the adsorbent which also enable diffusion of the Au(III) ions into the pores of the particles.

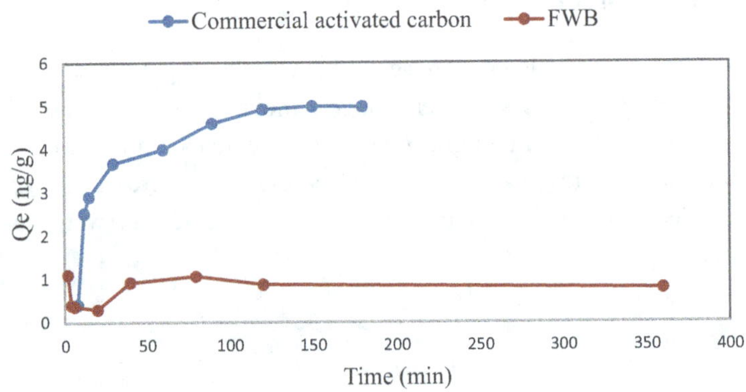

Figure 3.2.8: Effect of contact time in adsorption of [AuCl4]- complex

This is further clarified as a physical process resulting in an increased concentration gradient between the sorbate in aqueous solution and that in the sorbent. The fast sorption process might have also been due to strong interaction forces between the molecules adsorbed and the adsorbent hence a fast diffusion into the intraparticle matrix to reach rapid equilibrium [14]. The decrease in the adsorption of the metal at 4-20 mins might be due to diffusion-controlled kinetics within the microporous structures in the beads hence more time is required for diffusion to take place, this is also supported by Torres et al. [37]. From 120-360 mins there is a slow adsorption rate, this phase resembles the establishment of an equilibrium between the adsorption speeds and desorption [86]. The graph of Commercial Activated Carbon shows an increase in gold adsorption as the time proceeds and it reaches equilibrium. This shows that the Commercial Activated Carbon is better than that of FWB.

3.2.4.4.4 Effect of temperature

3.2.4.4.5 Effect of ionic strength

The effect of ionic strength as illustrated in Figure 3.2.9 showed that as the ionic strength increases the adsorption of gold from solution decreases. This shows that the adsorption process is directly affected by ionic strength since the maximum adsorption was observed were there was no ionic strength. The charge density of metal oxides is affected by ionic strength hence the decrease in the concentration of gold uptake [63].

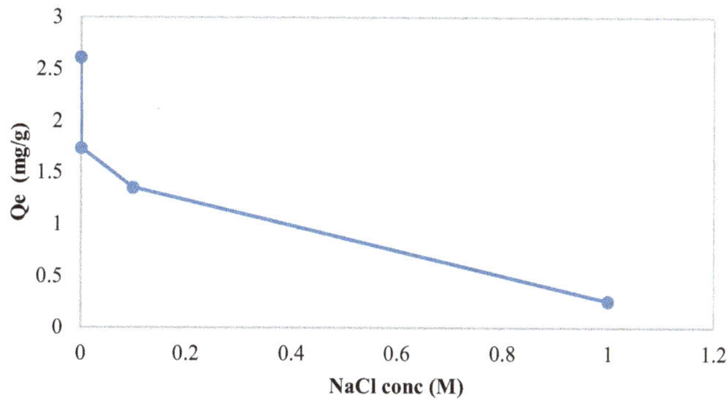

Figure 3.2.9: Effect of ionic strength in adsorption of [AuCl4]- complex

This influence might be due to competition between Cl- ions from NaCl which was used for background ionic strength and the [AuCl4]- complex during adsorption.

3.2.4.4.6 Effect of initial concentration

There is a gradual increase in adsorption as the initial concentration increases (Figure 3.2.10). This increase is as a result of vacant sites on the sorbent were the gold ions move from the bulk solution to the sorbent since there is an increase in diffusion gradient as the sorbate initial concentration increased from 3 mgL-1 to 100 mgL-1 [37]. The graph of Commercial Activated Carbon shows no significant difference in the adsorption of gold hence the FWB is better in the adsorption capacity.

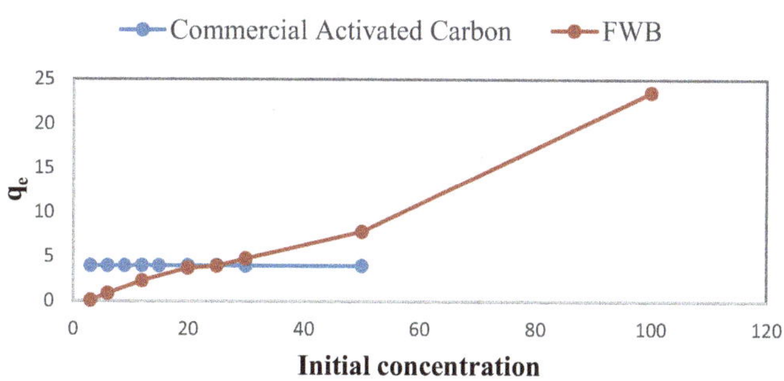

Figure 3.2.10: Effect of initial concentration in adsorption of [AuCl4]- complex

3.2.4.5.0 Adsorption isotherms

Adsorption isotherms are significant in the study and design of adsorption systems. In this study four commonly used isotherm models which are Langmuir, Freundlich, Temkin and Dubunin-Radushkevich were used.

3.2.4.5.1 Langmuir isotherm model

This model assumes that the sorption sites are identical and energetically equivalent and also maximum adsorption corresponds to a saturated monolayer of solute molecules on the adsorbent surface [69].

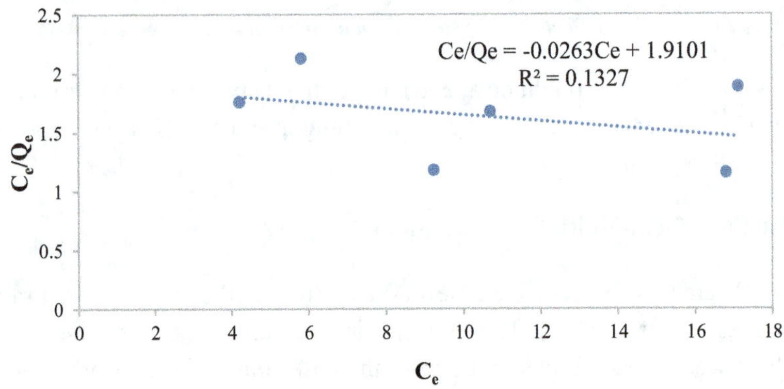

Figure 3.2.11: Langmuir Isotherm

The correlation coefficient, R^2, from the graph plotted above is 0.1327 showing that the equilibrium data did not suit the Langmuir isotherm hence not possible to determine the maximum sorption capacity in this way (figure 3.2.11). Torres et al. [37] also produced similar results in the adsorption of gold (III) chloride using alginate beads and concluded that a reduction of [AuCl4]- to Au0 after adsorption took place.

3.2.4.5.2 Freundlich isotherm model

Adsorption is described by the Freundlich isotherm model as a multilayer adsorption mechanism onto a heterogeneous surface that has different sorption energies (figure 3.2.12). It goes on to explain that the presence of different functional groups and various adsorbent–adsorbate interactions make the adsorption heterogeneous [23]. A plot of In Ce was done and a correlation coefficient of 0.9575 was produced.

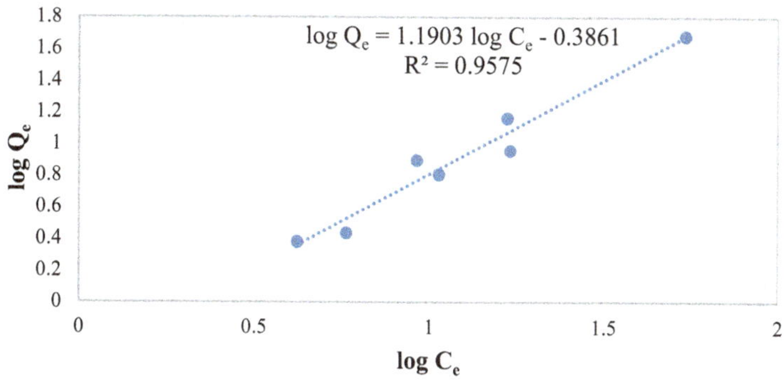

Figure 3.2.12: Freundlich Isotherm model

From equation (9) the value of 1/n is 1.1903 which does not fall in the heterogeneity range of $0.1 < 1/n < 1$. The value of n (0.840124) does not fall in the expected range of physisorption which is $1 < n < 10$ hence it is classified as chemical adsorption process. The value of KF was found to be 0.4133 [87].

3.2.4.6 Temkin Isotherm model

This model takes into consideration the adsorbent-adsorbate interactions and also the change of heat of adsorption during adsorption on the surface of the adsorbent [72]. The assumption is that there is a linear decrease in heat of adsorption with increase monolayer surface coverage owing to the adsorbent-adsorbate interactions and also that adsorption is characterized by a uniform distribution of binding energies up to some binding energy (figure 3.2.13).

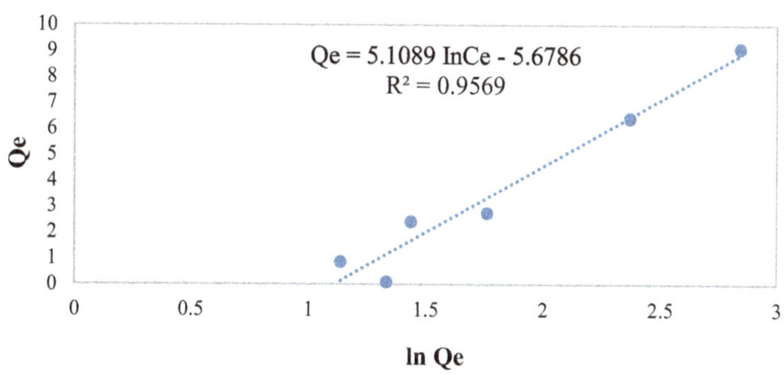

Figure 3.2.13: Temkin Isotherm model

The correlation coefficient as found in the plotted graph is 0.95569. According to equation (10) the heat of sorption, BT, and the intensity of sorption, KT, were 5.1089 KJ mol-1 and 0.329061 Lg-1. The value of bt does not fall within the expected range of bond energy for ion-exchange (6-15 KJ mol-1) hence there was no ionic exchange involved.

3.2.4.6.1. Dubunin-Radushkevich isotherm model

The Dubinin-Radushkevich isotherm model is an empirical model which describes the adsorption process with respect to pore filling mechanism [72]. It is usually applied to both homogeneous and heterogeneous surfaces to show the adsorption process that is taking place (Figure 3.2.14).

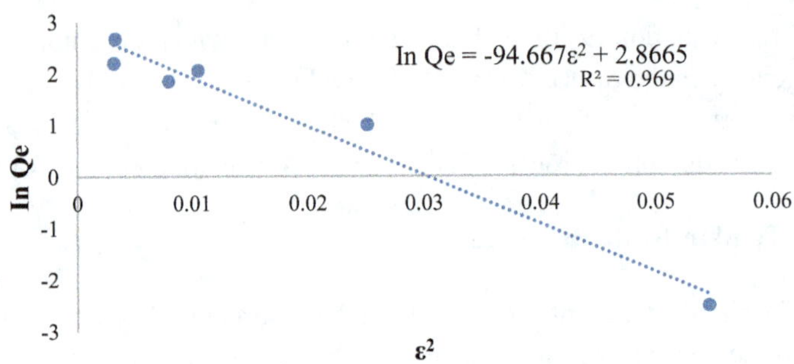

Figure 3.2.14: Dubunin-Radushkevich isotherm model

The graph plotted produced an R^2 value of 0.969 (fig 3.2.14). The value of β (adsorption energy constant) was 94.667. The value of qm was calculated from the graph and found to be 17.5754 mg g-1. 4.5.0 Kinetic studies Kinetics studies have two models which pseudo first and second kinetics which are employed to study the phenomena of adsorption process.

3.2.4.6.2 Pseudo first order kinetics model

This model is used for the adsorption kinetics of liquids on the solid phase. It is harmonized with the adsorption rate based on adsorption capacity. The correlation coefficient was obtained from the graph which is 0.9243, the value of qe was 1.278792 mg/g which is smaller than the experimental qe (4.175 mg/g) and k1 was found to be 0.000461 min-1. The SSE value was found to be 5.0000 x 10-5.

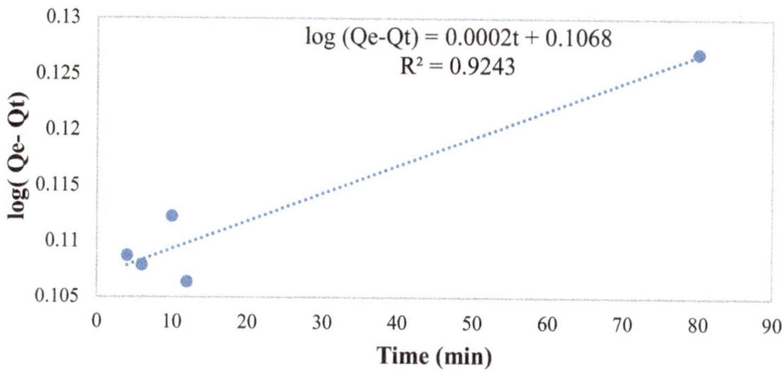

Figure 3.2.15: Pseudo first order kinetics model

3.2.4.6.3. Psuedo second order kinetics model

The model is used to give the kinetics of absorption based on absorption capacity of the solution. The correlation coefficient was 0.9996 from the graph plotted above. The constant of the pseudo second order rate constant, k2, was 4.07904 (mg min)-1 and the maximum adsorption capacity (qe) was 8.4246 mg/g which was double that of experimental qe (4.175 mg/g). The SSE value was found to be 4.8963 x 10-5. Basing on the qe values and the correlation values, this model fits the experimental.

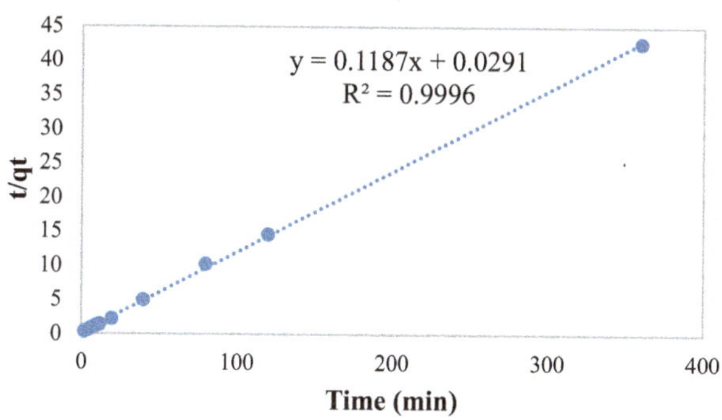

Figure 3.2.16: Pseudo second order kinetics model

3.2.4.7 Thermodynamic Studies

The R^2 value was found to be 0.9936, the values of ΔH° and ΔS° were calculated from the slope and intercept of fig 4.2.17, and the results were 29.42491 kJ/mol and 11.103 J/(mol-1 K) respectively.

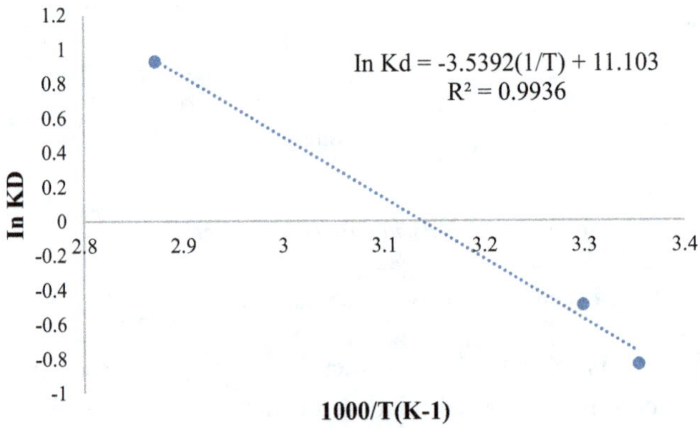

Figure 3.2.17: The Van't Hoff plot of lnKd versus 1/T

The positive value of ΔH° shows that the adsorption process was endothermic and Jawad et al. [74] explains the positive ΔS° as an increase in the degree of freedom to the adsorption species due to adsorption disorder and/ or loss of water upon binding of [AuCl4]- ions to the FWB surface (figure 3.2.8). The values of ΔG° calculated from equation (16) for temperatures 25, 30 and 70°C are 2075.755, 1241.232, and -2687.1 kJ/mol respectively hence this shows that at temperatures 25 and 30°C the adsorption process is non-spontaneous and at 70°C the process is spontaneous. For the positive and negative values of ΔS° and ΔG° respectively at 70°C the process shows that there was affinity between the adsorbent and adsorbate and for both positive values at 25 and 30°C it explains that there was no affinity between adsorbent and adsorbate [87].

Table 3.2.1: Summary of Adsorption isotherms

Isotherm model	Parameter	Value
Langmuir	R^2	0.1327
freundlich	R^2	0.9575
	n	0.840124
	K_F (mg g^{-1})(L $mg^{-1})^n$	0.4133

Temkin	R^2	0.95569
	K_T (Lg/g)	0.329061
	b_t (KJ /mol)	5.1089
Dubunin-Radushkevich	R^2	0.969
	β	0.329061
	q_m (mg g^{-1})	17.5754

Table 3.2.8: Summery of Kinetic studies

kinetic model	parameter	value
Psuedo first order kinetics	R^2	0.9243
	Q_e (mg/g)	1.278792
	K1 $(min)^{-1}$	0.000461
Psuedo second order kinetics	R^2	0.9996
	K_2 $(mg\ min)^{-1}$	4.07904
	Q_e (mg/g)	8.4246

Table 3.2.9: Adsorption Thermodynamics

ΔH^0 (kJ/mol)	ΔS^0 (J/(mol-1K))	ΔG^0 (kJ/mol)		
		25	30	70
29.42491	11.103	2075.755	1241.232	-2687.1

3.2.5. Conclusion and Recommendation

3.2.5.1 Introduction

This chapter gives summarized conclusion, recommendations and areas of further studies regarding this research.

3.2.5.2 Conclusion

• FWB were synthesized using the co-precipitation method, characterized and used for gold (III) ion recovery.

• The functional groups of FWB were hydroxyl, ethers, metal-oxygen and carboxyl groups.

• Optimum conditions were found to be pH 8, 0.005g dosage, 2 minutes contact time at ionic strength of 0.0 M.

• The highest percentage removal was found to be 99.97 %.

• The maximum adsorption capacity of the FWB was 17.5754 mg g-1 from the Dubunin-Radushkevich isotherm.

• Equilibrium data was better described by the Dubunin-Radushkevich, Freundlich and Temkin isotherm model basing on the R^2 value of 0.969, 0.9575 and 0.95569 respectively showing that the reaction was chemisorption and endothermic in nature.

• The equilibrium data fitted best the Pseudo second order kinetics in terms of the R2 value of 0.9996 showing that the adsorption process was chemisorption.

• The values of the thermodynamic parameters, ΔG°, ΔH° and ΔS° show that the adsorption process was spontaneous and endothermic and that there is a high affinity between the adsorbent and the adsorbate hence process was favorable and occurs via physical adsorption whereas at lower temperatures of 25 and 30 it is non-spontaneous and there is no affinity between adsorbent and adsorbate.

• The experimental qe of CAC (4.98 mg/g) resulted to be higher than that of FWB (4.175 mg/g) hence CAC is a better adsorbent compared to FWB

• Basing on the percentage removal of the adsorbents, there is no significant difference between FWB (99.97 %) and CAC (100 %) hence can be used as an alternative for the recovery of gold.

3.2.5.3 Recommendations

• FWB can be used for the removal of other heavy metals.

• Investigations on the effects of interferences, column studies desorption and regeneration studies should be explored.

• Further studies should be carried out on comparison of the effectiveness of gold cyanide and gold chloride removal with the FWB.

3.2.6 References

[1] Ricci BC, Ferreira CD, Aguiar AO, Amaral MCS. Integration of nanofiltration and reverse osmosis for metal separation and sulfuric acid recovery from gold mining effluent 2015;154:11–21. doi:10.1016/j.seppur.2015.08.040.

[2] Abidin MAZ, Jalil AA, Triwahyono S, Adam SH, Kamarudin NHN. Recovery of gold (III) from an aqueous solution onto a durio zibethinus husk. Biochem Eng J 2011;54:124–31. doi:10.1016/j.bej.2011.02.010.

[3] Morcali MH, Zeytuncu B, Ozlem E, Aktas S, Engineering M, Technology P, et al. Studies of Gold Adsorption from Chloride Media. Mater Res 2015;18:660–7.

[4] Gajigo O, Kayizzi-mugerwa S, John C. Gold Mining in Africa: Maximizing Economic Returns for Countries 2012.

[5] Finotelli P V, Sampaio DA, Morales MA, Rossi AM, Rocha-Leao MH. Ca Alginate as scaffold for iron oxide nanopartiocle synthesis. Brazilian J Chem Eng 2008;25:759–64.

[6] Murakami H, Nishihama S, Yoshizuka K. Hydrometallurgy Separation and recovery of gold from waste LED using ion exchange method. Hydrometallurgy2015;157:194–8. doi:10.1016/j.hydromet.2015.08.0 14.

[7] Bagal MR, Shaikh UP, Dhokte AO, Lande MK, Arbad BR, Aurangabad MS. Rapid solvent extraction of gold (III) with high molecular weight amine from malonate media 2013;5:131–7.

[8] Rossier M, Koehler FM, Athanassiou EK, Grass RN. Gold adsorption on the carbon surface of C / Co nanoparticles allows magnetic extraction from extremely diluted aqueous solutions 2009:1–11.

[9] The fast recovery of gold (III) ions from aqueous solutions using raw date pits: Kinetic, thermodynamic and equilibrium studies. J SAUDI Chem Soc 2013. doi:10.1016/j.jscs.2013.06.002.

[10] Soleimani M, Kaghazchi T. Activated Hard Shell of Apricot Stones: A Promising Adsorbent in Gold Recovery 2008;16:112–8.

[11] Hastuti S, Kuncaka A. L-ARGININE-MODIFIED SILICA FOR ADSORPTION OF GOLD (III) 2015;15:108–15.

[12] Afanas EY. Adsorption of Gold on Oxidized Tungsten 2013;58:793–8. doi:10.1134/S1063784213060029.

[13] Salarirad MM, Behnamfard A. Hydrometallurgy The effect of fl otation reagents on cyanidation , loading capacity and sorption kinetics of gold onto activated carbon. Hydrometallurgy 2010;105:47–53. doi:10.1016/j.hydromet.2010.07.009.

[14] Sathishkumar M, Mahadevan A, Vijayaraghavan K, Pavagadhi S, Balasubramanian R. Green Recovery of Gold through Biosorption, Biocrystallization, and Pyro-Crystallization 2010:7129–35.

[15] Wang X, Guo Y, Yang L, Han M, Zhao J, Cheng X. Nanomaterials as Sorbents to Remove Heavy Metal Ions in Wastewater Treatment. Environ Anal Toxicol 2012; 2:1–7. doi:10.4172/2161-0525.1000154.

[16] Bhargavi RJ, Maheshwari U. Synthesis and use of alumina nanoparticles as an adsorbent for the removal of Zn (II) and CBG dye from wastewater 2015:31–41. doi:10.1007/s40090-014-0029-1.

[17] Harikumar PS, Joseph L. Kinetic and thermodynamics studies of As (III) adsorption onto iron nanoparticles entrapped ca-alginate beads. Int J Plant, Anim Environ Sci 2012;2:159–66.

[18] Mohan D, Pittman CU. Arsenic removal from water / wastewater using adsorbents — A critical review 2007;142:1–53. doi:10.1016/j.jhazmat.2007.01.006.

[19] Ali Mansoori G, Bastami TR, Ahmadpour A, Eshaghi Z. Environmental Application of Nanotechnology 2008;2:439–93. doi:10.1142/9789812790248_0010.

[20] Chen L, Xin H, Fang Y, Zhang C, Zhang F, Cao X, et al. Application of Metal Oxide Heterostructures in Arsenic Removal from Contaminated Water. J Nanomater 2014;2014:1–10. doi:10.1155/2014/793610.

[21] Giraldo L, Moreno-piraján JC. Syntehsis of magnetite nanoparticles and exploring their application in the removal of Pt3+ and Au3+ ions from aqueous solutions. Eur Chem Bull 2013;2:445–52.

[22] Chen L, He B, He S, Wang T, Su C, Jin Y. Fe — Ti oxide nano-adsorbent synthesized by co-precipitation for fl uoride removal from drinking water and its adsorption mechanism. Powder Technol 2012;227:3–8. doi:10.1016/j.powtec.2011.11.030.

[23] Zhang G, Ren Z, Zhang X, Chen J. Nanostructured iron (III) -copper (II) binary oxide : A novel adsorbent for enhanced arsenic removal from aqueous solutions. Water Res 2013;47:4022–31. doi:10.1016/j.watres.2012.11.059.

[24] Bezbaruah AN, Kalita H, Jacob DL, Ugrinov AG, Payne SA. Ca – alginate-entrapped nanoscale iron : arsenic treatability and mechanism studies. J Nanopart Res 2014;16. doi:10.1007/s11051-013-2175-3.

[25] Bezbaruah AN, Sharanya S, Á CÁB. Encapsulation of iron nanoparticles in alginate biopolymer for trichloroethylene remediation 2011:6673–81. doi:10.1007/s11051-011-0574-x.

[26] Marsden JO, House CO. The Chemistry of Gold Extraction. Second. Colorado: Society for Mining, Metallurgy, and Exploration; 2006.

[27] Soni A, Tiwari A, Bajpaia K. Adsorption of o-nitrophenol onto nano iron oxide and alginate microspheres: Batch and column studies. African J Pure Appl Chem 2012;6:161–73. doi:10.5897/AJPAC11.071.

[28] Meteku BE. Production of activated carbon from palm kernel shell from gold adsorption using leachates from cocoa husk ash (crude potash) as activated agent. Kwame Nkrumah University of Science and Technology, 2013.

[29] Sze LP. Removal and Recovery of Heavy Metals by Amorphous TiO 2 Nanoparticles and Ca-alginate Immobilized TiO 2 Beads n.d.

[30] Malhotra P, Jain A. Role of Nanotechnology As A Tool for Sustainability: Potential of Zerovalent Metal Naoparticles (ZVN) and Their Metal Composites in Environmental Remediation 2016;3:143–50.

[31] Wang F, Zhao J, Pan F, Zhou H, Yang X, Li W, et al. Adsorption Properties toward Trivalent Rare Earths by Alginate Beads Doping with Silica. Ind Eng Chem Res 2013;52:3453–61.

[32] Eng JAC, Taman R, Me O, Ms M, Ha F. Metal Oxide Nano-particles as an Adsorbent for Removal of Heavy Metals. Adv Chem Eng 2015;5.

[33] Rdriguez A, R C. Removal of cadmium (II), lead (II) and chromium (VI) in water with nanomaterials. Universitat Autonoma de Barcelona, 2015.

[34] Xu P, Zeng GM, Huang DL, Feng CL, Hu S, Zhao MH, et al. Use of iron oxide nanomaterials in wastewater treatment: A review. Sci Total Environ 2012;424:1–10. doi:10.1016/j.scitotenv.2012.02.023.

[35] Nicomel N, Leus K, Folens K, Van Der Voort P, Du Laing G. Technologies for Arsenic Removal from Water: Current Status and Future Perspectives. Int J Environ Res Public Health 2015;13:62. doi:10.3390/ijerph13010062.

[36] Mishra SP. Adsorption of Cu and Zn on calcium alginate immobilized Penicillium sp . Indian J Chem Technol 2013;20:21–5.

[37] Torres E, Mata YN, Bla ML, Gonza F, Ballester A. Gold and Silver Uptake and Nanoprecipitation on Calcium Alginate Beads 2005:7951–8.

[38] Hidayat D, Purwanto A, Wang W, Okuyama K. Preparation of size-controlled tungsten oxide nanoparticles and evaluation of their adsorption performance. Mater Res Bull 2010;45:165–73. doi:10.1016/j.materresbull.2009.09.025.

[39] Marsden JO, House CO. The chemistry of gold extraction. Miner Eng 2006;5:1069–70.

[40] Yin C, Ng M, Saunders M, Goh B, Senanayake G, Sherwood A, et al. New Insights into the Adsorption of Aurocyanide Ion on Activated Carbon Surface: Electron Microscopy Analysis and Computational Studies Using Fullerene-like Models. Langmuir 2014;30:7703–9.

[41] Buah WK, Williams PT. Agricultural waste biomass converted to activated carbon as a material for gold processing 2012:396–402. doi:10.1007/s10163-012-0083-8.

[42] Nsimba EB. Cyanide and cyanide complexes in the gold mine pollted land in the East and Central Rand Goldfields, South Africa. 2009.

[43] Wai W, Eugene L, Mujumdar AS. Gold Extraction and Recovery Processes. Singapore: 2009.

[44] Prabhu N. Effect of tepmerature on the structural and optical properties of WO3 nanoparticles prepared by solvo thermal method. Dig J Nanomater Biostructure 2013;8:1483–90.

[45] Kamil F, Hubiter KA, Abed TK. Synthesis of Aluminum and Titanium Oxides Nanoparticles via Sol-Gel Method: Optimization for the Minimum Size. J Nanosci Technol 2016;2:37–9.

[46] Pudukudy M, Yaakob Z, Rajendran R. Visible light active novel WO 3 nanospheres for methylene blue degradation. Sch Res Libr 2013;5:208–12.

[47] Murugesan K, Sivakumar P, Palanisamy PN. An Overview on synthesis of metal oxide nanoparticles. South Asian J Eng Technol 2016;2:58–66.

[48] Lim J, Majetich SA. Composite magnetic — plasmonic nanoparticles for biomedicine: Manipulation and imaging. SciVerse Sci 2013;8:98–113.

[49] Wu W, Wu Z, Yu T, Jiang C. Recent progress on magnetic iron oxide nanoparticles: synthesis, surface functional strategies and biomedical applications. Sci Technol Adv Mater 2015;16:23501. doi:10.1088/1468-6996/16/2/023501.

[50] Co-precipitation method of synthesis and characterization of iron oxide nanoparticles 2014;73:87–90.

[51] Laurent S, Forge D, Port M, Roch A, Robic C, Elst L Vander, et al. Magnetic Iron Oxide Nanoparticles: Synthesis, Stabilization, Vectorization, Physicochemical Characterizations , and Biological Applications 2008:2064–110.

[52] Ramimoghadam D, Bagheri S, Bee S, Hamid A. Journal of Magnetism and Magnetic Materials Progress in electrochemical synthesis of magnetic iron oxide nanoparticles. J Magn Magn Mater 2014;368:207–29. doi:10.1016/j.jmmm.2014.05.015.

[53] Mishra D, Arora R, Lahiri S, Amritphale SS, Chandra N. Synthesis and Characterization of Iron Oxide Nanoparticles by Solvothermal Method 1 2014;50:628–31. doi:10.1134/S2070205114050128.

[54] Mohapatra BPM, Verma SAHC. Mössbauer studies of nano phase Ce – Fe oxide composites 2008:123–8. doi:10.1007/s10751-008-9739-0.

[55] Hidayat D, Purwanto A, Wang W, Okuyama K. Preparation of size-controlled tungsten oxide nanoparticles and evaluation of their adsorption performance n.d.:19.

[56] Voo W, Lee B, Idris A, Islam A, Tey DB. Production of ultra-high concentration calcium alginate beads with prolonged dissolution profile†. RSC Adv 2015;5:36687–95. doi:10.1039/C5RA03862F.

[57] Taylor P, Han Y, Lee W, Lee C, Park S, Kim K. Entrapment of Mg-Al layered double hydroxide in calcium alginate beads for phosphate removal from aqueous solution. Desalin Water Treat 2011;36:178–86. doi:10.5004/dwt.2011.2254.

[58] Tous S, Fathy M, Fetih G, Gad SFS. Preparation and Evaluation of Ketoprofen-loaded Calcium alginate beads. Int J PharmaTech Res 2014;6:1100–12.

[59] Pal P, Banat F. Contaminants in Industrial Lean Amine Solvent and their Removal using Biopolymers: A New Aspect. Phys Chem Biophys 2014;4:1–5. doi:10.4172/2161-0398.1000135.

[60] Dorati R, Genta I, Ferrari M, Vigone G, Merico V, Garagna S, et al. Formulation and stability evaluation of 3D alginate beads potentially useful for cumulus – oocyte complexes culture 2016;2048. doi:10.3109/02652048.2015.1134691.

[61] Farahmand E. Adsorption of Cerium (IV) from Aqueous Solutions Using Activated Carbon Developed from Rice Straw 2016;2016:189–200.

[62] Berger AH, Bhown AS. Comparing Physisorption and Chemisorption Solid Sorbents for use Separating CO 2 from Flue Gas using Temperature Swing Adsorption. Energy Procedia 2011;4:562–7. doi:10.1016/j.egypro.2011.01.089.

[63] Lim S-F, Zheng Y-M, Zou S-W, Chen P. Characterization of Copper Adsorption onto an Alginate Encapsulated Magnetic Sorbent by a

Combined FT-IR , XPS, and Mathematical Modeling Study. Environ Sci Technol 2008;42:2551–6.

[64] Mahmood Z, Nasir S, Jamil N, Sheikh A, Akram A. Adsorption studies of phosphate ions on alginate- calcium carbonate composite beads 2015;9:274–81. doi:10.5897/AJEST2014.1784.

[65] Wanghmare SS, Tanvir A. Flouride removal from water by mixed metal oxide adsorbent materials: A state-of-the-art review. Int J Eng Sci Res Technol 2015;4:519–36.

[66] Hua M, Zhang S, Pan B, Zhang W, Lv L, Zhang Q. Heavy metal removal from water/wastewater by nanosized metal oxides: A review. J Hazard Mater 2012;211-212:317–31. doi:10.1016/j.jhazmat.2011.10.016.

[67] Terrones SC. Volatile-organic-compound optic fiber sensor using a gold-silver vapochromic complex 2006;45:1–7. doi:10.1117/1.219 2470.

[68] Chen X. Modeling of Experimental Adsorption Isotherm Data 2015:14–22. doi:10.3390/info6010014.

[69] Brouers F, Al-musawi TJ. On the optimal use of isotherm models for the characterization of biosorption of lead onto algae 2016:1–16.

[70] Asthana A, Verma R, Kumar A, Bin A, Susan H. Journal of Environmental Chemical Engineering Glycine functionalized magnetic nanoparticle entrapped calcium alginate beads: A promising adsorbent for removal of Cu (II) ions 2016;4:1985–95.

[71] El-latif MMA, Ibrahim AM, Showman MS, Hamide RRA. Alumina / Iron Oxide Nano Composite for Cadmium Ions Removal from Aqueous Solutions 2013;2013:47–62.

[72] Kaygusuz H, Erim FB. Aluminum Alginate Beads. Clean - Soil, Air, Water 2014;42:1–7. doi:10.1002/clen.201300632.

[73] Bhaumik R, Kumar N. Adsorption of fluoride from aqueous solution by a new low-cost adsorbent: thermally and chemically activated coconut fibre dust. Clean Technol Environ Policy 2015:2157–72. doi:10.1007/s10098-015-0937-6.

[74] Jawad AH, Rashid RA, Azlan M, Ishak M, Wilson LD, Jawad AH, et al. Adsorption of methylene blue onto activated carbon developed from biomass waste by H 2 SO 4 activation : kinetic, equilibrium and thermodynamic studies. Desalin Water Treat 2016;3994:1–13. doi:10.1080/19443994.2016.1144534.

[75] Ren Z, Zhang G, Chen JP. Journal of Colloid and Interface Science Adsorptive removal of arsenic from water by an iron – zirconium binary oxide adsorbent. J Colloid Interface Sci 2011;358:230–7. doi:10.1016/j.jcis.2011.01.013.

[76] Li G, Gao S, Zhang G, Zhang X. Enhanced adsorption of phosphate from aqueous solution by nanostructured iron (III)– copper (II) binary oxides. Chem Eng J 2014;235:124–31. doi:10.1016/j.cej.2013.09.021.

[77] Cheng W, Zhang W, Hu L, Ding W, Wu F, Li J. RSC Advances. R Soc Chem 2016:1–39. doi:10.1039/C5RA26143K.

[78] Li LH, Xiao J, Liu P, Yang GW. amorphousization of metal oxide 2015:1–6. doi:10.1038/srep09028.

[79] Subhan MA, Monim SA, Bhuiyan MBR, Chowdhury AN, Islam M, Hoque MA. Synthesis, characterization of a multi-component metal oxide solution 2011;2011:9–15. doi:10.4236/ojic.2011.12002.

[80] Lakay EM. Superparamagnetic iron-oxide based nanoparticles for the separation and recovery of precious metals from solution Master of Science. university of Stellenbosch, 2009.

[81] Daemi H, Barikani M. Sharif University of Technology Synthesis and characterization of calcium alginate nanoparticles, sodium homopolymannuronate salt and its calcium nanoparticles. Sci Iran 2012;19:2023–8. doi:10.1016/j.scient.2012.10.005.

[82] Moreno-Barbosa JJ, Lopez-Velandia C, Maldonado A del P, Giraldo L, Moreno-piraja JC. Removal of lead (II) and zinc (II) ions from aqueous solutions by adsorption onto activated carbon synthesized from watermelon shell and walnut shell. Adsorption 2013;19:675–85. doi:10.1007/s10450-013-9491-x.

[83] Zhou Y, Nie H, Branford-white C, He Z, Zhu L. Removal of Cu 2 + from aqueous solution by chitosan-coated magnetic nanoparticles modified with α -ketoglutaric acid. J Colloid Interface Sci 2009;330:29–37. doi:10.1016/j.jcis.2008.10.026.

[84] Zhang YU, Yang MIN, Dou X. Arsenate Adsorption on an Fe - Ce Bimetal Oxide Adsorbent : Role of Surface Properties. Environ Sci Technol 2005;39:7246–53.

[85] Patrick de BM. Deposition of Au/Pt on Co/SiO2 for Fischer-Tropsch synthesis. University of Cape Town, 2014.

[86] Kamarchou A, Bebba AA, Douadi A. Elimination of phosphorus by activated carbon prepared from algerian dates stones. Res J Pharm Biol Chem Sci 2015;6:1271–8.

[87] Dias NC, Steiner PA, Cristina M, Braga B. Characterization and Modification of a Clay Mineral Used in Adsorption Tests. J Miner Mater Charact Eng 2015;3:277–88.

[88] Onundi YB, Mamun AA, Khatib MF Al, Ahmed YM. Adsorption of copper, nickel and lead ions from synthetic semiconductor industrial wastewater by palm shell activated carbon. Int J Environ Sci Tech 2010;7:751–8.

[89] Zheng H, Wang L. Banana Peel Carbon Containing Functional Groups Applied to the Selective Adsorption of Au (III) from Waste Printed Circuit Boards. Sci Res 2013;2013:29–36.

Part 4: Molecular Epidemiology of SARS-COV-2 Genetic Variants in Africa

4.1 Genetic diversity and spread dynamics of SARSCoV-2 variants present in African populations

Royal Society Open Science

Desire Mtetwa[1], Tafadzwa Manjengwa[1] and Zedias Chikwambi[2]

[1]Department of Biotechnology, School of Health Science and Technology, Chinhoyi University of Technology, Private Bag 7724, Chinhoyi, Zimbabwe

[2] African Institute of Biomedical Science and Technology, 911 Chiedza Park, Boronia, Harare, Zimbabwe

4.1.1 Abstract:

The dynamics of coronavirus disease-19 (COVID-19) have been extensively researched in many settings around the world, but little is known about these patterns in Africa. A total of 7540 complete nucleotide genomes from 51 African nations were obtained and analysed using the National Center for Biotechnology Information and Global Initiative on Sharing Influenza databases to examine the genetic diversity and spread dynamics of the severe acute respiratory syndrome coronavirus 2 (SARS-CoV-2) lineages circulating in Africa. Using various clade and lineage nomenclature schemes, we examined their diversity and used maximum parsimony inference methods to reconstruct the evolutionary hypotheses about the spread of the virus in Africa. According to this study, only 465 of the 2610 Pango lineages found to have existed in the world circulated in Africa three years after the COVID-19 pandemic, with five different lineages dominating at various points during the outbreak. We identified South Africa, Kenya and Nigeria as key sources of viral transmission among sub-Saharan African nations. These findings provide insights into the viral strains that circulate in Africa and their evolutionary patterns.

4.1.2. Introduction

In December 2019, severe acute respiratory syndrome coronavirus 2 (SARS-CoV-2) and the disease caused by the infection, named COVID-19, were originally identified in Wuhan, China, as a novel virus that produced a cluster of atypical pneumonia cases [1]. The outbreak soon became a global pandemic, resulting in a global public health emergency [2]. COVID-19 has affected all seven continents, with Africa being the least affected [3]. The first confirmed case in Africa occurred on 14 February 2020, in Egypt, while the first case reported in sub-Saharan Africa occurred on 27 February 2020, in Nigeria [4]. The first cases in other African countries, including Ghana, were recorded in March 2020 [5] on 12 March 2020. Globally, as of 21 February 2023, there have been 757.26 million confirmed cases of COVID-19, including 6.85 million deaths, reported to the World Health Organization (WHO) [6]. SARS-CoV-2 is the third novel

4.1.3 Methodology

4.1.3.1: Research Methods

Dataset mining and workflow SARS-CoV-2 genome sequences collected from Africa were obtained from the National Center for Biotechnology Information (NCBI) [11] and Global Initiative on Sharing Influenza Data (GISAID) [12] databases on 26 February 2023. A total of 24 415 African sequences were retrieved from both databases to examine the number of lineages circulating within Africa. The two databases had only 8044 complete genome sequences combined from Africa, and these sequences, excluding those with low coverage using NextClade [13], were retrieved to determine spread dynamics. A total of 5908 sequences from 23 African countries are available in the NCBI database and 2137 sequences from 41 African countries are available in the GISAID database. The sequences were aligned using the online version of MAFFT [14] multiple sequence alignment tool [15] with Wuhan-Hu-1 (MN 908947.3) as the reference sequence, and sequences with more than 5.0% ambiguous letters were removed. Duplicates were removed using the goalign dedup software, and only high-quality African complete sequences remained (n = 7540).

4.1.3.2 Phylogenetic reconstruction

Using IQ-TREE multicore software version v1.6.12 [16] and NextClade [13], phylogenetic reconstruction of the dataset was performed. IQ-TREE was used in its default mode with Wuhan-Hu-1 (MN908947.3) as the reference genome as an outgroup; the command 'iqtree -s alignment_filename.fasta –mGTR+G' was used. NextClade was also used in its default mode. coronavirus associated with significant outbreaks in the twenty-first century and the seventh coronavirus known to infect humans [7].

The first was SARS-CoV in 2003, which also appeared in China [8], and the second was the Middle East respiratory syndrome coronavirus (MERS-CoV) in 2012, which appeared in Saudi Arabia [8,9]. SARS-CoV-2 is more contagious but appears to have a lower-case fatality rate (CFR) than SARS-CoV and MERS-CoV, both of which cause serious illnesses in humans [1,10]. The emergence of SARS-CoV-2 on the African continent necessitates a thorough examination of its genomic and evolutionary patterns. Comparative analysis of viral genome sequences is an extremely useful method for gaining insights into pathogen emergence and evolution. As a result, this study conducted an in-depth investigation into the epidemiology and evolution of SARS-CoV-2 in Africa to shed light on pandemic dynamics, investigate the number of variants circulating within Africa, and access the past deployment of control measures efficiency and inform future interventions on the African continent.

4.1.3.2 Lineage classification

PANGOLin, a Web application [17], was used to classify sequences into their lineages. The objective was to determine the most important SARS-CoV-2 lineages circulating in Africa from an epidemiological perspective, as well as the lineage dynamics within and across the African continent, as this naming system integrates genetic and geographic data on SARS-CoV-2 dynamics [17].

4.1.3.3. Phylogeographic reconstruction

Variants of Concern (VOC), Variants of Interest (VOI) and Variants Under Monitoring (VUM) were designated based on the WHO framework as of 20 January 2022. We included one lineage, Variant A, and labelled it the VOI for the purposes of this analysis. This lineage was included because it demonstrates the continued evolution of African lineages into potentially more transmissible variants. The VOI, VOC and VUM that emerged on the African continent were marked. These were Variant A (VOI), Beta and Omicron (VOC) and IHU Variant and Eta (VUM). Genome sequences of the five lineages were extracted from the NCBI database for phylogeographic reconstruction. An approach similar to that described above (including alignment using online MAFFT) was employed. Phylogeographic reconstruction for all variants circulating in Africa and all VOI, VOC and VUM were conducted using PASTML [18] and maps using QGIS. PASTML was used for coalescent analysis and/or ancestral state reconstruction.

4.1.3 Results

4.1.3.1 Genetic diversity and lineage dynamics in Africa,

The NextClade results showed that 465 of the 2610 Pango lineages circulated worldwide in Africa. Of the 8044 complete genomes retrieved from NCBI and GISAID, 7540 were retained after removing duplicates and sequences with more than 5% ambiguous letters from the dataset. We determined the import and export of viruses between African nations using ancestral location state reconstruction of dated phylogeny (figure 4.1.1). The early stage of the pandemic was marked by the dominance of lineage B.1, a large European lineage that caused most deaths in Italy (it approximately correlated with the northern Italian breakout in early 2020). Following the emergence of Beta in South Africa, it became the most common SARS-CoV-2 lineage in Africa during the second wave of the pandemic. Owing to the relaxation of travel restrictions, other variants of concern, such as Alpha, the UK lineage, and Delta, an Indian lineage, made their way into Africa and became dominant during the third wave of the pandemic. Eta and Gamma are some notable lineages that dominated during different waves of the pandemic. Omicron preceded the fourth pandemic wave and was first reported in South Africa and Botswana. Egypt reported the first case of SARS-CoV-2 infection in Africa (figure 4.1.1), which was of lineage B. This lineage was exported from other continents, primarily Europe, to other African countries.

There were distinct waves over time, according to a time series of case incidences in Africa (figure 4.1.2). The virus entered Africa in March 2020, marking the beginning of the first wave, with May 2020 seeing the highest number of cases. In July/August 2020, the Beta variant first appeared in South Africa, signalling the start of the second wave. August saw the greatest number of cases from this wave. The arrival of Delta, Alpha and Gamma variants marked the beginning of the third wave (figure 4.1.2). Owing to the easing of travel restrictions, the variants entered Africa in October 2020, November 2020 and January 2021, respectively. The third wave's peak number of cases was noted in January 2021. Before the number of instances began to decline in February 2021, there was a noticeable surge from October 2020 to January 2021 (figure 4.1.2). In November 2021, Omicron emerged in South Africa marking the start of the fourth wave. The waves appeared to be defined by either newly introduced variants in populations from other populations or novel arising variants as a result of mutations.

NextClade detects mutations in other sequences from a reference sequence and uses these differences to classify sequences into clades or lineages. Figure 3a shows the genetic diversity of clades in the reference sequences. Different colours represent different clades that originate from different mutations, and the branches of the tree represent the distance between these clades from the original sequence. GISAID separates the clades of SARS-CoV-2 into S, O, L, V, G, GH, GR, GV and GRY (figure 3b). When the pandemic first started, the S and L clades circulated. At first, S remained common, whereas L divided into G and V. G went on to break into GR, GH, and eventually GV. After July 2020, GR separated into GRY. The mutations that led to their branching are represented by the letters [12].

Alternatively, Nextstrain divides the SARS-CoV-2 strains into 19A, 19B, 20A, 20B, 20C, 20D, 20E, 20F, 20G, 20H, 20I and 20J (figure 4.1.3a). The initial reference strain in these clades is 19B. 20J/501Y.V3 denotes the P.1 strain that originated and spread from Brazil; 20H/501Y.V2 denotes the B.1.351 strain that began in South Africa; and 20I/501Y.V1 is the B.1.1.7 variation that originated in Britain [13]. From figure 4.1.3, we can clearly elaborate that GISAID clade S (figure 4.1.3b) is the same as Nextclade 19B (figure 4.1.3a), while GISAID clades L, O and V correspond to Nextclade 19A. GISAID clades G, GH and GR correspond to Nextclade 20A, 20C and 20B, respectively.

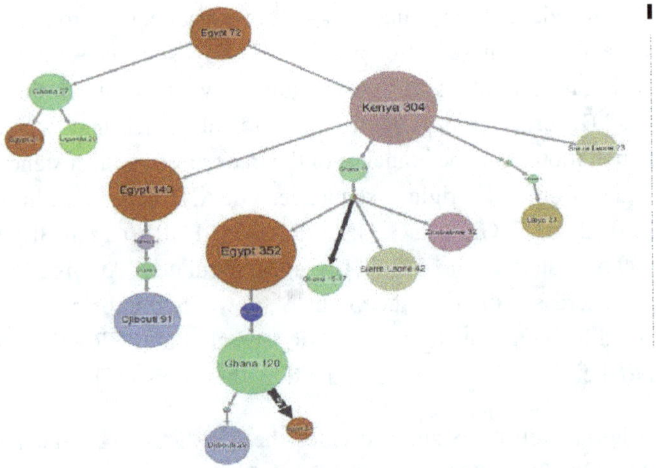

Figure 4.1.1. Spread dynamics reconstruction within Africa irrespective of the lineages. The figure shows compressed tree visualizations created by PastML using an F81-like model and MPPA. According to the map, various colours denote various countries. The sizes of the circles represent samples inside and the thickness of the arrow relates to a number of sequence IDs making up the circle. After entering Egypt, the virus spread to Kenya and Ghana. From there, there were imports and exports of the virus between African countries, and diversification events occurred across the continent.

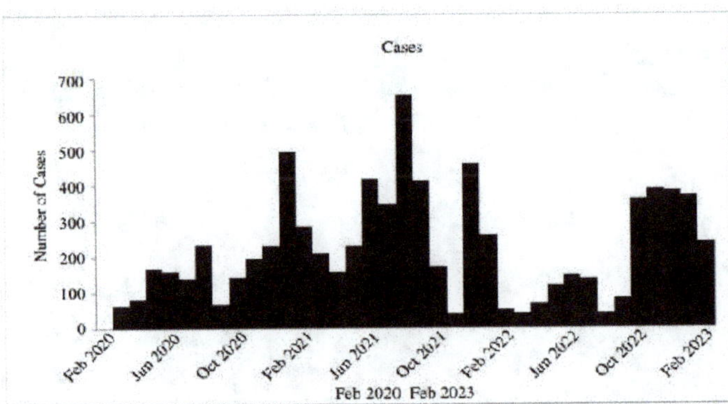

Figure 4.1.2 Visual presentation of epidemic waves and case rates through time in Africa.

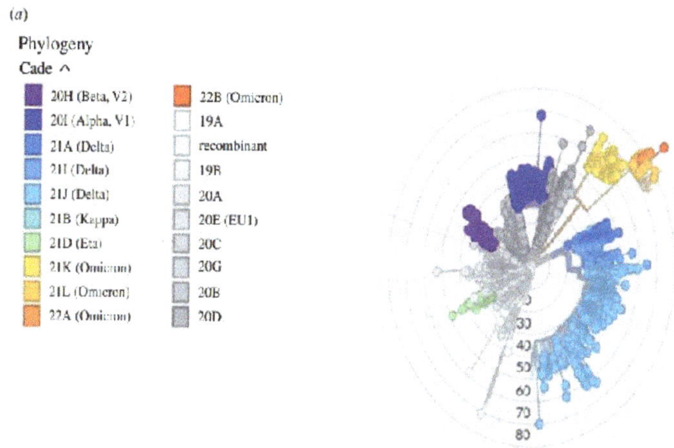

(b)

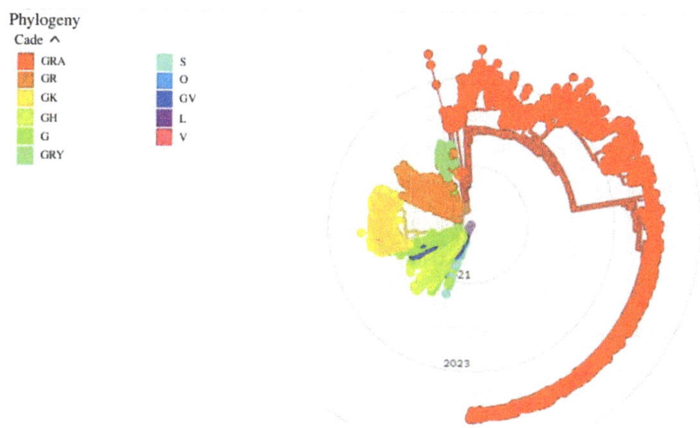

Figure 4.2.3 Phylogenetic tree reconstruction of clades circulating in Africa produced by NextClade (a) and GISAID (b). Different colours correspond to different clades as shown in the key (left).

4.1.3.2 Emergence and spread of new SARS-CoV-2 variants

We carried out a phylogeographic analysis of VOC Beta and Omicron, VUM Eta and IHU variant and one additional variant that arose and that we identified as VOI for this analysis as variant A to understand how some of the important SARS-CoV-2 mutations are spreading throughout Africa.

4.1.3.4 Variants of concern

4.1.3.4.1 Beta

Beta (B.1.351) was first sampled and reported in South Africa (figure 4a) in October 2020. When it first appeared in the Eastern Cape, the virus spread rapidly throughout South Africa, as the variant spread faster than previous variants of the virus (figure 4.1.4b). This variant was responsible for the second wave of the pandemic in South Africa. The variant had spread to bordering countries by November 2020 and to Djibouti, Malawi and Seychelles by December 2020. The predominant lineage in the majority of southern and eastern African populations by March 2021 was Beta. Beta also moved directly from southern Africa to eastern and central Africa (figure 4.1.4a).

4.1.3.4.2. Omicron

South Africa notified the WHO on 24 November 2021 that a new SARS-CoV-2 variant, Omicron, was discovered. Omicron was initially discovered in specimens collected in Botswana on 11 November 2021, and in South Africa on 14 November 2021. Since then, Omicron has been detected in samples collected in South Africa on 8 November 2021. VOC Omicron expanded into Botswana and South Africa's neighbouring countries (figure 5a), then into northern and western Africa (Morocco and Nigeria), as well as into their neighbouring countries. Other transmissions from Botswana to Mali and Ghana have been reported. Obviously, in-between transmission and re-introduction of the variant to the southern countries occurred (figure 5b).

4.1.4. Variants under monitoring

4.1.4.1. Eta

The spread dynamics of Eta demonstrate the movement of Eta directly from Nigeria to neighbouring countries such as Ghana, Mali, Togo, Sierra Leone and East Africa such as Kenya and Djibouti, and northern Africa such as Libya (figure 4.1.6a). Eta was exported to other continents and was reintroduced back to Africa (figure 4.1.6b). 3.2.2.2.

4.1.4.2 IHU Variant

The IHU Variant originated in central Africa and was first reported in the Congo (figure 4.1.7a). It was directly transmitted to Ghana, Kenya and Reunion (figure 4.1.7b).

4.2.4.3 Variant of interest

4.1.4.4. Variant A

For the sake of this investigation, we identified Variant A as a VOI because it offers excellent examples of the virus's ongoing evolution in Africa. Since its discovery in September 2020, Variant A has regionally extended to neighbouring Rwanda and Kenya, as well as to the Democratic Republic of Congo (DRC), South Sudan and Zambia in the south (figure 4.1.8a). Figure 4.1.8b shows that Variant A was also exported outside Africa.

4.1.5 Discussion

The first SARS-CoV-2 case in Africa was reported in Egypt, followed by Nigeria, South Africa, Ghana and Kenya on the same day (12 March 2020). To date, Africa has contributed approximately 9.49 million cases and 175 000 deaths, which are 1.25 and 2.56% of the total cases and deaths reported worldwide, respectively [6]. A minimum of 84.88 million cases and 1.97 million deaths were recorded in 2020. In 2021, 205.88 million cases and 3.49 million deaths were reported while 442.53 million cases and 1.23 million deaths were reported in 2022 [6]. The highest numbers of cases were recorded in 2022, whereas 2021 recorded the highest number of deaths. This makes COVID-19 the number one infectious disease, surpassing tuberculosis. Although it seems that people have adapted to coexist with the virus and uplifting lockdown restrictions, these results show that SARS-CoV-2 is still circulating and killing.

Egypt, Kenya and South Africa appear to be key suppliers of imports to other African countries. Wilkinson et al. [19] found that this is likely owing to the fact that these three nations have the most deposited sequences and are among the first to report cases of SARS-CoV-2 in Africa. Lineage B was the first SARS-CoV-2 case reported in Egypt, indicating that its early introduction in Africa was an Asian lineage reported by China. The early stage of the pandemic was distinguished by the dominance of lineage B.1, a large European lineage (the origin of which roughly corresponds to the northern Italian outbreak early in 2020) rather than lineage B, which caused the establishment of the pandemic in China. According to Kamara and Essien [20], lineage B.1 was introduced multiple times to African countries in the early phase of the pandemic; thus, most introductions were predominantly from Europe rather than from Asia.

(a)

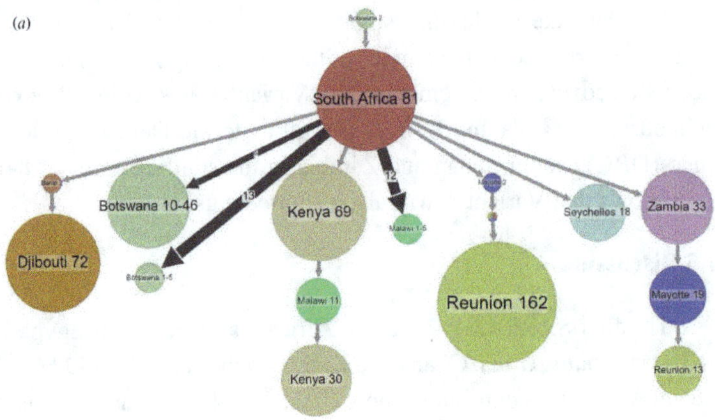

(b)

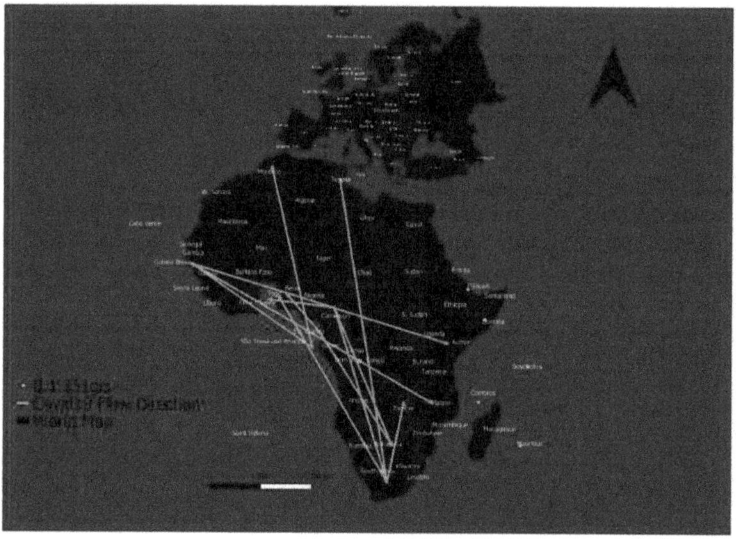

Figure 4.1.4. Spread dynamics reconstruction of Beta. (a) Compressed visualizations produced by PastML using MPPA with an F81-like model. The different countries are represented in the tree with different colours. The sizes of the circles represent samples inside and the thickness of the arrow relates to a number of sequence IDs making up the circle. (b) Spread dynamics of Beta on a geographical map.

Our phylogeographic analysis of African countries revealed that after entering Egypt, the lineage developed into the B.1 lineage before extending to Kenya, Morocco and Ghana. Subsequently, the B.1 lineage was imported and exported across African countries, and the lineage diversified into numerous variants (diversification events) across the continent. Lineage A also entered the picture from other continents as air travel restrictions were relaxed, allowing for intra-country and occasional international viral movements between neighbouring countries, presumably via road and rail ties. Although some border posts between countries were closed during the initial lockdown period, others remained open to allow trade to continue. As a result, variants have evolved within and spread across continents.

(a)

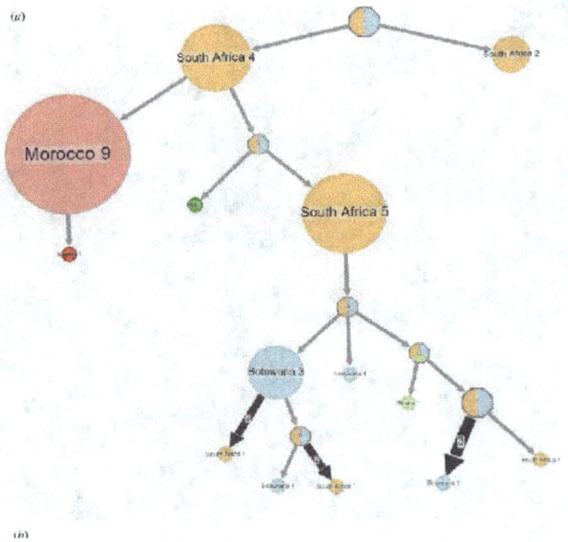

(b)

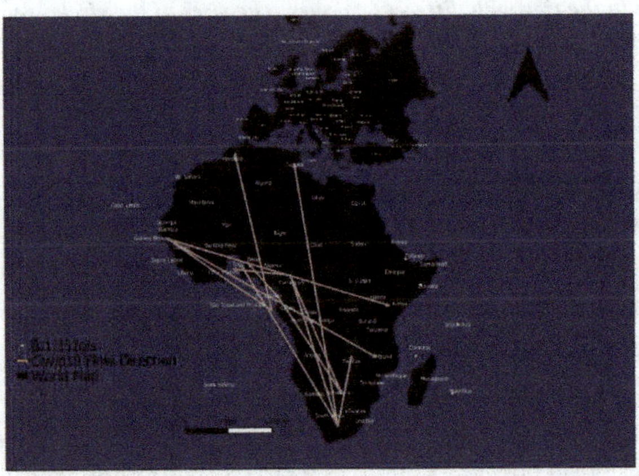

Figure 4.1.5. Spread dynamics reconstruction of Omicron. (a) Compressed visualizations produced by PastML using MPPA with an F81-like model. Different countries are represented in the tree with different colours. The sizes of the circles represent samples inside and the thickness of the arrow relates to a number of sequence IDs making up the circle. The Joint and MAP predictions are shown for the uncertain nodes. Although MAP and Joint disagree on the tree root, MPPA projections for all nodes are included in predictions (including those with unique MPPA prediction). (b) Spread dynamics of Omicron on a geographical map.

(a)

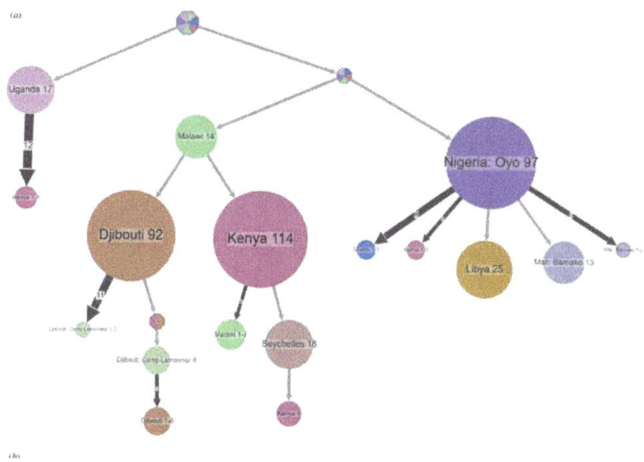

(b)

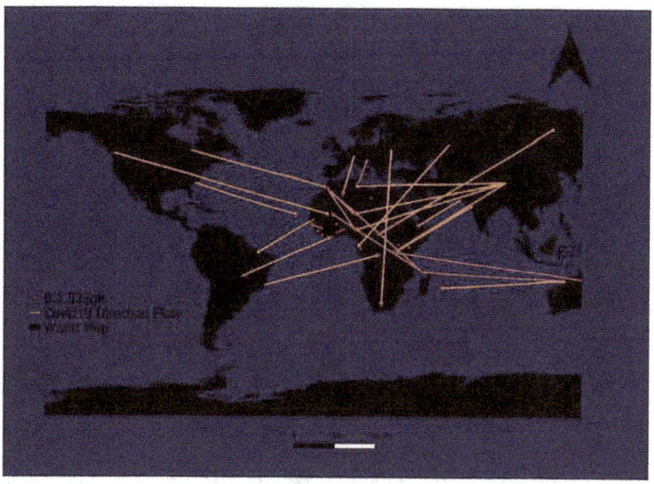

Figure 4.1.6 Spread dynamics reconstruction of Eta. (a) Compressed visualizations produced by PastML using MPPA with an F81-like model. Different countries are represented in the tree with different colours. The sizes of the circles represent samples inside and the thickness of the arrow relates to a number of sequence IDs making up the circle. The Joint and MAP predictions are shown for the uncertain nodes. MAP and Joint disagree on the tree root, but their predictions are included in MPPA predictions for all nodes (including those with unique MPPA prediction). (b) Spread dynamics of Eta on a geographical map.

(a)

(b)

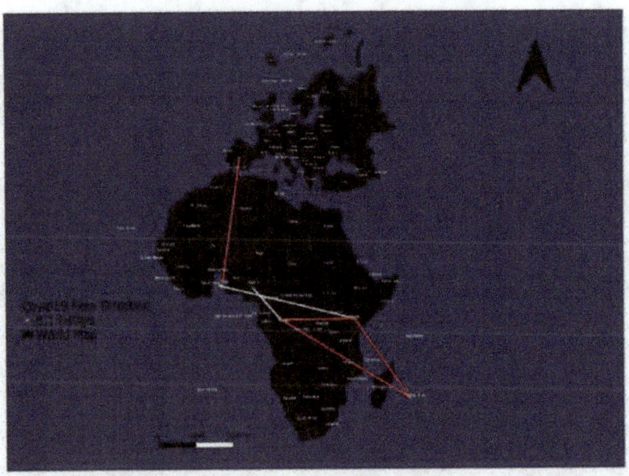

Figure 4.1.7. Spread dynamics reconstruction of IHU Variant. (a) Compressed visualizations created by PastML using a F81-like model and MPPA. Different countries are represented by different colours. (b) Spread dynamics of IHU Variant on a geographical map.

When lineage Beta first appeared in South Africa, it quickly spread throughout the country because the variant was thought to propagate faster than previous virus versions. By March 2021, Beta had surpassed Variant B (B.1) as the dominant lineage in most southern and East African countries. During the second wave of the epidemic, it was the most frequently discovered SARS-

CoV-2 lineage in Africa. This occurred because of the loosening of air and land travel limitations, which resulted in the arrival of various variants and lineages from other continents. VOC, such as Alpha of the UK lineage and Delta of the Indian lineage, made their way to Africa and became prominent during other pandemic waves. According to our phylogeographic reconstruction, Beta moved directly from southern Africa to eastern and central Africa. A discrete phylogeographic examination of a larger sample of viruses by Katella [21] implied that transmission to West Africa may have occurred via East Africa, with a probable European intermediary. BA.1, an alias of Omicron, was initially detected in South Africa and preceded the fourth pandemic wave. South Africa notified the WHO on 24 November 2021 that a new SARS-CoV-2 variant, Omicron, was discovered. Omicron was initially discovered in specimens collected in Botswana on 11 November 2021, and in South Africa on 14 November 2021. Since then, South Africa has discovered Omicron in samples taken on 8 November 2021 [22]. The spike protein of the Omicron variant is characterized by at least 30 amino acid substitutions, three small deletions, and one small insertion. Owing to these mutations, there was a marked reduction in vaccine effectiveness, enhanced transmissibility and increased risk for reinfection and immune escape as compared with other variants like Delta. Consequently, Omicron was designated as a VOC. Omicron extended into Botswana and South Africa's neighbouring countries, then into northern and western Africa (Morocco and Nigeria), as well as into their neighbouring countries (figure 4.1.5). Other transmissions from Botswana to Mali and Ghana have been reported. Clearly, the variety was reintroduced to southern and other countries during the time between transmissions. As a result, most southern African countries were placed on the blacklist and travel to other countries was prohibited.

Eta is a VOI defined by six substitutions in the spike protein and two deletions in the N-terminal domain. According to Katella [21], this variant was first sampled in the UK in mid-December 2020, but the phylogeographic reconstruction suggested that it originated in Nigeria in November 2020. Since then, it has spread throughout much of Nigeria and neighbouring Ghana. Our phylogeographic reconstruction also demonstrated the movement of Eta directly from Nigeria into neighbouring countries such as Ghana, Mali, Togo, Sierra Leone and eastern Africa such as Kenya and Djibouti and northern Africa, such as Libya (figure 4.1.6). The scope of this VOI distribution in the region is unknown because of poor sampling from neighbouring nations in western and Central Africa.

(a)

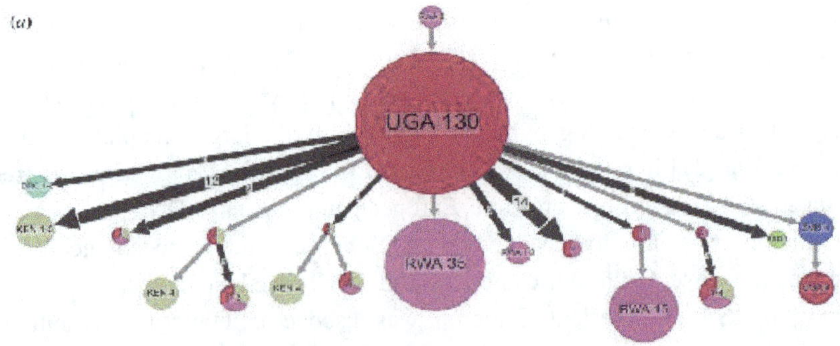

(b)

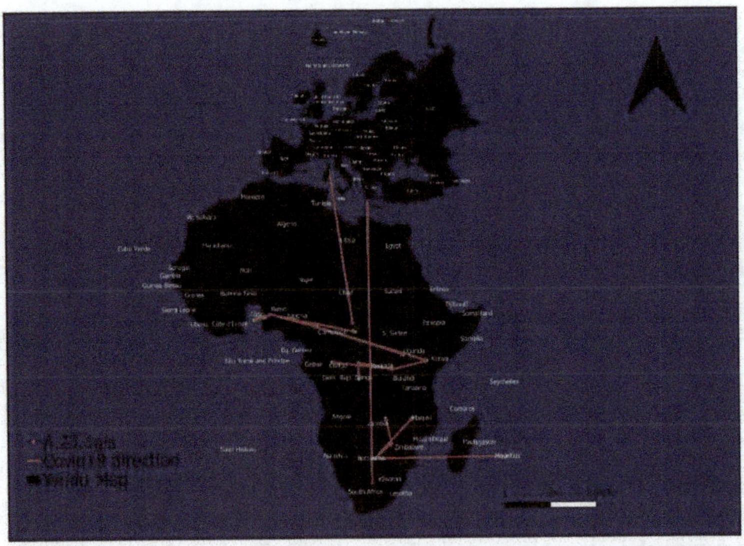

Figure 4.1.8. Spread dynamics reconstruction of Variant A. (a) Compressed visualizations produced by PastML using MPPA with an F81-like model. Different countries are represented in the tree with different colours. The sizes of the circles represent samples inside and the thickness of the arrow relates to a number of sequence IDs making up the circle. The Joint and MAP predictions are shown for the uncertain nodes. MAP and Joint disagree on the tree root, but their predictions are included in MPPA predictions for all nodes (including those with unique MPPA prediction). (b) Spread dynamics of Variant A on a geographical map.

The IHU Variant is a variant that originated in central Africa. According to our phylogeographic reconstruction, it was first reported in Congo and directly transmitted to Ghana, Kenya and Reunion (figure 4.1.7). Given the small sample size of all African countries, the extent of the VUM spread in this region is unknown. What we know is that it has two sub-lineages, B.1.640.1 and B.1.640.2, and as of 17 February 2022, B.1.640.2 was a VOC because of its 46 mutations and 37 deletions in its genetic code, many affecting the spike protein. This sub-lineage was also dubbed as the IHU rose in France [23].

We designated Variant A as a VOI because it presents good examples of the continued evolution of the virus within Africa. According to Haseltine [24], the variant contains several mutations observed in variants of concern, as well as six unique substitutions. However, the variant does not have a similar origin to any of the variants of interest or concern, including Alpha, Beta, Gamma, Delta and Mu [24]. All variants shared a mutation that identified their common ancestry; Variant A, however, does not [25]. It has more in common with the A.30 variety, which was first discovered in Angola and may have originated in Tanzania. According to Haseltine [24], neither of these variants has a common ancestor with the other major strains, and the identification of two separate yet distantly related variations in East Africa is troubling in and of itself. The fact that these variations originate separately from all other variants worldwide, despite the fact that they lack the characteristic trio of mutations that link all other variants, illustrates the adaptability of SARS-CoV-2 to local conditions.

Lineage Variant A, characterized by three spike mutations (F157L, V367F and Q613H), was first detected in July 2020 in a Ugandan prison in Amuru [25]. The lineage was then transmitted to the Kitgum prison, possibly aided by the transfer of the prisoners. Subsequently, the A.23 lineage infiltrated the general population and spread to Kampala, introducing new spike mutations (R102I, L141F, E484K and P681R) as well as additional mutations in nsp3, nsp6, ORF8 and ORF9, prompting the creation of a new lineage classification, Variant A [25]. Since its discovery in September 2020, Variant A has spread regionally into neighbouring Rwanda and Kenya, as well as to the DRC, South Sudan and Zambia in the south (figure 8). However, according to Wilkinson et al. [19], the phylogeographic reconstruction of Variant A suggests that the introduction into Ghana may have occurred via Europe, whereas the introduction into southern Africa likely occurred directly from East Africa. This is consistent with epidemiological data indicating that the case discovered in South Africa was the result of contact with a person who had recently visited Kenya.

High rates of COVID-19 testing and consistent genomic surveillance in the south of the continent have resulted in early detection of VOCs, such as Beta and Omicron [26]. Since the discovery of these southern African variants, several other SARS-CoV-2 VOIs have emerged around the world, including on the African continent, including B.1.525 in West Africa and Variant A in East Africa. Strong evidence suggests that both VOIs are becoming more common in regions where they have been identified, implying that they may be more suitable than other variants in these regions. In the later months of 2020, the Beta variant moved from South Africa to neighbouring nations, reaching as far north as the DRC by February 2021, according to our phylogenetic analysis of the Beta branch.

This spread could have been aided by rail and road networks that connect South Africa's ocean ports to economic and industrial centres in Botswana, Zimbabwe, Zambia and the DRC's southern regions. The rapid and seemingly unhindered penetration of the virus into these countries implies that present land border regulations aimed at limiting the virus's international transmission are ineffective and much needs to be done to implement and improve African land borders as far as epidemiology is concerned to contain such outbreaks in the future. From a public health standpoint, genetic surveillance is the only instrument in the pandemic preparedness toolset. However, genetic surveillance has not yet been effectively performed in Africa. The usefulness of molecular surveillance as a technique for monitoring pandemics relies heavily on ongoing and consistent sampling, rapid viral genome sequencing and timely reporting.

When this is accomplished, molecular surveillance can be used to detect the shifting pandemic traits early. Molecular monitoring data can inform public health interventions when such changes occur. In this context, molecular surveillance data collected by most African countries are less useful than they could be. For example, in some circumstances, the time lag between when virus samples are acquired and when sequences for these samples are stored in sequence repositories is so great that the primary utility of the genomic surveillance data is lost. Depending on the laboratory or country, numerous variables contribute to this lag: (i) lack of reagents owing to global supply chain disruptions, (ii) lack of equipment and infrastructure within the source country, (iii) lack of technical skills in laboratory methods or bioinformatics support, and (iv) reluctance of health officials to share data. To determine the genetic properties of currently circulating viruses in these nations, more recent sampling and fast reporting are required.

As a result, our study's fundamental flaw is the patchiness of the African genetic monitoring data. It is worth noting that phylogeographic reconstruction of viral propagation is strongly reliant on sampling, with the caveat that the exact paths of viral migration between nations cannot be deduced if connecting countries are not sampled. Furthermore, uneven sampling between African countries is certainly biasing our efforts to recreate SARS-CoV-2 migration dynamics across the continent. It is no surprise that we identified South Africa, Kenya and Nigeria as key sources of viral transmission between sub-Saharan African nations because they had the most SARS-CoV-2 genomes sampled and sequenced. The strength of national public health programmes to embrace genomic surveillance as a tool for preventing the appearance and spread of dangerous variants determines their reliability as a strategy for preventing the emergence and spread of dangerous variants. Ishikawa et al. [18] stated that, as in most other parts of the world, the success of genomic surveillance in Africa requires more samples to be tested for COVID-19, with a higher proportion of positive samples being sequenced within days of sampling.

The degree of dissemination of specific variations in the region is unknown because of limited sampling across all African countries. Some variants were sequenced to exhaustion and were deposited in databases, whereas others, such as IHU Variant and Eta, had relatively few sequences deposited in databases. Some countries have a considerable number of sequences deposited in sequence repositories, whereas others have few or no sequences. As a result, the results of our study were skewed. PASTML, which we used to perform phylogeographic reconstruction, also required caution with sampling bias, and it was necessary to subsample sequences to better represent the number of declared cases by location and/or use of additional data, such as travel history, or phylogeographic predictions would be incorrect. Complete genome sequencing of SARS-CoV-2 is important for obtaining useful information about the viral lineages, variants of interest, and variants of concern; however, only approximately 32% of the nucleotide sequences deposited in the NCBI database from Africa are complete, with 68% partial nucleotide sequences.

As a continent, we need to be able to deposit complete genomic sequences because most researchers only take complete genomes for these types of studies. Perhaps because of this, we discovered that there are fewer lineages circulating throughout Africa than actually.

4.1.5 Conclusion

Africa has been the least-affected continent for several reasons. This may be owing to skewed statistics being reported or the fact that many people have turned to using ethnobotanical treatments to treat COVID-19 symptoms because Africa has a strong tradition of using plants. This might be because African populations have lower rates of comorbid conditions and are older than those of other continents. Our findings suggest that SARS-CoV-2 variants circulating in Africa follow unique patterns of evolution and transmission. We also found that all variants shared a mutation that identified common ancestry. However, Variant A does not, it having more in common with the A.30 variety, which was first discovered in Angola, and may have originated in Tanzania. Neither of these variants had a common ancestor with other major strains. The fact that these variations originate separately from all other variants worldwide, despite the fact that they lack the characteristic trio of mutations that link all other variants, illustrates the adaptability of SARS-CoV-2 to local conditions. Since the discovery of the southern African variants Beta and Omicron, several other VOIs have emerged worldwide, including those on the African continent, specifically in West Africa (Eta) and East Africa (Variant A). According to the available data, both VOIs are clearly increasing in frequency in the areas where they have been detected, implying that they may be more suitable than other variants in these regions, thus suggesting that SARS-CoV-2 variants are specific to geographic conditions. Routine genomic surveillance of the variants, employing phylogenetic and phylodynamic analyses with wider coverage and higher resolution, is essential to provide insight into the effectiveness of different intervention strategies against the pandemic. The effectiveness of molecular surveillance as a method for tracking pandemics strongly depends on continuous and reliable sampling, rapid viral genome sequencing and prompt reporting, all of which must be improved in Africa. Additionally, the pandemic outbreak revealed that current land border regulations aimed at limiting the virus's international transmission are ineffective, and much needs to be done to implement and improve African land borders as far as epidemiology is concerned to contain such outbreaks in the future. We also need universal or common criteria for sample selection, library preparation and sequencing platforms, bioinformatics workflows and data interpretation to ensure consistent data quality standards across public databases.

Ethics. No human or animal subjects were included in this study and ethical approval was not required.

Data accessibility. The datasets generated and analysed during the current study are publicly available at [27]. Data is available in the supplementary material [28].

Declaration of AI use. We have not used AI-assisted technologies in creating this article.

Authors' contributions. D.M.: conceptualization, data curation, formal analysis, investigation, methodology, project administration, resources, validation, visualization, writing—original draft, writing—review and editing; T.M.: data curation, validation, visualization, writing—review and editing; Z.C.: conceptualization, supervision, validation, writing—review and editing.

All authors gave final approval for publication and agreed to be held accountable for the work performed therein.

Conflict of interest declaration. We declare we have no competing interests.

Funding. The authors have received no financial support for the research, authorship, or publication of this paper.

Acknowledgements. We thank all those who have contributed sequences to the GISAID and NCBI databases and Wallace Gara for helping with constructing GIS maps.

4.1.6 References

1. Zhang YZ, Holmes EC. 2020 A genomic perspective on the origin and emergence of SARS-CoV-2. Cell 181, 223–227. (doi:10.1016/j.cell.2020.03.035)

2. Zukurov JP, do Nascimento-Brito S, Volpini AC, Oliveira GC, Mario Janini LR, Antoneli F. Estimation of genetic diversity in viral populations from next generation sequencing data with extremely deep coverage. See http://tanden.url.ph/.

3. Lone SA, Ahmad A. 2020 COVID-19 pandemic—an African perspective. Emerg. Microbes Infect. 9, 1300–1308. (doi:10.1080/22221751.2020.1775132)

4. Olarenwaju B et al. 2021 Evolution and genetic diversity of SARS-CoV-2 in Africa using whole genome sequences. Int. J. Infect. Dis. 103, 282–287. (doi:10.1016/j.ijid.2020.11.190)

5. Africa Centres for Disease Control and Prevention (Africa CDC). 2021 Annual progress report 2020. See https://africacdc.org/covid-19/.

6. World Health Organization. WHO coronavirus (COVID-19) dashboard. See https://covid19.who.int (accessed 28 February 2023).

7. Corman V et al. 2020 Detection of 2019 novel coronavirus (2019-nCoV) by RT-PCR. Euro Surveill. 25, 2000045. (doi:10.2807/1560-7917.ES.2020.25.3.2000045)

8. Alanagreh L, Alzoughool F, Atoum M. 2020 The human coronavirus disease covid-19: its origin, characteristics, and insights into potential drugs and its mechanisms. Pathogens 9, 331. (doi:10.3390/pat hogens9050331)

9. Cui J, Li F, Shi ZL. 2019 Origin and evolution of pathogenic coronaviruses. Nat. Rev. Microbiol. 17, 181–192. (doi:10.1038/s41579-018-0118-9)

10. Kumar S, Nyodu R, Maurya VK, Saxena SK. 2020 Morphology, genome organization, replication, and pathogenesis of severe acute respiratory syndrome coronavirus 2 (SARS-CoV-2). In Coronavirus disease 2019 (COVID-19) (ed. S Saxena), pp. 23–31. Singapore: Springer. (doi:10.1007/978-981-15-4814-7_3)

11. National Center for Biotechnology Information (NCBI), National Library of Medicine (US). See https://www.ncbi.nlm.nih.gov/ (accessed 25 February 2023).

12. Khare S et al. 2021 Perspectives: GISAID's role in pandemic response. China CDC Weekly 3, 1049–1051. (doi:10.46234/ccdcw2021.255)

13. Aksamentov I, Roemer C, Hodcroft EB, Neher RA. 2021 Nextclade: clade assignment, mutation calling and quality control for viral genomes. J. Open Source Softw. 6, 3773. (doi:10.21105/joss.03773)

14. Katoh K, Rozewicki J, Yamada KD. 2019 MAFFT online service: multiple sequence alignment, interactive sequence choice and visualization. Brief. Bioinform. 20, 1160–1166. (doi:10.1093/bib/b bx108)

15. Katoh K, Standley DM. 2013 MAFFT multiple sequence alignment software version 7: improvements in performance and usability. Mol. Biol. Evol. 30, 772–780. (doi:10.1093/molbev/mst010)

16. Nguyen LT, Schmidt HA, von Haeseler A, Minh BQ. 2015 IQ-TREE: a fast and effective stochastic algorithm for estimating maximum-likelihood phylogenies. Mol. Biol. Evol. 32, 268–274. (doi:10.1093/molbev/msu300)

17. O'Toole Á et al. 2021 Assignment of epidemiological lineages in an emerging pandemic using the pangolin tool. Virus Evol. 7, 1–9. (doi:10.1093/ve/veab064)

18. Ishikawa SA, Zhukova A, Iwasaki W, Gascuel O, Pupko T. 2019 A fast likelihood method to reconstruct and visualize ancestral scenarios. Mol. Biol. Evol. 36, 2069–2085. (doi:10.1093/molbev/msz131)

19. Wilkinson E et al. 2021 A year of genomic surveillance reveals how the SARS-CoV-2 pandemic unfolded in Africa. Science 374, 423–431. (doi:10.1126/science.abj4336)

20. Kamara J, Essien U. 2022 COVID-19 in Africa: supply chain disruptions and the role of the Africa continental free trade agreement. J. Glob. Health 12, 03085. (doi:10.7189/jogh.12.03085)

21. Katella K. 2023 Omicron, Delta, Alpha, and more: what to know about the coronavirus variants. Yale Medicine. See https://www.yalem edicine. org/news/covid-19-variants-of-concern-omicron (accessed 3 February 2023).

22. Viana R et al. 2022 Rapid epidemic expansion of the SARS-CoV-2 Omicron variant in southern Africa. Nature 603, 679–686. (doi:10.1038/s41586-022-04411-y)

23. Cella E. 2021 Chemotherapy. SARS-COV-2 lineages and sub-lineages circulating worldwide: a dynamic overview. See https://www.karger .com/Article/FullText/515340 (accessed 28 February 2023).

24. Haseltine WA. 2021 A second east African variant: A.23.1. Forbes. See https://www.forbes.com/sites/williamhaseltine/2021/09/16/a-secondeast african-variant-a231/?sh=43790f106ed2 (accessed 22 June 2022).

25. Lubinski B, Frazier L, Phan MVT, Bugembe DL, Tang T, Daniel S, Cotten M, Jaimes J, Whittaker GR. 2021 Spike protein cleavage-activation mediated by the SARS-CoV-2 P681R mutation: a case-study from its first appearance in variant of interest (VOI) A.23.1 identified in Uganda. SSRN. (doi:10.2139/ssrn.3966642)

26. European Centre for Disease Prevention and Control. SARS-CoV-2 variants of concern as of 23 February 2023. See https://www.ecdc.europa.eu/en/covid-19/variants-concern (accessed 24 February 2023).

27. Mtetwa D. 2024 Genetic diversity and spread dynamics of SARS-CoV-2 variants present in African populations. [Dataset] Dryad (doi:10.5061/dryad.1c59zw42d)

28. Mtetwa D, Manjengwa T, Chikwambi Z. 2024 Supplementary material from: Genetic diversity and spread Dynamics of SARS-Cov-2 variants present in African populations. FigShare (doi:10.6084/m9.figshare.c.7184008).

4.2 Identification of SARS-CoV-2 mutation hotspots within the spike protein Region

Desire Mtetwa mtetwadesire@gmail.com

School of Health Science and Technology, Department of Biotechnology, Chinhoyi University of Technology, Private Bag 7724, Chinhoyi, Zimbabwe.

Abstract

The objective of this integrated mini review was to find mutation hotspots of SARS-CoV-2 variants circulating within African populations and document them. SARS-CoV-2 is composed structurally of four structural proteins: the spike (S), envelope (E), membrane (M), and nucleocapsid (N). Among these structural proteins, the virus enters host cells through the S protein. Therefore, in order to comprehend viral transmission and create viral vaccines, sequence analysis of the genomic region encoding the viral S protein is required, thus attention was given to nonsynonymous mutations within the spike protein region. From all the variants circulating within the African continent we only selected and focused on the five variants of concern as these were the major variants that seem to affect vaccine efficacy. We found out that All five Variants of Concern (VOCs) exhibited three clusters of S gene mutations, one at the NTD, one at the RBD, and one at CTD 2, two of which are regions that neutralizing antibodies are designed to neutralize and D614G is a mutation found in all five VOCs.

Keywords: Mutation hotspot, spike protein, SARS-CoV-2 variants

4.2.0 Introduction

A detailed analysis of the virus's genetic and evolutionary trends is required in regard to the appearance of SARS-CoV-2 on the African continent. Understanding how pathogens develop and evolve can be learned via comparative examination of viral genome sequences. To provide insight into the dynamics of pandemics and assist in the creation and execution of control measures on the African continent, this project will perform all-encompassing research of the SARS-CoV-2 epidemic and its progression in Africa.

Due to constant mutation, the SARS-CoV-2 virus has swiftly mutated and now affects millions of people worldwide (Wu, Ye, et al., 2020). Since the SARS-CoV-2 genome sequence became public, changes at various locations in the genome have been discovered, raising worries about the virus's increased transmissibility (Wu et al., 2020). In 2020, new coronavirus variants expanded significantly. First, a more contagious strain (Alpha) was discovered by British researchers in the London region. (FinancialNewsMedia.com, n.d.). Then South African researchers found one (Beta) that appeared to evade the immune system (FinancialNewsMedia.com, n.d.). Next, yet another variant (Gamma) was detected in Brazil because it appeared to have the potential to re-infect individuals who had previously been exposed to it (FinancialNewsMedia.com, n.d.) and following the discovery of one variant (Delta) in India, another variant (Omicron) with additional mutations and a higher transmission potential than Delta was discovered initially in South Africa and Botswana. Concerns about the ability of the present vaccines to control the virus's ever-evolving strains have increased as a result of vaccine development to combat and contain the worldwide pandemic (FinancialNewsMedia.com, n.d.). It has been further shown that the SARS-CoV-2 variants exhibit great implication on vaccine efficacy.

The virus's propensity for mutation has motivated researchers worldwide to make active efforts to monitor and comprehend the development of variants of concern (Wu et al., 2020). One of the initial mutations that quickly spread over the globe, D614G, was initially noted in April 2020 (Hoffmann et al., 2020). The S protein's D614G mutation, is a result of an A-to-G base change at the position 23,403 in the Wuhan reference strain, and has been associated with the emergence of a more transmissible form of SARS-CoV-2 (Thomson et al., 2021). Other mutations in the S gene area, besides the D614G, have been found in isolates from various geographical origins (Thomson et al., 2021). Globally, the coronavirus disease (COVID-19) pandemic has a high incidence of transmission and mortality (Thomson et al., 2021).

4.2.1 Taxonomic classification

SARS-CoV-2 is classified as follows:

Kingdom: Orthornavirae

Phylum: Pisuviricota

Class: Pisoniviricetes

Order: Nidovirales

Family: Coronaviridae

Subfamily: Orthocoronavirinae

Genus betacoronavirus

subGenus: Embecovirus

Species: Human coronavirus HKU1

Abbreviations HCoV-HKU1

4.2.2 SARS-CoV-2 Virus

An entirely novel betacoronavirus called SARS-CoV-2 was discovered for the first time in China in December 2019. (cmay, 2021). Three groups of SARS-CoV-2 variants are defined by a variant categorization scheme that was created by an interagency group inside the US government: Variant of interest (VOI) a genetic variant with particular genetic markers linked to altered receptor binding, decreased neutralization by antibodies produced against prior infection or vaccination, decreased treatment effectiveness, potential diagnostic impact, or anticipated rise in disease severity; Variant of concern (VOC) a variant for which there is evidence of an increase in transmissibility, more severe disease (such as an increase in hospitalization and deaths), a notable reduction in neutralization by antibodies produced during prior infection or vaccination, decreased effectiveness of treatments or vaccines, or failures in diagnostic detection; and Variant of high consequence, has convincing evidence that medical countermeasures (MCMs) or preventative measures (PCMs) are much less effective than previously circulating variants. ("Next IAS," n.d.). SARS-CoV-2 is composed structurally of four structural proteins: the spike (S), envelope (E), membrane (M), and nucleocapsid (N) (Thoutam, Breitzig, Lockey, & Kolliputi, 2020). The S protein, one of these structural proteins, facilitates the virus's entrance into host cells (Thoutam, Breitzig, Lockey, & Kolliputi, 2020). Therefore, in order to comprehend viral transmission and create viral vaccines, sequence analysis of the genomic region encoding the viral S protein is required (Thoutam, Breitzig, Lockey, & Kolliputi, 2020).

4.2.3 SARS-CoV-2 in Africa

Africa accounts for quite a low number of cases and deaths – by the mid-May 2023, ~9.52 million cases and ~175 000 deaths were reported on the continent, translating to under 8% of the global burden. Seroprevalence surveys and autopsy studies, however, suggests that the correct statistics may be several times greater than reported in some African nations (Wilkinson et al., 2021). Between the middle of February and the beginning of March 2020, Nigeria, Egypt, and South Africa reported the initial COVID-19 case on the continent. By the end of March 2020, the majority of nations had reported cases (Wilkinson et al., 2021).

The majority of these early cases were found in air travellers returning from places with high rates of community transmission (Wilkinson et al., 2021). To contain the spread of the virus and enable health services time to prepare, many African nations implemented early public health and social measures (PHSM), such as foreign travel restrictions, quarantines for travellers who returned, and internal lockdown measures (Wilkinson et al., 2021). Different countries have implemented different lockdown and travel restrictions, which has led to the emergence of local breakouts (Tegally et al., 2021).

4.2.4 Early Phylogenetic analysis of the virus in Africa

Genomic surveillance has been central to the COVID-19 response on the continent since the beginning of the pandemic (Wilkinson et al., 2021). Genomic data from the earliest imported cases may be obtained and shared owing to the diverse laboratories in Africa's rapid adoption of SARS-CoV-2 sequencing. The SARS-CoV-2 sequences that have been shared throughout this epidemic, notably from one of the initial patients in Wuhan, China (MN908947.3), are helping to characterize and comprehend this newly developing virus (Tegally et al., 2021). To comprehend the emergence of SARS-CoV-2 and monitor the dynamics of transmission around the world, genomic epidemiology is crucial (Tegally et al., 2021). The monitoring of ongoing SARS-CoV-2 evolution and the detection of new SARS-CoV-2 variants has both depended heavily on genomic surveillance (Wilkinson et al., 2021).

Lineage A and lineage B are the two primary phylogenetic lineages into which SARS-CoV-2 is commonly divided. Despite the fact that both lineages began in China, lineage A spread over the world from Asia whereas lineage B mostly spread from Europe (Tegally et al., 2021). The preponderance of lineage B.1 during the pandemic's early stages was one of its main characteristics. African nations were repeatedly introduced to this (Wilkinson et al., 2021). As air travel virtually stopped in March/April 2020, the epidemic reached a phase defined in sub-Saharan Africa by persistent low levels of intra-country migrations as the number of detectable virus imports into Africa dropped and B.1.351 became the most regularly identified SARS-CoV2 lineage found in Africa due to irregular international viral transfers between nearby nations, most likely through road and rail (Wilkinson et al., 2021).

Despite being introduced into a number of African nations, lineage A viruses only make about 1.3% of the continent's analysed genomes. The earliest SARS-CoV-2 lineage is Lineage A, which corresponds to the initial Wuhan virus samples from December 2019 (Wilkinson et al., 2021). Despite the fact that lineage A viruses first caused numerous small clustered outbreaks in a number of nations (such as Burkina Faso, Cote d'Ivoire, and Nigeria), lineage B viruses eventually mainly superseded them as the pandemic progressed. However, there is proof that lineage A viruses are becoming more common in several African nations (Wilkinson et al., 2021).

4.2.5 Early Phylogeographic analysis of the virus in Africa

Early SARS-CoV-2 introductions were mostly from Europe, but as the pandemic advanced, there was an increased transmission between African nations. While imports from outside of Africa, primarily from Europe, predominated in the initial phases of the epidemic, the dynamics then changed as the pandemic spread, with an increase in imports from other African nations (Wilkinson et al., 2021). Despite the fact that these three nations have the most deposited sequences, South Africa, Kenya, and Nigeria stood out as significant import sources into other African nations. The southern African region is particularly worth noting, because South Africa is the origin of a significant portion (about 80%) of imports to other nations in the region (Wilkinson et al., 2021).

In contrast to the rest of the continent, the Northern African region has a dissimilar pattern, with most viral introductions coming from Europe and Asia (especially the Middle East) than from other African nations. (Wilkinson et al., 2021). With a minimum of 356 exportation episodes from Africa to the rest of the world being identified in the dataset, the spread of the virus globally has also been facilitated by Africa. The majority of exports went to Europe (41%), Asia (26%) and North America (14%), which is consistent with the origin of imports. Exportation events were more uniformly spaced out across time than importation events were (Wilkinson et al., 2021).

4.2.6 Vaccines

Following the coronavirus's isolation in late 2019 (Wikipedia, 2020), the genetic sequence of the virus was published on January 11, 2020, prompting an immediate international response to get ready for an outbreak and speed up the creation of a COVID-19 vaccine (Wikipedia, 2020). Since the beginning of 2020, unprecedented government and global pharmaceutical industry partnership has sped up the development of vaccines (Wikipedia, 2020). In order to create lots of vaccine candidates and get ready for international immunization campaigns to protect against SARS-CoV-2 virus, corporations, governments, international health organizations, and university research groups committed tens of billions of dollars by June 2020 (Wikipedia, 2020). There are four vaccine types namely: Whole Virus (conventional inactivated vaccines); Protein Subunit vaccines; Viral Vector vaccines and Nucleic Acid vaccines. Whole viruses are conventional vaccines that use whole viruses to trigger an immune response ("There Are Four Types of COVID-19 Vaccines: Here's How They Work," n.d.). There are two primary methods. There are two types of vaccines: inactivated vaccines, which utilize viruses whose genetic material has been destroyed so they cannot multiply but can still elicit an immune response, and live attenuated vaccines, which employ a virus that has been weakened but can still replicate without causing illness. Both employs tried-and-true technology and formal regulatory approval procedures, but live attenuated ones pose a danger of disease transmission to individuals with weakened immune systems and frequently in need of careful cooling storage, making their adoption in countries with limited resources more challenging. People with weakened immune systems can receive inactivated viral vaccines, although they may also require cold storage ("There Are Four Types of COVID-19 Vaccines: Here's How They Work," n.d.). In order to elicit an immune response, protein subunit vaccines generally use portions of the pathogen, such as protein fragments. The danger of adverse effects is reduced, but the immunological response might be less robust as a result ("There Are Four Types of COVID-19 Vaccines: Here's How They Work," n.d.). This is why adjuvants are frequently needed to aid in enhancing the immune response. Hepatitis B vaccination is a prime example of an existing subunit vaccine. To give cells the instructions they need to create the antigen in nucleic acid vaccinations, genetic material, such as RNA or DNA is needed. This is typically the viral spike protein, as it is in the case of COVID-19. Once inside human cells, this genetic material employs the protein-making machinery to create the antigen that will set off an immune response. Additionally, viral vector vaccines function by instructing cells genetically to manufacture antigens ("There Are Four Types of COVID-19 Vaccines: Here's How They Work," n.d.). However, they vary from nucleic acid vaccines because they convey these instructions into the cell using a virus that isn't harmful and isn't the target of the vaccination. The adenovirus, which causes the common cold, is one virus type that has frequently been utilized as a vector ("COVID-19 VACCINES," n.d.). Similar to nucleic acid vaccines, the antigen is produced by our own cells using those instructions to stimulate an immune reaction. Since viral vector vaccines can imitate real viral infection, they ought to strongly elicit an immune response ("COVID-19 VACCINES," n.d.).

By April 2021, National regulatory authorities have granted emergency use of 17 vaccines for public use. These vaccines are Pfizer-BioNTech and Moderna (Nucleic Acid vaccines), Sinopharm-BBIBP, Sinopharm-WIBP, CoronaVac, Covaxin, CoviVac, Minhai, COVIran Barakat and QazCovid-in (Whole Virus vaccines), Sputnik light, Sputnik V, OxfordAstraZeneca, Convidecia and Johnson & Johnson (Viral Vector vaccines), EpiVacCorona and RBD-Dimer (Protein subunit vaccine) (Wikipedia, 2020). Nine of these vaccines namely Oxford-AstraZeneca, Pfizer-BioNTech, Moderna, Johnson&Johnson Sputnik light, Sinopharm (BBIBP), CoronaVac, Covaxin, and Sputnik V were being administered in Africa with OxfordAstraZeneca dominating in the continent in April 2021. There are many concerns that these vaccines are malfunctioning, South Africa reported that Oxford-AstraZeneca was causing blood clots and has since been discontinued in South Africa.

4.2.7 Antibodies produced by vaccines

Important immunological molecules known as antibodies defend their hosts by attaching to antigens, which are their infectious targets, and generating a focused immune response (Weitzner, 2015). Immune responses brought on by spontaneous infections or vaccinations mostly involve serum immunoglobulin G (IgG). The most common antibody, IgG, is produced by leukocytes, or white blood cells, particularly B lymphocytes. These IgGs are structurally different depending on which vaccine has been administered or if it's a natural infection. Literature shows that the Pfizer vaccine elicits RBD-binding IgGs and anti-S IgG among others (Kim et al., 2023). AstraZeneca elicits IgG generated against S1, RBD, and full-length spike proteins. Sinopharm triggers an immune response against nucleocapsids, membranes, envelope proteins and spike proteins. These immune responses can validate vaccine efficacy (Kim et al., 2023).

4.2.8 Methodology

4.2.8.1 Dataset mining and workflow From the National Center for Biotechnology

Information (NCBI) database and GISAID database ("GISAID - Gisaid.org," n.d.), SARS-CoV-2 genomic sequences gathered from Africa were retrieved (NCBI, 2020) on February 26, 2023. 24415 African sequences were retrieved from both databases to examine the number of lineages circulating within Africa. The two databases had only 8044 complete genome sequences combined from Africa, and these sequences excluding those with low coverage using NextClade ("Nextclade," n.d.) were retrieved to determine spread dynamics. There were 5908 sequences from 23 different African nations in the NCBI database and 2137 sequences from 41 African countries from the GISAID database. With the help of the MAFFT's online version, a multiple sequence alignment tool ("MAFFT – a Multiple Sequence Alignment Program," n.d.) the sequences were aligned (Katoh et al., 2018) with the Wuhan-Hu-1 (MN 908947.3) as the reference sequence, and sequences with more than 5.0% ambiguous letters were removed. Duplicates were removed using goalign dedup software and only high-quality African complete sequences remained (n=7540).

4.2.8.2 Identification of mutation hotspots

A fasta file with 7540 Spike protein sequences was uploaded on Galaxy Europe platform ("Galaxy | Europe," n.d.). We chose NextClade to analyze for mutation hotspots. 4156 sequences were retained after quality filtering with NextClade in a tsv file.

4.2.9 Results

4.2.9.1 Identification of mutation hotspots

It is worth noting that D614G is a mutation found in all five Variants of Concern (VOCs). All five VOCs have three clusters of S gene mutations as shown in (Table 4.2.1), these three clusters are at the NTD, RBD and CTD 2, two of which are regions that neutralizing antibodies are designed to neutralize. Omicron has a greater overall number of mutations than any other variant, by a wide margin (Table 4.2.1).

Table 4.2.1: Affected regions of the spike protein in SARS-CoV-2 variants (Alpha, Beta, Delta, Gamma and Omicron).

	Sub-unit 1				Sub-unit 2				
	N-terminal Domain NTD (14-305)	**Receptor-Binding Domain RBD (311-527)**	**CTD 1 (528-590**	**CTD2 (591-680)**	**C**	**Prefusion peptide (686-816)**	**834-909**	**HR1 (910-984)**	**1068-1162**
Alpha	**69/70▲, Y144▲**	**N501Y**	A570D	D614G	**P681H**	T716I		S982A	D1118H
Beta	**L18F,** D80A, D215G, ▲242-244, R246I	**K417N, E484K, N501Y**		D614G		A701V			
Gamma	**L18F,** T20N, P26S, D138Y, R190S	**K417T, E484K, N501Y**		**D614G, H655Y**					T1072I
Delta	T19R, Q107H,	L452R, **E484Q, T478K**		**D614G**	**P681R**			D950N	

	▲156/157, R158G								
Omicron	A67V, **▲69/70**, T95I, G142D, **▲143/145,** ▲211, L212I, ins214EPE	G339D, S371L, S373P, S375F, **K417N**, N440K, G446S, S477N, **T478K**, **E484A**, Q493R, G496S, Q498R, **N501Y**, Y505H	T547K	**D614G,** **H655Y,** N679K	**P681H**	N764K, D796Y	N856K	Q954H, N969K, L981F	

4.2.10 Discussion and Conclusion

4.2.10.1 Discussion

All five VOCs have three clusters of S gene mutations (Table 4.2.1), these three clusters are at the NTD, RBD and CTD 2, two of which are regions that neutralizing antibodies are designed to neutralize. Alpha variant was first detected in United Kingdom (Mtetwa et al., 2022) and carries a significant number of mutations (28 in total), of the 28, 20 (3 deletions and 17 nonsynonymous substitutions) are amino acid-altering, 9 are S gene mutations (Table 4.2.1), 4 of which have established pharmacological effects. The RBD contains the mutation N501Y, which has been demonstrated to impact the ACE2 receptor's binding affinity in the prior studies (McLean et al., 2022).

This mutation is shared among all VOCs except delta variant. Numerous other SARSCoV-2 lineages, including omicron, have the deletion at position 69-70 (69/70▲), which has been observed in conjunction with RBD alterations and may be related to immune evasion or infectivity (McLean et al., 2022) (Table 4.2.1). The Alpha variant shows a deletion at position 144 which is in consistence with other lineages such as the Omicron, and this alteration may affect the coordination and stability of the spike glycoprotein (McLean et al., 2022), and provide resistance to neutralizing antibodies that are directed towards NTD region (McLean et al., 2022). The spike protein's alteration P681H is close to the furin cleavage site at the intersection of the S1 and S2 domains and has been demonstrated to enhance transmission in an animal model by encouraging entrance into human lung cells (McLean et al., 2022), alpha variant shares this mutation with omicron variant.

The Beta variant was discovered initially in South Africa in October 2020 (Mtetwa et al., 2022). The S gene has 8 nonsynonymous changes and variable alterations at position L242 are among the 20 mutations found in Beta (McLean et al., 2022). 3 of these mutations are in crucial RBD locations linked to immunological evasion: the N501Y, E484K mutation shared with Gamma, Delta and Omicron plus K417N shared with Gamma and Omicron (Table 4.2.1). Brazil made the initial discovery of the Gamma variant in December 2020 (Mtetwa et al., 2022). Gamma has 31 mutations, 21 of which alter the functionalities of the amino acids (McLean et al., 2022). Of the 21 mutations, 20 are nonsynonymous changes and 1 is a deletion. Further, the spike protein is affected by 11 mutations, including 2 that are shared with Beta (N501Y and E484K) and a distinct mutation at position 417 (K417T) (McLean et al., 2022).

The Delta variant was first stated in India in October 2020 (Mtetwa et al., 2022). 21 onsynonymous mutations, 1 deletion, and 5 synonymous mutations make up this variant. P681R, a point mutation that also affects Alpha and Omicron and is located next to the furin cleavage site, is one of 9-point mutations that impact the spike protein, and the RBD's L452R mutation has been associated with enhanced ACE2 binding and neutralizing antibody resistance. Additionally, the spike protein has a deletion at position 156/157 and the Omicron-shared mutation T478K. The Omicron variant was recorded in several nations in November 2021 including South Africa and Botswana (Mtetwa et al., 2022). The aggregate number of mutations is substantially higher than has ever been observed with any prior variation, despite the fact that some of them are shared with the other VOCs. Omicron carries 45 nonsynonymous mutations, 10 synonymous mutations, 7 deletions, 1 insertion, and 7 deletions. The majority of the nonsynonymous alterations originate in the NTD and RBD of the S gene (Table 4.2.1). Key mutual mutations among different VOCs are of the deletion at position 69–70 common with Alpha variant, K417N mutual with Beta variant, N501Y common with Alpha, Beta and Gamma variants, and P681H common with Alpha and Delta variants (Table 1). These five VOCs have spread over the world and have taken over as the predominant variations in the areas where they were initially discovered.

4.2.10.2 Conclusion

Due to whole-genome sequencing, we have the ability to watch in fine detail the investigation of SARS-CoV-2 mutations throughout different genetic environments. The emergence of the Alpha variant in United Kingdom was awe-inspiring and depressing. Data from the genome sequence showed that it had picked 23 mutations, 9 of which are in the spike protein region all at once. The majority of these alterations have minimal evolutionary significance; however, some are in charge of this variant's faster rate of dissemination. Similar surges of evolution have occurred before like the emergence of a variant of interest in Tanzania It included 34 distinct mutations, as well as instances of recombination, in which various SARS-CoV-2 genomes merge to create a hybrid, have also been identified. These occurrences might be uncommon, but their possible evolutionary relevance shouldn't be overlooked. A comparatively sluggish regular mutation rate does not necessarily indicate a slow rate of evolution. Even though we may not fully understand the molecular causes, because potential mutations have repeatedly appeared during the pandemic, we can simply detect them from the genome sequence data. Despite the fact that natural selection has repeatedly chosen the same mutations, mutation is a blind process. The term "evolutionary convergence" describes this. It is likely that several mutations rather than a single mutation alone determine the characteristics of the virus. We become aware of the risks of unchecked spread when we recognize how a few uncommon incidents can change the course of the entire pandemic. As more people contract the disease, the risk of unusual evolutionary occurrences producing different varieties of concern will rise.

4.2.11 References

cmay. (2021, July 16). COVID-19 Updates: Globally COVID deaths and cases rising again. Retrieved June 16, 2023, from MRIGlobal website:

https://www.mriglobal.org/covid-19-updates-globally-covid-deaths-and-cases-rising-again/

COVID-19 VACCINES. (n.d.). Retrieved June 16, 2023, from www.slideshare.net website: https://www.slideshare.net/imaging_70/covid19-vaccines-242698369

FinancialNewsMedia.com. (n.d.). A US Government Interagency Group Defines Three Classes Of SARS-Cov-2 Variants. Retrieved June 16, 2023, from www.prnewswire.com website: https://www.prnewswire.com/news-releases/a-us-government-interagency-group-definesthree-classes-of-sars-cov-2-variants-301254674.html

Galaxy | Europe. (n.d.). Retrieved from usegalaxy.eu website: https://usegalaxy.eu

GISAID - gisaid.org. (n.d.). Retrieved from gisaid.org website: https://gisaid.org

Hoffmann, M., Kleine-Weber, H., Schroeder, S., Krüger, N., Herrler, T., Erichsen, S., … Pöhlmann, S. (2020). SARS-CoV-2 Cell Entry Depends on ACE2 and TMPRSS2 and Is Blocked by a Clinically Proven Protease Inhibitor. Cell, 181(2), 271–280. https://doi.org/10.1016/j.cell.2020.02.052

Kim, J. W., Cho, A. H., Shin, H. G., Jang, S. H., Cho, S. Y., Lee, Y. R., & Lee, S. (2023). Development and Characterization of Phage Display-Derived Monoclonal Antibodies to the S2 Domain of Spike roteins of Wild-Type SARS-CoV-2 and Multiple Variants. Viruses, 15(1), 174. https://doi.org/10.3390/v15010174

Mtetwa D., Mutetwa T. F., Manjengwa T., Mazadza A., Chikwambi Z., 2022. Genetic diversity and spread dynamics of SARSCoV-2 variants present in African populations. Royal Society Open Sciencedoi: 10.1098/ros231326.

NCBI. (2020). National Center for Biotechnology Information. Retrieved from Nih.gov website: https://www.ncbi.nlm.nih.gov

Nextclade. (n.d.). Retrieved from clades.nextstrain.org website: https://clades.nextstrain.org

Next IAS. (n.d.). Retrieved June 16, 2023, from www.nextias.com website: https://www.nextias.com/current-affairs/10-07-2021

Tegally, H., Wilkinson, E., Lessells, R. J., Giandhari, J., Pillay, S., Msomi, N., … Sigal, A. (2021). Sixteen novel lineages of SARS-CoV-2 in South Africa. Nature Medicine, 27(3), 440–446. https://doi.org/10.103 8/ s41591-021-01255-3

There are four types of COVID-19 vaccines: here's how they work | Gavi, the Vaccine Alliance. (n.d.). Retrieved June 16, 2023, from www.gavi.org website: https://www.gavi.org/vaccineswork/there-are-four-types-covid-19-vaccines-heres-howtheywork?gclid=EAIaIQobCh MI_Inqx5ft7 wIVlJ1LBR1CIwCHEAAYASAAEgInWvD_Bw

Thomson, E. C., Rosen, L. E., Shepherd, J. G., Spreafico, R., da Silva Filipe, A., Wojcechowskyj, J. A., … Pinto, D. (2021). Circulating SARS-CoV-2 Spike N439K Variants Maintain Fitness While Evading antibody-mediated Immunity. Cell, 184(5). https://doi.org/10.1016/j.c ell.2021. 01.037

Thoutam, A., Breitzig, M., Lockey, R., & Kolliputi, N. (2020). Coronavirus: a shift in focus away from IFN response and towards other inflammatory targets. Journal of Cell Communication and Signaling, 14(4), 469–470. https://doi.org/10.1007/s12079-020-00574-3

Weitzner, B. (2015). N E X T-G E N E R AT I O N A N T I B O D Y M O D E L I N G. Retrieved from https://jscholarship.library.jhu.edu/bit stream/ handle/1774.2/37835/WEITZNERDISSERTATION-2015.pdf

Wikipedia. (2020, March 12). COVID-19 vaccine. Retrieved from Wikipedia website: https://en.wikipedia.org/wiki/COVID-19_vaccine

Wilkinson, E., Giovanetti, M., Tegally, H., & San, J. E. (2021). A year of genomic surveillance reveals how the SARS-CoV-2 pandemic. MedRxiv : The Preprint Server for Health Sciences, 27.

Wu, C., Ye, D., Mullick, A. E., Li, Z., Danser, A. H. J., Daugherty, A., & Lu, H. S. (2020). Effects of Renin-Angiotensin Inhibition on ACE2 and TMPRSS2 Expression: Insights into COVID-19. Hypertension. https://doi.org/10.1161/hypertensionaha.120.15782

Wu, F., Zhao, S., Yu, B., Chen, Y.-M., Wang, W., Song, Z.-G., … Zhang, Y.-Z. (2020). A new coronavirus associated with human respiratory disease in China. Nature, 579(7798). https://doi.org/10.1038/s41586-020-2008-3

www.ingramcontent.com/pod-product-compliance
Lightning Source LLC
LaVergne TN
LVHW052220200926
848481LV00001B/161